Topics in Lightwave Transmission Systems

Optical Fiber Communications

Edited by

Tingye Li
AT&T Bell Laboratories
Crawford Hill Laboratory
Holmdel, New Jersey

A table of contents for Volume 1 appears at the end of this volume.

Topics in Lightwave Transmission Systems

Edited by

TINGYE LI

*AT&T Bell Laboratories
Crawford Hill Laboratory
Holmdel, New Jersey*

Published by arrangement with AT&T

ACADEMIC PRESS, INC.
Harcourt Brace Jovanovich, Publishers

Boston   San Diego   New York
London   Sydney   Tokyo   Toronto

This book is printed on acid-free paper.

Copyright © 1991 by American Telephone and Telegraph Company
ALL RIGHTS RESERVED.
NO PART OF THIS PUBLICATION MAY BE REPRODUCED OR
TRANSMITTED IN ANY FORM OR BY ANY MEANS, ELECTRONIC
OR MECHANICAL, INCLUDING PHOTOCOPY, RECORDING, OR
ANY INFORMATION STORAGE AND RETRIEVAL SYSTEM, WITHOUT
PERMISSION IN WRITING FROM THE PUBLISHER.

ACADEMIC PRESS, INC.
1250 Sixth Avenue, San Diego, CA 92101

United Kingdom Edition published by
ACADEMIC PRESS LIMITED
24-28 Oval Road, London NW1 7DX

Library of Congress Cataloging-in-Publication Data
(Revised for volume 2)

Optical fiber communications

    Vol. 2-   has imprint: Boston : Academic Press.
    Includes bibliographies and index.
    Contents: v. 1. Fiber fabrication—[v. 2] Topics
in Lightwave Transmission Systems
    1. Optical communications. 2. Optical fibers.
I. Li, Tingye.
TK5103.59.O675 1985   621.382'7   84-9232
ISBN 0-12-447301-6 (v. 1 : alk. paper)
ISBN 0-12-447302-4 (v. 2: alk. paper)
PRINTED IN THE UNITED STATES OF AMERICA

91 92 93 94   9 8 7 6 5 4 3 2 1

CONTENTS

CONTRIBUTORS vii
PREFACE ix

1. Optical Transmitter Design
   M. Dixon and J.L. Hokanson

   I. Introduction 1
   II. System Constraints on Transmitter Design 3
   III. Optical Properties of Sources for PCM Transmitters 18
   IV. Circuit Strategies for PCM Transmitters 26
   V. Practical PCM Transmitters and Their Performance 44
   VI. Multigigabit-per-Second Transmission Systems 66
   VII. Conclusion 71
   References 72

2. Lightwave Receivers
   Gareth F. Williams

   I. Introduction 79
   II. Receiver and Device Requirements of Lightwave Systems 83
   III. Receiver System and Noise Considerations 85
   IV. First and Second-Generation Lightwave Receivers 112
   V. Active-Feedback Lightwave Receiver Circuits 121
   References 148

3. Frequency and Phase Modulation of Semiconductor Lasers
   Soichi Kobayashi, Yoshihisa Yamamoto, and Tatsuya Kimura

   I. Introduction 151
   II. Direct Frequency Modulation 153

	III.	Phase Modulation by Injection Locking	186
	IV.	Summary	197
		Acknowledgments	199
		References	199

4. Coherent Optical Fiber Transmission Systems
   Shigeru Saito, Yoshihisa Yamamoto, and Tatsuya Kimura

	I.	Introduction	203
	II.	System Operations and Configurations	206
	III.	Advantages of Optical Heterodyne or Homodyne Detection	211
	IV.	System Applications	225
	V.	Essential Technology for Developing Coherent Systems	233
	VI.	System Experiments	251
	VII.	Conclusion	257
		References	258

5. Nonlinear Effects in Optical Fibers
   Andrew R. Chraplyvy

	I.	Introduction	267
	II.	Nonlinear Gain and System Parameters	268
	III.	Stimulated Raman Scattering	269
	IV.	Carrier-Induced Phase Modulation	278
	V.	Stimulated Brillouin Scattering	281
	VI.	Four-Photon Mixing	285
	VII.	Multiplexing Effects	288
	VIII.	Scaling	290
	IX.	Conclusions	292
		References	293

INDEX 297

CONTRIBUTORS

Numbers in parentheses indicate the pages on which the authors' contributions begin.

Andrew R. Chraplyvy (267), AT&T Bell Laboratories, Crawford Hill Laboratory, Holmdel, New Jersey 07733

M. Dixon (1), AT&T Bell Laboratories, Breinigsville, Pennsylvania 18103

J. L. Hokanson (1), AT&T Bell Laboratories, Breinigsville, Pennsylvania 18103

Tatsuya Kimura (151, 203), NTT Basic Research Laboratories 3-9-11, Musashino-shi, Tokyo, 180 Japan

Soichi Kobayashi (151), Phototonic Integration Research, Inc. (PIRI), 1375 Perry Street, Columbus, Ohio 43201-2693

Shigeru Saito (203), NTT Basic Research Laboratories, Musashino-shi, Tokyo, 180 Japan

Gareth F. Williams (79), NYNEX Corporation, 500 Westchester Avenue, Room 2F7, White Plains, New York 10604

Yoshihisa Yamamoto (151, 203), NTT Basic Research Laboratories 3-9-11, Musashino-shi, Tokyo, 180 Japan

PREFACE

Twenty years have passed since the attainment of the first low-loss optical fiber intended for communication applications. The rapid research progress in the lightwave field that ensued has led to the widespread application of optical fiber communication throughout the telecommunications industry. Optical fiber transmission systems are now in commercial use under the ocean to link continents, on land to join cities, and in metropolitan areas to connect telephone central offices and to distribute broadband services to the customers. Indeed, optical fiber has emerged as the undisputed transmission medium of choice in almost all areas of telecommunication, mainly because it offers unrivaled transmission capacity at lower cost.

Research continues today on many fronts to explore the vast potential bandwidth of low-loss single-mode fibers for long-distance transport as well as for local networks and local distribution. Monolithic integration of photonic and electronic devices is being pursued for both transmission and switching, to enhance subsystem performance and functionality, and to reduce overall system cost. The continuing interest in increasingly higher speed devices and systems is shown by demonstrations of transmission experiments at data rates of up to 20 Gb/s. Coherent detection offers high receiver sensitivity and efficient means for channel selection in a densely packed wavelength-multiplexed system, but noncoherent techniques involving optical amplifiers and tunable optical filters may pose a serious challenge. Various coherent and noncoherent experimental systems have been demonstrated for transmission and networking. Recent developments in erbium-doped fiber amplifiers are evoking renewed excitement in the field as optical amplifiers, operated as non-regenerative repeaters, make wavelength multiplexing attractive and therefore promise significant benefits in operational flexibility and overall cost for both terrestrial and undersea long-haul transmission. Nonlinear effects in the transmission fiber are therefore of great interest, as these effects may impair signal transmission when distances between regenerative repeaters become very large.

This book is a second volume of a treatise on optical fiber communications that is devoted to the science, engineering, and application of information transmission via optical fibers. The first volume, published in 1985, dealt exclusively with fiber fabrication. The present volume contains topics that pertain to subsystems and systems. Transmitters and receivers, which are basic to systems now operating in the field, are treated in the first two chapters. The next two chapters cover topics relating to coherent systems: frequency and phase modulation of the optical carrier, and systems considerations and experiments. The last chapter reviews the fundamentals of nonlinear effects in optical fibers and considers how systems are affected by various nonlinear phenomena.

Tingye Li

Optical Transmitter Design

M. DIXON AND J. L. HOKANSON

AT&T Bell Laboratories
Solid State Technology Center
Breinigsville, Pennsylvania

I. Introduction 1
II. System Constraints on Transmitter Design 3
   A. Modulation Properties of Sources 4
   B. Modulation Format 8
   C. Source Wavelength and Linewidth 10
   D. Modulation Rate 14
III. Optical Properties of Sources for PCM Transmitters 18
   A. Optical Pulse Requirements 18
   B. Direct Current Modulation of Lasers and LEDs 19
   C. Summary of Source Properties 25
IV. Circuit Strategies for PCM Transmitters 26
   A. Modulator Circuits 27
   B. Biasing Circuits for LEDs 29
   C. Biasing Circuits for Lasers 31
   D. Source Temperature Control 40
   E. Equivalent Circuit Models 41
V. Practical PCM Transmitters and Their Performance 44
   A. LED Transmitters 44
   B. Laser Transmitters 48
   C. Testing of Transmitters 64
VI. Multigigabit-per-Second Transmission Systems 66
   A. System Overview 66
   B. System Performance Limits 69
   C. FT Series G 1.7 GB/s 1.55-μm DFB WDM Channel 69
VII. Conclusion 71
   A. Summary of Presenter Transmitter Design 71
   References 72

I. INTRODUCTION

The initial driving force for lightwave technology has been in the interoffice and intercity trunking and loop feeder network of the telecommunications system. The technology has expanded to capture both the terrestrial and the subcable long-haul transmission networks, and it is moving rapidly to capture the last few kilometers to the home markets. Exciting advances in optical technology have been made over the last decade, with performance as measured by the bit rate × distance product almost doubling each year. The development of very high frequency, long-wavelength, single-frequency semiconductor lasers, wideband photodetectors, and microwave electronic circuits in both silicon and gallium arsenide, along with the concurrent reduction of fiber loss toward the Rayleigh scattering limit, has brought about this phenomenal growth. As a result, large quantities of digitized information can be transmitted over long distances and provide the foundation for the envisioned future broadband integrated services digital networks, B-ISDN (Catania, 1986). Such lightwave communication systems are configured presently, as shown in Fig. 1, to be compatible with existing communication networks. It is seen immediately that optical transmitters play an integral role in such systems. However, as a result, the choice of optical sources, modulation schemes, and even optical performance are not dictated solely by the transmitter designer. At a time when optical transmitter design is itself evolving, transmitter design decisions are often made to accommodate advances in, or limitations of, other parts of a rapidly changing system.

For these reasons we shall initially discuss in Section II those system considerations that have limited the majority of viable present-day optical transmitter designs for long-haul communications to the direct-current modulation of semiconductor light-emitting diodes (LEDs) and lasers. This will include a brief description of optical sources, their available power, wavelengths, linewidths, bandwidth, transmission characteristics of the fiber medium, modulation format (analog or digital), the physical properties of the sources available for modulation, and the modulation rate. These will be discussed in the context of the technical considerations affecting their choice.

Following that, Section III will review the properties of optical sources that are important for the most widely adopted of the modulation schemes,

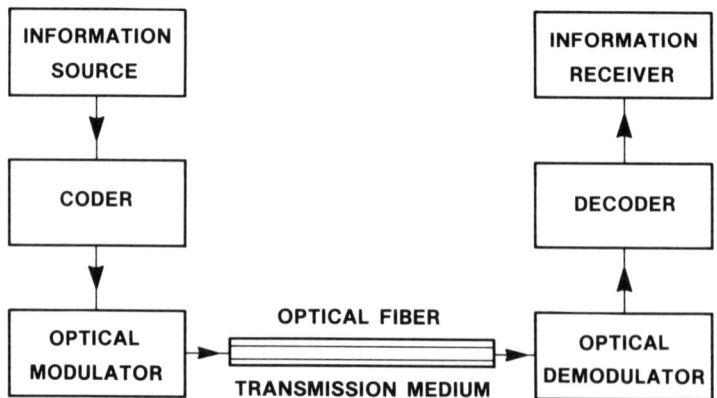

FIG. 1. Schematic of an optical fiber transmission system.

viz., pulse code modulation. This will cover, in a general manner, the desirable optical pulse characteristics imposed by lightwave system constraints. It will be followed by a description of the dynamic pulse response characteristics of semiconductor LEDs and lasers, and their limitations in meeting such system objectives. As the modulation rates cascade into the multigigabit range and large fiber chromatic dispersion is encountered, system limitations brought about by the effect of direct modulation on the dynamic spectral characteristics of the laser source (Linke *et al.*, 1985) has led to successful experiments with external modulators (Korotky and Alferness, 1988). In this scheme, the laser is operated in a cw mode, and its light is modulated using the external modulator.

Section IV will discuss the modulation and feedback control strategies that have been suggested to produce and to retain adequate transmitter performance with both time and environmental changes of the direct current modulated semiconductor sources and their packaging. Topics include new ICs in Si-bipolar or GaAs MESFET for electrical-to-optical conversion; microwave interconnections between components in the data path; and for long-wavelength laser sources, thermoelectric heat pump temperature control. The section will conclude with a discussion of computer modeling of laser diodes.

Section V will present a more detailed discussion of practical transmitter design. No attempt at completeness will be made, as this section represents only our combined transmitter design experiences for real-world systems. However, it will serve to illustrate the problems encountered, circuit strategies adopted, and performance obtainable with pulse modulated transmitters up to a few hundred Mb/s with LED sources and extending into the multigigabit range with laser sources. Both short-wavelength (0.8–0.9 μm) and long-wavelength (1.3 μm) devices will be discussed.

Section VI will discuss the successful laboratory demonstrations aimed at the next generation of very large capacity systems with long repeater spans. The maximum bit rate attainable in present experimental systems is limited by laser modulation speeds and dynamic optical spectra; by external modulator capabilities; by detector and receiver bandwidths; and finally, by the performance of electronic amplifiers and circuits (Li and Linke, 1988).

Finally, the concluding Section VII will summarize the status of present-day transmitter design.

II. System Constraints on Transmitter Design

The optical transmitter in a lightwave system, of the type shown in Fig. 1, performs the function of electro-optic transducer, as well as that of signal modulator. Some choices in the transmitter design are strongly influenced by the system environment in which it is embedded. These include (1) modulation properties of the source, (2) modulator format, (3) source wavelength and linewidth, transmission characteristics of the fiber medium, and (4) modulation rate. We will outline the significance of these properties for present transmitter design.

A. Modulation Properties of Sources

A number of physical properties of optical sources potentially could be modulated (Kaminow and Li, 1979). They can be seen clearly in Fig. 2. However, our current choices from this list are reduced by limitations in the state of the art of fibers and detectors, as well as of the sources themselves.

1. Fibers

Optical loss and its wavelength dependence, optical dispersion in its three forms of material, waveguide, and intermodal dispersion, and optical depolarization caused by birefringence and fiber dimension variations are all encountered during the transmission of a modulated signal. The difficulty of producing fibers that retain a fixed polarization state (Rashleigh and Marrone, 1983) has so far eliminated modulation of the polarization vector.

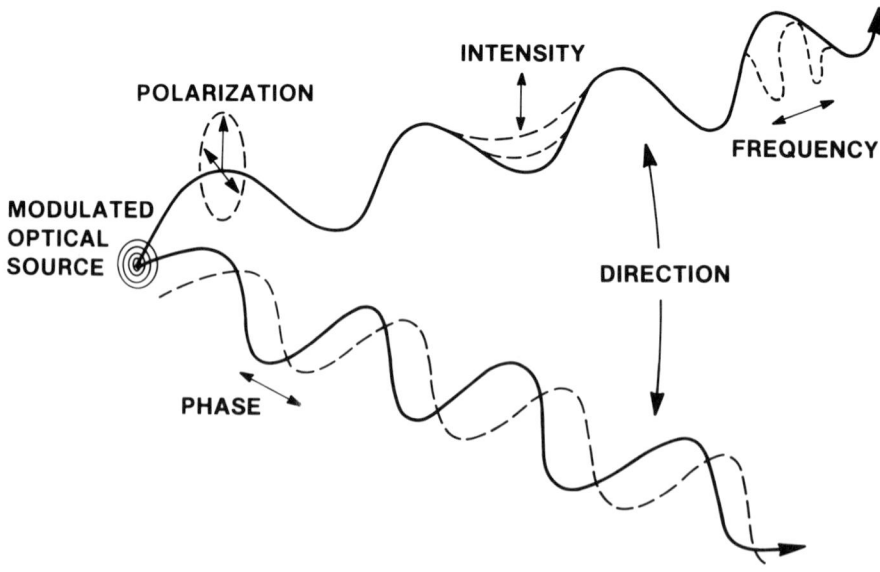

FIG. 2. Optical wave properties for modulation.

Fiber development has, however, improved tremendously in recent years, opening up low loss and dispersion windows around 0.83 μm, and especially at 1.3 μm, as well as at the very attractive wavelength of 1.55 μm.

Multimode (MM) fiber was introduced into manufacture ahead of SM fiber because of difficulties encountered in coupling light into single-mode (SM) fiber, and because of the complexities of splicing and interconnections between SM fiber cables (Kaiser and Keck, 1988). Today, MM fiber systems are increasingly being used for short-distance (a few kilometers) transmission applications such as data links, because of higher loss and significantly lower bandwidths than those of SM fiber systems. In addition, surface-emitting LEDs require large fiber core diameters (>50 μm) to couple sufficient light into the fiber to produce a viable system. Figure 3a shows the values of loss and fiber laser bandwidth achievable in MM fiber using MCVD (Nagel et al., 1982) and VAD (Kuwahara et al., 1981) growth technologies. The MM fiber bandwidth is limited by both intermodal and chromatic (sum of material and waveguide) dispersion mechanisms. For broad optical spectral width devices such as LEDs, both the chromatic and intermodal dispersion mechanisms determine the MM fiber bandwidth (Refi, 1986). For the much narrower optical spectral width lasers, the MM fiber bandwidth is dominated by intermodal dispersion effects that are ≤ 3 Gb/s km (Henry et al., 1988).

Single-mode fiber development, with core diameters of less than 10 μm, is concentrated between 1.2 and 1.6 μm because of the availability of zero chromatic dispersion at these wavelengths (Iwahashi, 1981). In these fibers, modal dispersion is nonexistent. Figure 3b shows the values of loss and chromatic dispersion for standard SM fiber, dispersion-shifted (DS) fiber, and dispersion-flattened (DF) fiber (Kaiser and Keck, 1988). The losses are approaching the Rayleigh scattering limit. In the DS fiber, the zero chromatic dispersion point is shifted to 1.55 μm by compensating the material dispersion with increased waveguide dispersion. The DF fibers with a small dispersion over the 1.3 to 1.6 μm range make these fibers attractive for wavelength-division-multiplex (WDM) applications. These fibers are the lowest-loss, largest-transmission-bandwidth medium available having the potential of producing bandwidths of 900 Gb/s km (Henry et al., 1988). Virtually all of the installed transmission systems today are standard SM fiber with zero dispersion at 1.3 μm. The migration to DS fiber for future applications is uncertain with the possible exception of new, higher-data-rate subcable routes where long repeater spans are required.

2. Sources

The restrictions of available power, linearity, modulation bandwidth, dynamic optical spectral characteristics, and stability at the transmission

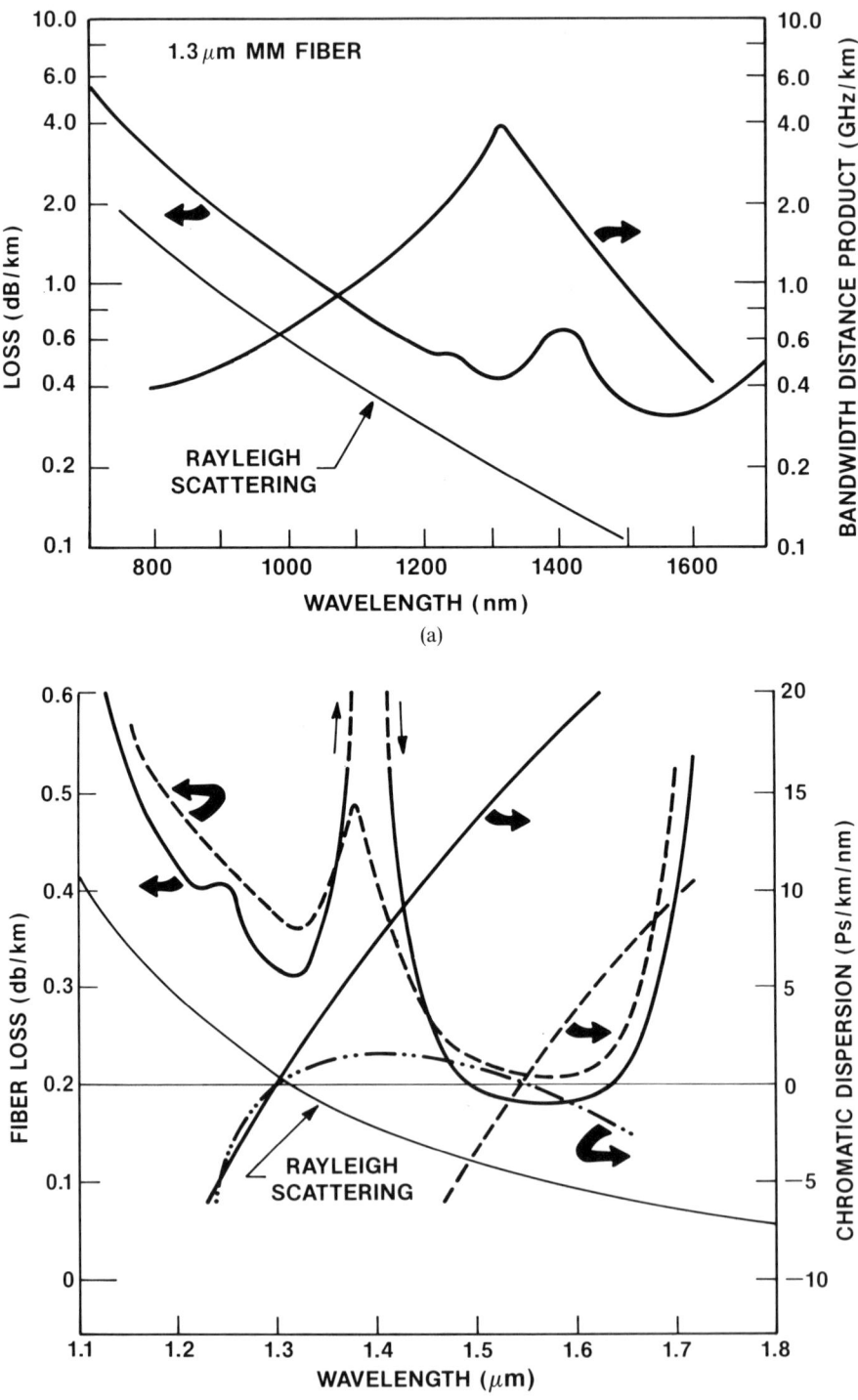

FIG. 3. (a) Spectral loss and laser bandwidth characteristics of multimode optical fiber as a function of wavelength. (b) Spectral loss and chromatic dispersion of standard single-mode fiber with zero dispersion at 1.3 μm (solid line), dispersion-shifted (DS) fiber with zero dispersion at 1.55 μm (dashed line), and dispersion-flattened (DF) fiber (dot/dashed line).

windows of 0.83, 1.3, and 1.55 μm must all be considered. Also to be considered is the need for a source emission characteristic that is compatible with coupling into fibers of less than 50 μm core diameter. The stringent requirements for stable wavelength sources, for carrier and local oscillators, has meant that phase and frequency modulation techniques, although very attractive (Favre *et al.*, 1981), have not been implemented.

3. Detectors

These devices influence the choice of source modulation property as a result of their wavelength-dependent quantum efficiencies, finite modulation bandwidths and limitations due to their characteristic noise generation (Lee and Li, 1979).

In Fig. 4 are shown the quantum efficiencies of the available detectors with wavelength. Fortunately, silicon photodetectors with excellent properties have been available for over a decade at the 0.83-μm wavelength, with modulation bandwidths greater than 2 GHz. At the longer wavelengths of 1.3 and 1.55 μm, the most commonly used detector is the InGaAs PIN photodiode, which is characterized by its relative ease of fabrication, extremely

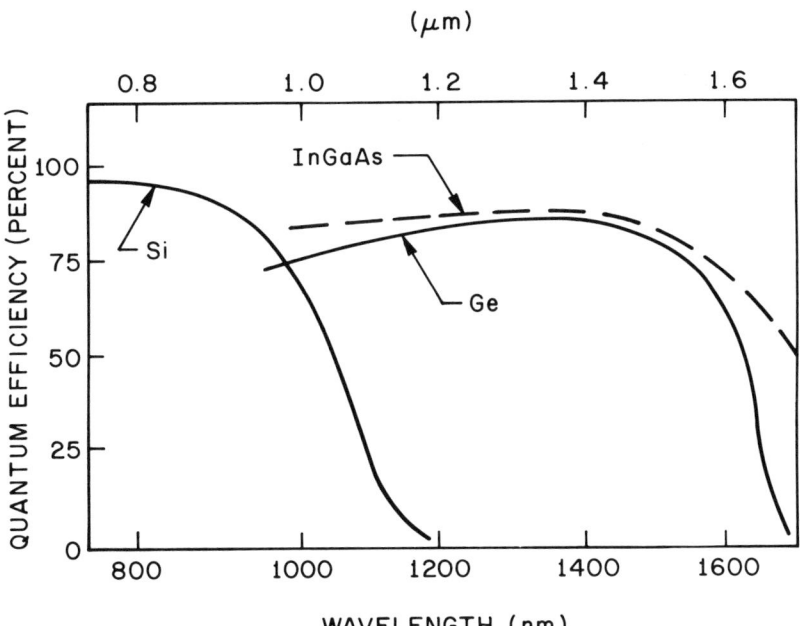

FIG. 4. Typical optical detector characteristics as a function of wavelength.

high reliability, excellent frequency response, and low noise. To enhance receiver sensitivities, particularly at higher data rates, much research has been carried out on InGaAs avalanche photodiodes (APD) (Forrest, 1988). The most recent devices, known as separate-absorption-grading-multiplication avalanche photodiodes (SAGM APD), have achieved a gain—bandwidth product as high as 70 GHz (Campbell et al., 1988).

Choice of source property to be modulated must then be matched to both the available detector properties and the available fiber transmission windows. All the restraints on source, fiber, and detectors we have briefly mentioned have meant that intensity modulation of optical sources and direct photon detection at the receiver has become the most widely used lightwave scheme.

B. Modulation Format

The electrically coded information applied to our intensity-modulated device could be either analog or digital. Two factors, however, have limited most practical, installed systems to a digital or pulse code modulated (PCM) format. First, optical fiber is basically a noise-limited transmission medium (Jones, 1982) and so requires signal-to-noise ratios of greater than 60 dB for analog transmission (Yasugi et al., 1982). Second, the high linearity of output signal needed to avoid distortion in analog systems is not readily obtained in optical sources.

Both these considerations presently limit analog transmission to short, unrepeatered links where the optical fiber costs are under the greatest pressure from conventional systems (Hanson, 1982; Mills, 1982). It should be pointed out, however, that an analog format does have appeal for video applications because of the high data rate needed for PCM transmission of TV signals. For example, the 5 MHz bandwidth of an FM color TV signal requires approximately 90 Mb/s of data in the PCM modulation format (Tseng and Chen, 1983). High definition color TV (HDTV) signals are even more demanding of bandwidth, requiring 26 MHz or approximately 400 Mb/s of digital data (Pinnow, 1983). Special attention, however, should be given to coding technologies. The progress made in the last 20 years of history shows that both audio and video coding bit rates have constantly decreased at a rate of greater than five times per decade. The most recent experimental results indicate that the standard TV signal can be transmitted with only 20 Mb/s data rate and HDTV with 140 Mb/s (Catania, 1986).

Linearization techniques, such as circuit predistortion of the signal (Pan, 1981; Shumate and DiDomenico, 1982), have been employed in LED transmitters to improve signal-to-noise characteristics in analog transmission systems where one of the advantages of optical fiber, such as immunity to interference, dielectric isolation, or small diameter cable size, makes it preferable to a system using either paired or coaxial cable. It is still not clear, though, whether analog or digital transmission of video will be best for telecommunications applications.

The restraints of high signal-to-noise and optical linearity are not applicable, however, to the PCM format. The extremely broad and flat bandwidth available with optical fibers (Gloge et al., 1979) makes pulse code modulation possible, but the limited power available from optical sources has meant that binary, or on–off keyed (OOK), PCM format is desirable. The lower optical power requirements of binary coding improve source reliability and can help reduce the practical difficulties encountered in obtaining high coupling efficiency into small-diameter fibers. However, in future systems requiring very high bit rates (Pan, 1983), which fully utilize the bandwidth capacity of the fiber, higher-order modulation schemes may be necessary.

As a result of these present limitations, we shall concentrate on transmitters driven by binary PCM signals, as they are of overwhelming importance in large-capacity, long-haul transmission systems utilized today or contemplated in the near future for voice, data, and video transmitters. We shall see in the next section that intensity modulation of sources with PCM signals can be readily obtained by direct current modulation of semiconductor lasers and LEDs. These two luxuries of direct current modulation of the source and direct photon detection by the receiver have meant that the simple system configuration shown in Fig. 5 has emerged as the easiest to implement.

Other chapters of this treatise will cover in much more detail the advantages and problems associated with using phase, frequency, gain, polarization, and direction modulation in both waveguide modulators and semiconductor lasers.

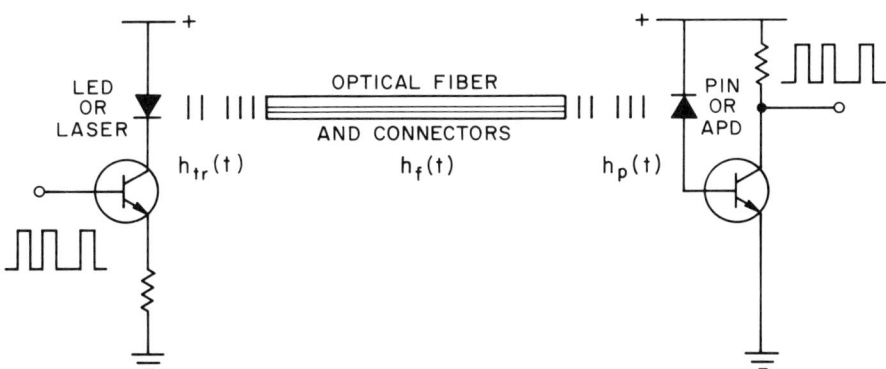

FIG. 5. A simple lightwave system.

C. Source Wavelength and Linewidth

Both the absolute and the spectral width of the optical radiation from semiconductor LEDs and lasers have had a major influence on the development of transmitters that use intensity modulation by direct current injection of PCM signals. Optical sources have proceeded through three generations of development and research to make better use of the fiber transmission capabilities.

1. First Generation, ~0.8 μm

Laser systems were developed at wavelengths between 0.8 and 0.9 μm using the GaAs and AlGaAs material systems (Burrus et al., 1979; Kressel et al. 1982). This was because fiber loss and laser lifetime were well controlled at these wavelengths (Chynoweth and Miller, 1979). Multimode fiber was used as the transmission medium, and silicon PIN or APD as detectors. Typical optical spectral properties of these short-wavelengths LEDs and lasers at this wavelength are compared schematically in Fig. 6a. The spectrum of the 0.83-μm laser is shown on an expanded scale in Fig. 6b. The RMS spectral with (σ) is typically less than 5 nm.

The LEDs have spectral widths that are an order of magnitude greater than those of lasers, while modulation bandwidths, BW, are an order of magnitude smaller. Also, the available power from the LED that can then be coupled into multimode fibers, Pc is significantly smaller than that achievable with lasers. It must also be traded off with the modulation bandwidth because of an inverse relationship between coupled power and modulation bandwidth.

At 0.83 μm, the wide spectral width of LEDs, when taken with the large chromatic dispersion of some 100 ps/km/nm in the optical fibers (Gloge et al., 1979), means that dispersive effects limit transmission of digital optical pulses to bit rates below 50 Mb/s. The narrower spectral width of the 0.83-μm lasers means that pulse spreading due to such dispersive effects can be suppressed until bit rates approach 400 Mb/s (Miller, 1979). The lasers were the multifrequency Fabry–Perot type, which lased in the fundamental transverse mode and exhibited a number of longitudinal modes. The optical and current confinement for this laser structure was dictated by "gain-guiding."

2. Second Generation, 1.3 μm

The transition to the 1.3 to 1.6 μm wavelength region saw a number of concurrent developments that took place in the late 1970s and early 1980s. These include surface-emitting LEDs (Dentai et al., 1977) and long-lived 1.3-μm "index-guided" Fabry–Perot-type multifrequency lasers (Nakamura,

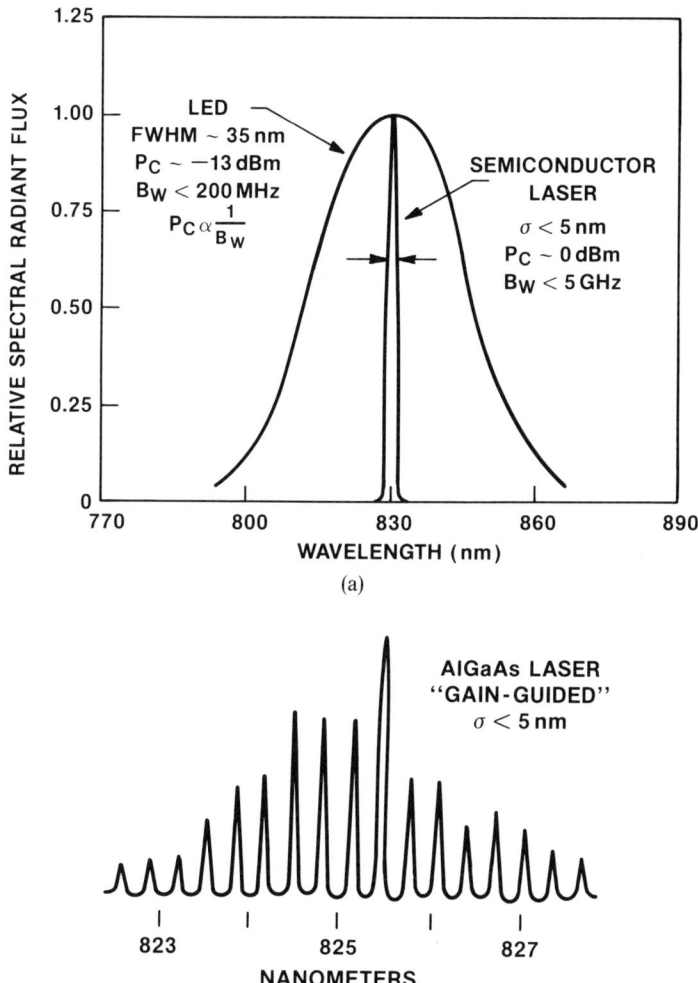

FIG. 6. (a) Typical LED and laser optical spectra. (b) Optical spectrum of 0.83-μm "gain-guided" AlGaAs laser. (c) Optical spectrum of 1.3-μm Fabry–Perot laser under 1.7 Gb/s, NRZ modulation $I_{bias} = 0.8 I_{th}$. Center wavelength and RMS spectral width (σ) indicated. (d) Optical spectrum of 1.55-μm DFB laser under 1.7 Gb/s, NRZ modulation $I_{bias} = 0.8 I_{th}$. Sided mode suppression ratio SMSR indicated. (e) Optical spectrum of 1.55-μm DFB laser under 1.7 Gb/s, NRZ modulation $I_{bias} = 0.8 I_{th}$. Center wavelength and maximum skew at -30 dB indicated.

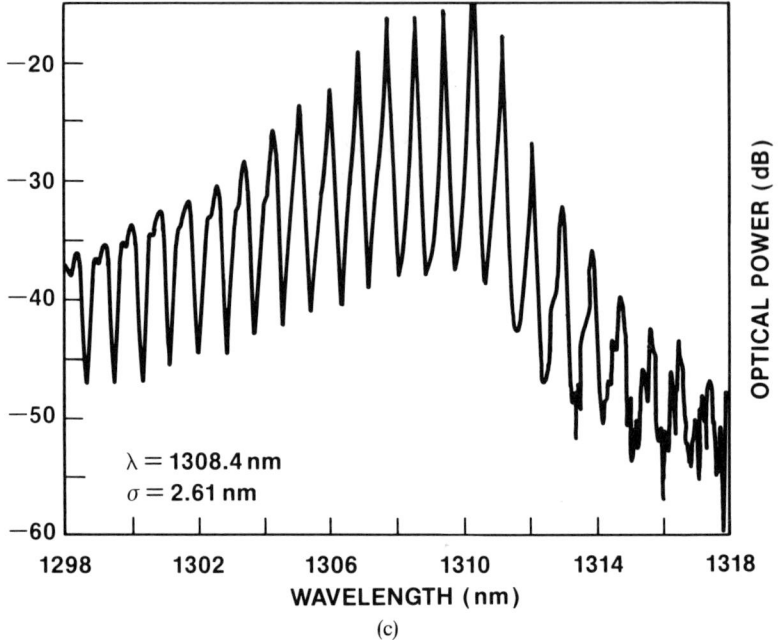

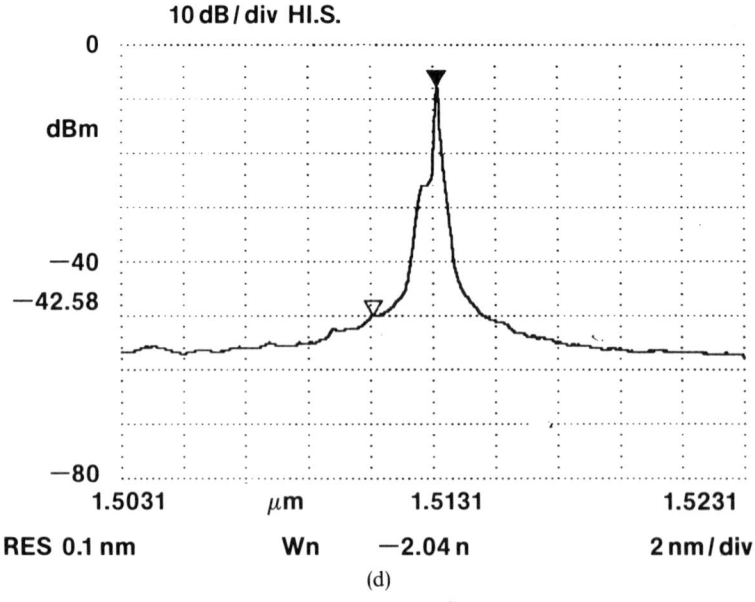

FIG. 6. (*continued*)

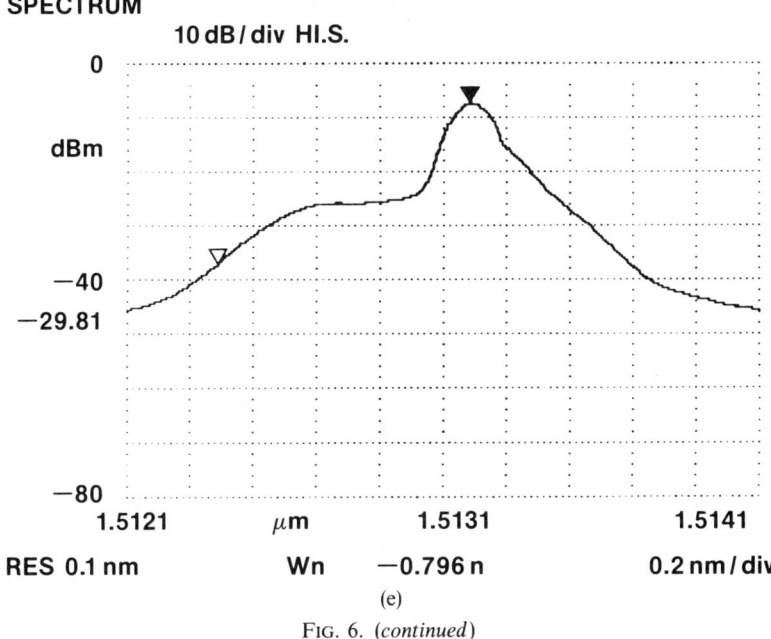

FIG. 6. (continued)

1983) and detectors (Smith et al., 1982). At the same time came the rapid development of single-mode fiber with cabled losses as low as 0.4 dB/km at 1.3 μm. Chromatic dispersion was reduced to zero, and rate-limiting modal dispersion was nonexistent. Laser sources could couple an appreciable amount of their power into a single-mode fiber, and virtually all laser-based lightwave systems today use single-mode fiber. The surface-emitting (SE) LED still requires the use of a MM fiber in order to couple sufficient energy to make a viable link. Edge-emitting (EE) LEDs (Kressel and Ettenberg, 1975) were developed to take advantage of SM fiber, but their low coupled power has so far has limited their acceptance. Recent experiments on standard SM fiber have achieve 1.2 Gb/s 10 km and 600 Mb/s 20 km (Ohtsuka et al., 1987) and 2 Gb/s 9.8 km (Fujita et al., 1987). Short-span subscriber loop systems for broadband ISDN and optical CATV are potential applications for these devices. The full-width half-maximum (FWHM) optical spectral width of the surface emitter is ~125 nm, and that of the edge emitter, ~75 nm. In order to take advantage of the zero chromatic dispersion at 1.3 μm, it is important that the center wavelength of the source be closely matched to this zero dispersion point. The temperature coefficient of wavelength of both SE LEDs and EE LEDs is +0.4 nm/°C. Temperature also affects the optical output power of LEDs. The temperature coefficient of output power for a SE LED is -0.025 dB/°C, and for a EE LED, -0.10 dB/°C.

Figure 6c shows a typical longitudinal mode spectrum of a multifrequency laser. The RMS spectral width (σ) is approximately 2.5 nm. This has allowed for the development of 40-km terrestrial links operating up to 1.7 Gb/s (Fishman et al., 1986; Nakagawa et al., 1986), and submarine cable spans of 60 km at 296 Mb/s (Runge and Trischitta, 1984). These systems are limited by the mode partition noise generated by the multifrequency laser and the small but finite dispersion in the fiber (Agrawal et al., 1988). Here it is extremely important to match the center wavelength to the zero dispersion wavelength of the fiber. The temperature coefficient of wavelength for these devices is $+0.4$ nm/°C. For this reason, a thermoelectric heat pump is used to heat and cool the laser diode to 20°C.

3. Single-Frequency Lasers at 1.3 and 1.55 µm

The 1980s have brought about the development of the single-frequency laser, which now permits the third-generation design of long-span multigigabit systems. A brief review of the research and development now taking place in this area will be given in Section VI. The solution to the dispersive impairments at 1.55 µm is to use a laser with an extremely narrow spectrum. The most successful approach today is the distributed feedback (DFB) laser, in which the Fabry–Perot structure has been modified to include a perturbed waveguide to provide a frequency-selective feedback mechanism (Sakai et al., 1983). The optical spectrum of a typical single-frequency DFB laser is shown in Fig. 6d and 6e. The linewidth of the DFB laser is considerably more narrow than that of its multifrequency laser predecessor. The suppression ratio of side modes to the DFB mode (SMSR) is greater than 33 dB. In order to obtain the most stable operation of these devices, the laser temperature is maintained at 20°C by a thermoelectric heat pump. The next obstacle to the realization of long-span multigigabit systems, laser chirping and side-mode oscillations, will be discussed in Section VI.

The research and development efforts directed to matching the source wavelength and linewidth properties to make best use of the transmission properties of the fiber medium are summarized in Fig. 7 (Henry et al., 1988). The improvements in system performance, as measured by bit rate × span product in each generation, appear to be quite rapid at the beginning and then tend to saturate after three or four years.

D. Modulation Rate

Although the ultimate modulation rate in a lightwave system is limited by the modulation bandwidths of the source, fiber, and detectors, it was economic

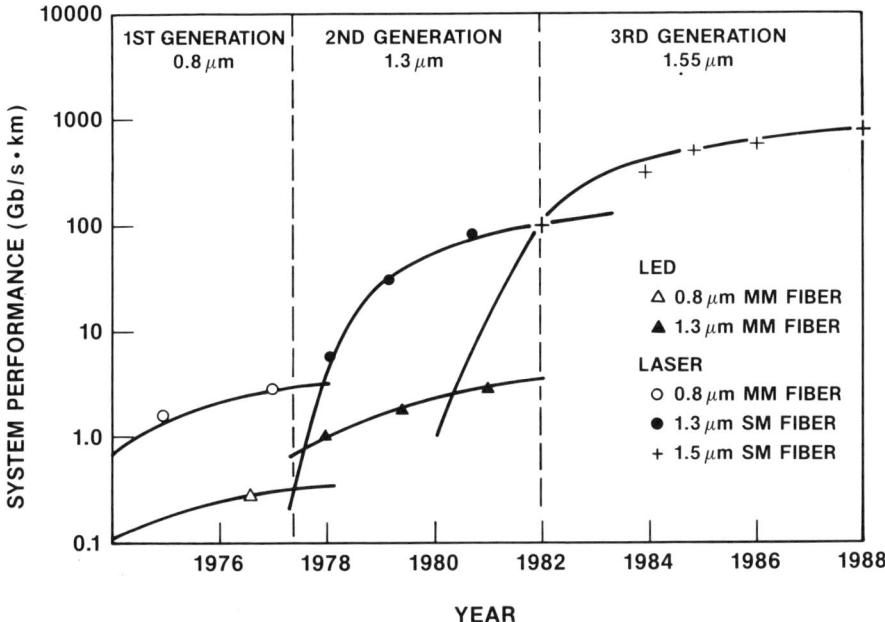

Fig. 7. Progress in lightwave transmission performance. Points correspond to benchmark laboratory demonstration and year. The curves illustrate the evolution of the three generations of lightwave transmitters (0.8 μm LED and laser, 1.3-μm LED and multifrequency Fabry–Perot laser, and 1.3-μm and 1.55-μm DFB single-frequency laser) to match source wavelength characteristics to the optimum transmission capabilities of the fiber medium.

considerations as much as technological factors that determined the earliest transmitter modulations rates in real-world systems.

Cook (1979) indicated that a natural "lightwave window" existed, in the interoffice trunking of the U.S. telecommunications hierarchy, around systems with 672 two-way, 4-kHz voice channels. This corresponds to the T3 rate of 44.7 Mb/s in the PCM format. This "lightwave window," illustrated in Fig. 8, was bounded below 30 Mb/s by the cost competitiveness of twisted pair cables, while the upper limit of 100 Mb/s was bounded by technological challenges of the emerging lightwave developments in long-wavelength (1.3 and 1.55 μm) lasers and LEDs and single-mode fiber. Thus, most initial transmitter development for optical transmission systems in the U.S. has concentrated at the FT3 and FT3C data rates of 44.7 and 90.5 Mb/s and an MM fiber transmission medium.

In 1979, the most advanced development was in multimode fibers with cabled losses of approximately 6 dB/km at 0.83 μm and 1 dB/km at 1.3 μm wavelength (Miller, 1979). These data, shown in Fig. 9 for repeater spacing with data rate and using fibers and LED and lasers then available, illustrated

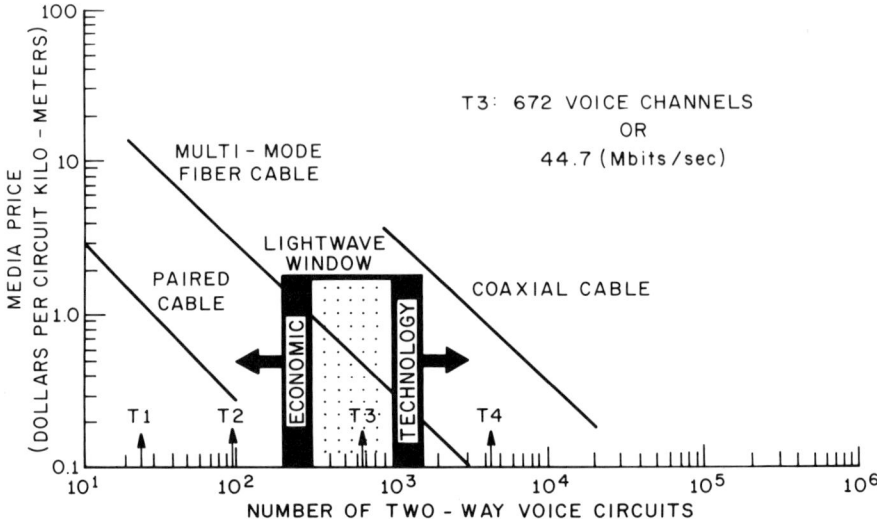

FIG. 8. A comparison of transmission and media costs (Jacobs, 1980).

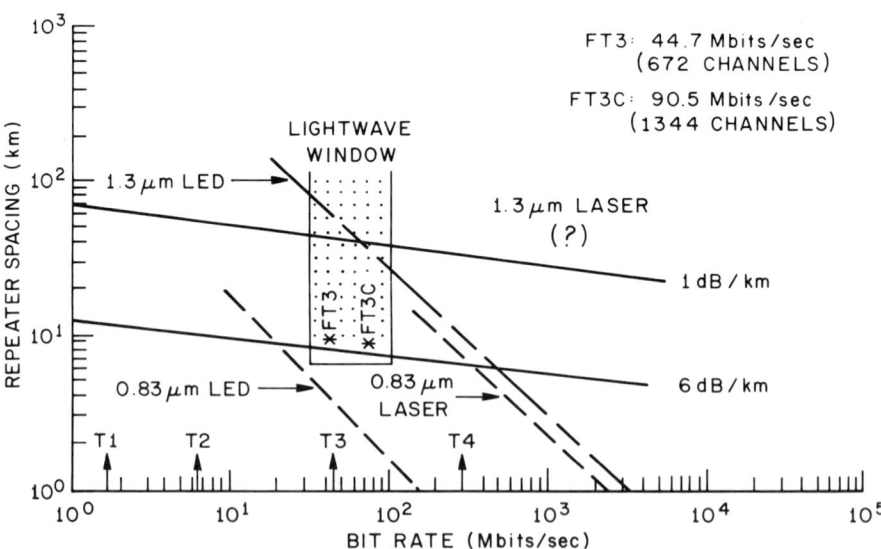

FIG. 9. Repeater spacing vs. bit-rate limitations due to MM fiber loss and dispersion (Miller, 1979).

that viable commercial system with link spacings up to 7 km were possible. However, only the 0.83-μm laser or 1.3-μm LED transmitter-based systems were within Cook's lightwave window.

The development of long-wavelength LED and laser sources designed to take advantage of fiber with zero chromatic dispersion was accompanied by the design of sources with a large modulation bandwidth. Surface-emitting LEDs typically have bandwidths of a few hundred megahertz and permit practical systems data rates of a few hundred megabits per second and MM repeater spans of tens of kilometers (Lee et al., 1988). EE-LEDs also have bandwidths of several hundred megahertz. Injection lasers are inherently high-speed devices, with modulation bandwidths bounded by the relaxation frequency, which increases with output power of $\sim 3-5 \text{ GHz}/\sqrt{(\text{mW})}$ (Bowers and Pollack, 1988). Typical bandwidths of long-wavelength lasers today are ~ 10 GHz, with resonant frequencies as high as 24 GHz reported

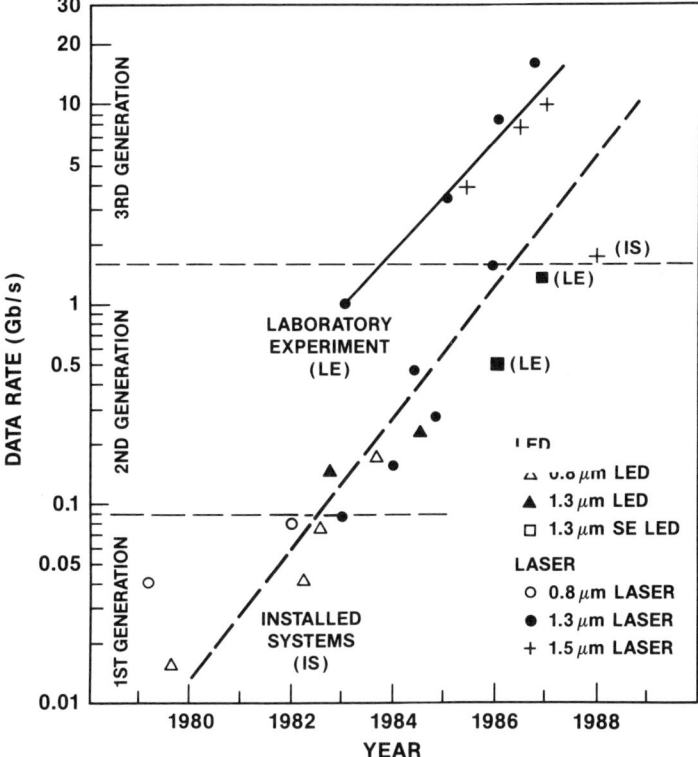

FIG. 10. Progress in lightwave transmitter data rate. Points show earliest installed optical system, benchmark laboratory demonstration, and year.

(Yausa et al., 1986). The impact of these efforts on source bandwidth may be seen in Fig. 10 (Goodfellow et al., 1985). Data transmission rates of lightwave systems versus installation rate show an increase in data rate of nearly two orders of magnitude in the last decade. Laboratory experiments show data rates almost an order of magnitude higher, with 16 GB/s operation being the fastest generated to data (Gnauck and Bowers, 1987). Circuits with the capabilities to provide modulation and bias control for these LED and laser sources will be discussed in Sections IV and VI.

III. Optical Properties of Sources for PCM Transmitters

We have discussed the system constraints reducing transmitter design in most cases to PCM of LEDs and lasers. This section will be a general treatment of the desirable characteristics of optical pulses for long-haul transmission systems, and how well they can be achieved by direct current modulation of such lasers and LEDs.

A. Optical Pulse Requirements

In the typical fiber optic link, shown in Fig. 5, the received pulse shape $h_p(t)$ is dependent upon both the transmitted pulse shape $h_{tr}(t)$ and the fiber impulse response $h_f(t)$:

$$h_p(t) = h_{tr}(t)h_f(t). \qquad (1)$$

The objective of the system designer, as we have seen, is to obtain an optical link with the highest (bit rate) × (fiber km) product at the lowest cost (i.e., the minimum dollar cost per Mb/s × km). A key step in this evaluation is the optimization of the receiver sensitivity of a chosen design to the convoluted pulse $h_p(t)$. Such analyses of the minimum received power detectable in digital receivers have been covered recently in excellent reviews (Personick, 1979; Smith and Personick, 1982).

The best pulse shape from the receiver point of view would be a high-power, narrow-width pulse. However, the transmitter designer usually limits the launched power, to ensure the reliability of the source, or to fit the constraints of the coupling efficiency of the source to the selected transmission fiber. Because optical transmitters are power limited, the optimal pulse width turns out to be full or 50% duty cycle (Smith and Personick, 1982). This compromise actually becomes an advantage in very high speed systems where the modulation bandwidth of the source can become a limiting factor.

Beyond power requirements, the transmitter designer, however, looks to the analyses of the receiver designer for detailed guidance on the other important launched optical pulse requirements. The desirable optical pulse characteristics to be specified for the launched optical data stream are (1) low pulse-to-pulse width variations, t_j; (2) fast rise, t_r, and fall, t_f, times; (3) high extinction ratio, ε (i.e., on/off ratio); (4) small amplitude variations and pattern dependence, ΔP_p; (5) low source noise, ΔP_n; (6) temperature and time stability of the optical power, P_{AV}, and of the parameters in (1) to (5). These parameters are shown schematically in Fig. 11.

However, before reviewing the digital circuits and strategies that have been suggested for producing pulses with these characteristics, we will review the direct current modulation characteristics of both lasers and LEDs.

B. Direct Current Modulation of Lasers and LEDs

A number of excellent books and review articles have covered the physical properties and device design of semiconductor lasers and LEDs (Burrus et al., 1979; Kressel et al., 1982) and their direct current modulation behavior (Arnold et al., 1982). Our objective here is merely to introduce those optical properties and the pulse response characteristics that have relevance to PCM optical transmitter design.

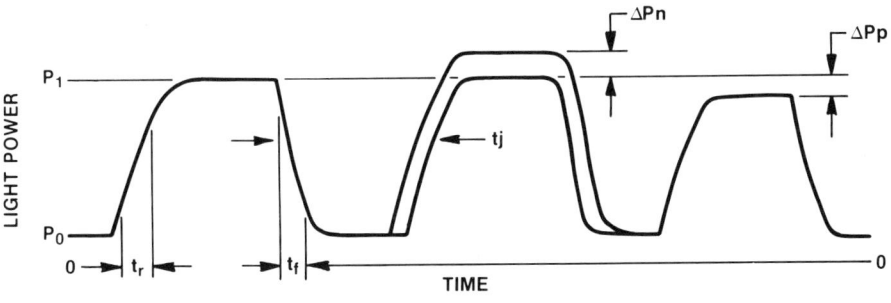

FIG. 11. Schematic representation of the objectives of a digital lightwave transmitter.

1. Light Emission in Semiconductors

Carriers excited to the conduction band of *direct* band-gap semiconductors, such as GaAlAs or InGaAsP, recombine by three competing processes: (1) nonradiative recombination, (2) spontaneous emission, and (3) stimulated emission of optical radiation.

In the case of LEDs, only the first two processes are operative, and the radiation is incoherent. This produces the wide source linewidth shown in Fig. 6a. The two types of LED geometries that have evolved for telecommunications use are the *surface* or *Burrus* emitter, and the *edge* emitter, with the latter structure being very similar in construction to double heterostructure lasers.

Although surface emitters can produce a peak power almost twice that of edge emitters, the greater directionality of the emitted beam of edge emitters produces radiance values almost an order of magnitude higher (Burrus *et al.*, 1979). This latter fact means that the optical power coupled into fibers is two to three times greater for the edge emitter. Typically, high-speed surface emitters of InGaAsP can only couple -13 dBm into 50-μm core multimode fibers (Conradi *et al.*, 1982), compared to -9 dBm for edge emitters (Botez and Ettenberg, 1979). Much work continues on increasing the coupling efficiency into fibers by such innovative techniques as growing an "integral lens" onto the emitting surface (Carter *et al.*, 1979).

Both LED structures produce almost linear electro-optic transfer functions, as shown in Fig. 12. Nonlinearities in the transfer function can occur at high currents because of both junction heating and carrier leakage (Kressel *et al.*, 1982). Also, edge emitters can develop a tendency to lase at low temperatures. However, the fact that these transfer functions are almost linear and have low temperature dependence and small variations with time (Lee, 1982) means that practical LED transmitter design is greatly simplified, as we shall see in Section IV.B and V.A.

In sharp contrast, double heterostructure lasers have device geometries with good carrier and photon confinement. When an inverted carrier population is obtained by carrier injection, and it is coupled with optical feedback from partially reflective facets, a Fabry–Perot cavity with optical gain and the ability to emit coherent stimulated radiation is produced. Lasers, then, have all three carrier recombination processes operative, which produces the high nonlinear electro-optic transfer functions shown in Fig. 12. This function is characterized by an almost linear region of incoherent spontaneous emission below the lasting threshold, I_{th}. Above I_{th}, the spontaneous emission is amplified, and the resulting coherent stimulated emission produces a second linear region of significantly greater slope, which is referred to as the differential or external quantum efficiency, η.

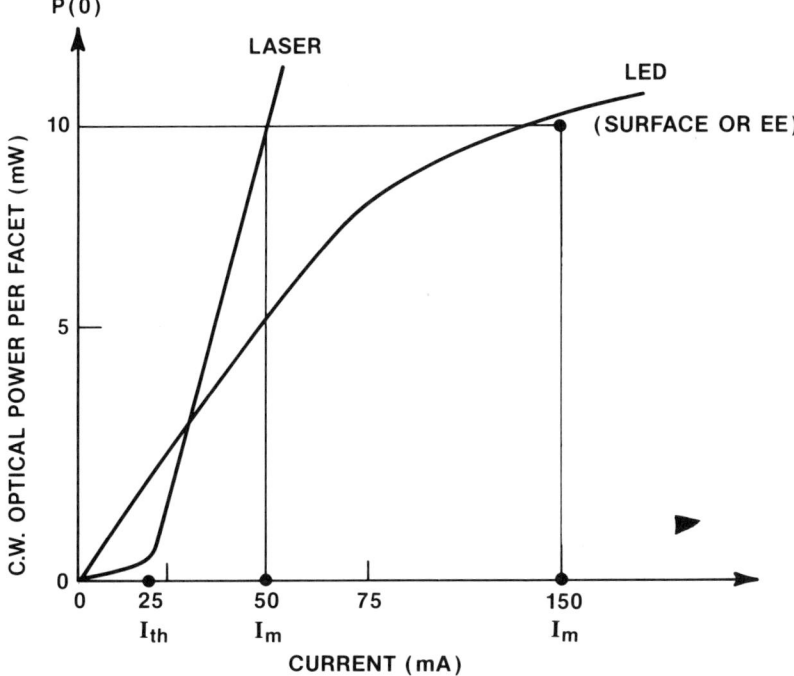

FIG. 12. Typical light–current characteristics of InGaAsP LEDs and lasers.

The lasting threshold transition is strongly dependent upon temperature and has been found to be represented by a characteristic temperature T_0 defined by

$$I_{th} = I_{th_0} \exp\left[-\frac{T}{T_0}\right]. \quad (2)$$

Not only is the threshold temperature-dependent, but I_{th} and η are subject to variations with both temperature and time. In fact, these variations limit the reliability of the device [Dixon, 1980).

Also, we shall see in Section IV.C that these changes strongly influence the transmitter biasing strategy. Much material and device design effort has been expended to minimize extrinsic nonradiative recombination processes, lower the room temperature I_{th}, increase T_0, and obtain η values independent of applied current, temperature, and time. As a result, two basic types of laser structures have emerged for communications applications, *gain-guided* and *index-guided* devices.

Practical first-generation lasers of AlGaAs have been gain-guided devices, in which there is only weak optical and current confinement in the lateral direction. Such stripe geometry lasers are simple to fabricate by growth techniques such as liquid phase epitaxy and with stripe delineation by proton bombardment (Dyment et al., 1972), oxide isolation (Dyment, 1967) or zinc diffusion (Yonezu et al., 1973). They are prone, however, to optical instabilities. Lasers with stripe widths greater than 10 μm exhibit kinks (Paoli, 1976), light jumps (Anthony et al., 1980), and self-pulsations (Paoli, 1977). These phenomena increase both the noise, ΔP_n, and t_j, and they can decrease the extinction ratio, ε (Tsang et al., 1983). Reducing the stripe width to below 5 μm suppresses these phenomena (Dixon, 1980). However, such lasers exhibit double-lobed far-field patterns (Petermann, 1981a) and often have asymmetric emission from the two lasing facets, creating problems for coupling into fibers (Marcuse and Nash, 1982). All these devices tend to lase with optical wavelength spectra containing multiple longitudinal modes, giving linewidths typically on the order of 2 nm (Streifer et al., 1982). The presence of source instabilities in gain-guided lasers, and their impact on system performance and reliability, has been one of the principal concerns of laser transmitter designers. These topics will be covered more fully in Section V.B.

The next generation of lasers have not only switched the material system to InGaAsP, but the device structure to *index-guiding*. In such devices, both current and optical confinement are created by having a real refractive index step in the lateral as well as the vertical direction. Device structures such as the buried heterostructure (Tsukada, 1974), channel substrate planar (Aiki et al., 1978), or buried cresent (Oron et al., 1983), although technologically more difficult to produce, have been grown with room temperature I_{th} in the range of 10 to 40 mA. In contrast, gain-guided AlGaAs lasers have I_{th} values of 60 to 100 mA. The search for the best long-wavelength laser structure and growth technique continues in many research laboratories around the world.

These index-guided lasers are free, however, from the optical nonlinearities and instabilities of the gain-guided devices and tend to lase in a small number of longitudinal modes (Streifer et al., 1982). However, their emission in a small number of longitudinal modes, and hence their greater degree of coherency than gain-guided lasers, raises the added complications of mode partition noise (Ito et al., 1977; Ogawa and Vodhanel, 1982) and modal noise if used in multimode fibers (Epworth, 1978). They also appear to be more sensitive to optical feedback (Hirota and Suematsu, 1979; Arnold, 1981). Modal noise and optical feedback effects both cause increases in ΔP_n. Mode partition noise at wavelengths with finite dispersion increases pulse width variation, t_j, and ΔP_n, in the received pulse.

A final added complication of the switch to the InGaAsP system is a significant decrease in T_0. For AlGaAs lasers, T_0 values as large as 200°C

(Ettenberg and Kressel, 1980) can be obtained, giving small values of dI_{th}/dt. However, for InGaAsP, T_0 values of 60 to 75 K are typical (Kressel and Ettenberg, 1982). It is believed that this is an intrinsic limitation caused by Auger recombination in these narrower band-gap semiconductors (Thompson and Henshall, 1980), although the role of carrier leakage, intervalence band absorption, and recombination at defects and impurities is still a strongly debated point (Sermage et al., 1982). Such competing nonradiative recombination processes produce a rapidly increasing I_{th} and decreased lasing efficiency above room temperature (Kressel and Ettenberg, 1982). Thermoelectric cooling of such devices must then be incorporated into laser transmitters used in practical systems, unless I_{th} can be made low ($\lesssim 5$ mA) and the slope efficiency made high (>0.1 mW/mA).

In general, however, room-temperature facet powers of 5 to 10 mW can be readily obtained from both types of laser structures and material systems. Also, since the emitting areas of these devices are less than 1 μm^2 ($<0.2 \times 5$ μm), and since the beam emission is highly directional, coupling efficiencies of approximately 50% can be attained in both multimode fibers and single-mode fibers. Thus, average launched power, in the range of -3 to 0 dBm, is readily available from laser transmitters without affecting the reliability of the laser source.

2. Pulse Response

There are significant differences in the pulse responses of LEDs and lasers, due primarily to the presence of the stimulated recombination processes in lasers.

The normalized output power, $P(\omega)$, of an LED, when modulated about a dc bias with a small ac signal of frequency ω, is given by

$$\frac{P(\omega)}{P(0)} \approx \frac{1}{[1 + (\omega t_c)^2]^{1/2}}. \tag{3}$$

The power suffers relaxation effects at $\omega \gtrsim 1/t_c$, as shown in Fig. 13a. This occurs at low frequencies because the carrier lifetime, t_c, is given by

$$\frac{1}{t_c} = \frac{1}{t_r} + \frac{1}{t_{nr}}, \tag{4}$$

where t_r and t_{nr} are the radiative and nonradiative lifetimes, respectively. When doping levels are low, t_c is dominated by the spontaneous radiative carrier lifetime $t_{sp} \cong 10^{-8}$ s $\ll t_{nr}$. The 3-dB bandwidth is thus limited to approximately 20 MHz. Attempts to improve the bandwidth significantly, by decreasing t_{nr} with heavy doping of the recombination region, causes a significant drop in radiant efficiency. Thus, increasing the bandwidth to 200 MHz

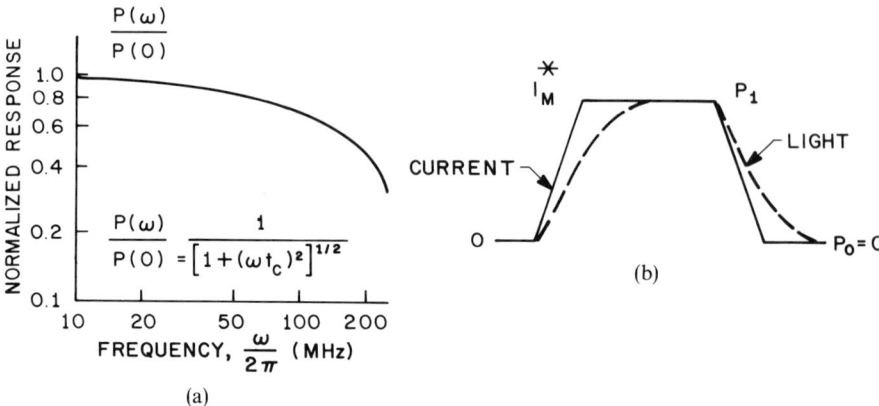

FIG. 13. Typical modulation characteristics of LEDs: (a) small-signal response; (b) pulse response.

in AlGaAs causes almost a five-fold decrease in output power available (Kressel et al., 1982). Similar behavior is observed in the InGaAsP system (Lee, 1982).

The corresponding pulse response is shown in Fig. 13b. The optical rise and fall times are seen to be limited by the bandwidth of the spontaneous emission characteristics, an obvious drawback for the LED transmitter designer.

In sharp contrast, lasers, when operated in the lasing region, have the small-signal response function shown in Fig. 14a. This suppression of relaxation effects to the gigahertz region is because of the very short carrier lifetime $t_c \approx 10^{-11}$ s, and photon lifetime, $t_p \leq 10^{-11}$ s, in the presence of stimulated radiation. Relaxation effects do occur beyond the resonance frequency of

$$\omega_r = \left[\frac{1}{t_c t_p}\left(\frac{I_m}{I_{th}} - 1\right)\right]^{1/2}, \tag{5}$$

where I_m is the dc current applied to bias the laser above its threshold, I_{th}. The details of the frequency response and the shape of the resonance can be quite different depending on the laser structure and the amount of spontaneous light coupled into the lasing modes (Figueroa et al., 1982; Tucker and Pope, 1983).

Under pulsed conditions, the result of the resonance effect causes the appearance of relaxation oscillations shown in Figure 14b, which die away as the carrier and photon populations come to equilibrium (Marcuse and Lee, 1983). As with the small signal resonance, these pulsed oscillations are damped

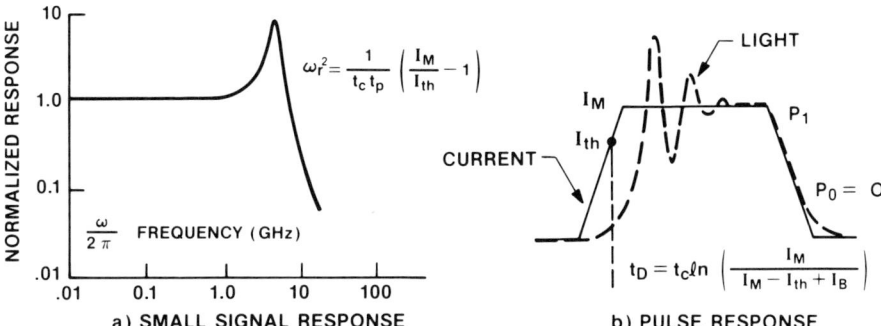

FIG. 14. Typical modulation characteristics of lasers: (a) small-signal response; (b) pulse response.

in lasers with large spontaneous emission contributions or high lateral carrier diffusion (Arnold et al., 1982). The variability of the modulation response of different laser types to a given current pulse is an added challenge to circuit designers. On the one hand, they desire fast-rising optical pulses, but on the other, they would like to avoid compromising the pattern dependence, ΔP_P, and the device reliability with large optical relaxation spikes. Relaxation oscillations become a limiting factor for the modulation rate of laser transmitters as the data rate approaches the relaxation frequency, ω_r.

In Fig. 14b, another effect is observed when the pulsed modulation is from a zero-bias condition, $I_B = 0$. The length of the optical pulse is shorter than the current pulse by a time typically of several nanoseconds; it is given by

$$t_d = t_{sp} \ln\left[\frac{I_M}{I_M - I_{th} + I_B}\right], \qquad (6)$$

where I_M is the amplitude of the current pulse. The "timing delay" gives rise to both pattern dependence, ΔP_P, and pulse jitter, t_j, effects (Garrett and Midwinter, 1980). This delay also disappears when the laser is dc biased at $I_B = I_{th}$. However, since I_{th} is strongly temperature dependent, the transmitter designer has to provide not only a circuit with dc bias to reduce timing delay, but also some form of feedback control. This is necessary to eliminate an average power decrease and to minimize changes in the timing delay with both time and temperature variations of I_{th}.

C. Summary of Source Properties

The AlGaAs and InGaAsP LEDs offer inexpensive, reliable sources requiring relatively simple transmitter circuitry, when compared to lasers.

These devices find applications in a rapidly expanding data link market. The 0.8-μm devices typically operate up to 50 Mb/s over a few kilometers span length of MM fiber. The 1.3-μm LEDs can operate with bit rates up to a few hundred megabits per second and transmission distances below 20 km.

The AlGaAs lasers provided the first generation of optoelectronic devices for telecommunications applications using MM fiber. Data rates of 90 Mb/s and 7 km span were achieved. The advent of single-mode fiber with low loss and zero chromatic dispersion at 1.3 μm, and index-guided 1.3 μm multifrequency lasers, extended the performance range of the second-generation systems to nearly 10 Gb/s km. The current single-frequency DFB lasers at 1.3 and 1.55 μm have a performance potential of 900 Gb/s km. The long-wavelength lasers have a somewhat limited temperature range of operation, and thermoelectric heat pumps are generally used to both cool and heat the laser in order to extend the operating temperature range.

IV. Circuit Strategies for PCM Transmitters

This section will outline the basic circuits and strategies to produce optical pulses by direct current modulation of the LED or laser source. Our review will be directed at circuits capable of producing pulses of the quality depicted in Fig. 11, at data rates up into the gigabit-per-second range. Emphasis will be placed on circuits that can retain the pulse properties with both time and temperature changes of the optical sources.

A schematic of a complete temperature-controlled laser transmitter is shown in Fig. 15. It serves to present an overview and show the complexities that the transmitter designer encounters. It consists of three sections. First, the modulator section (a) provides the high-speed current pulses to the optical source, (b) has a buffered data input stage to accommodate a range of data signal amplitudes, and (c) supplies a data reference signal that linearly adjusts the feedback circuit to compensate for duty cycle variations of the data signal. For the LED source, the entire transmitter consists only of parts (a) and (b) of the modulator section. Second, the feedback circuit provides automatic power control, laser safety to inhibit the laser bias during power-up, and band-gap-regulated current sources for modulation, temperature, and data reference stability over voltage and temperature variation. Third, the proportional temperature control circuit for the thermoelectric heat pump (TEHP) heats and cools the laser to 20°C.

This section is divided into five parts describing modulator circuitry, the simple biasing circuits for LEDs, the more complex biasing circuits for lasers, and source temperature control. The concluding part will describe recent developments in equivalent circuit models for lasers.

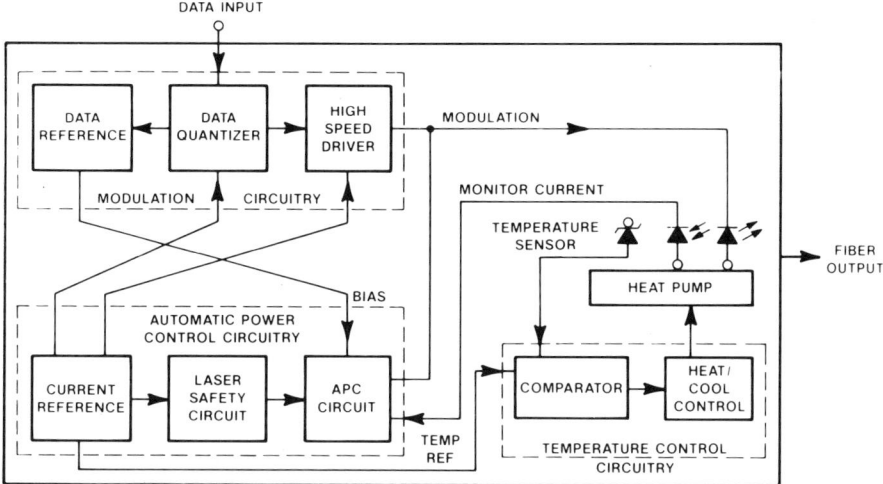

FIG. 15. Schematic of temperature-controlled laser transmitter.

A. Modulator Circuits

The requirements of a modulator are that it should be capable of delivering a current pulse to a laser up to 80 mA with rise/fall time of $T/3$, where T is the pulse width of a single bit. An LED requires a current pulse up to 150 mA. Several digital circuit designs have been used to produce such direct modulation of LEDs and lasers (Shumate and DiDominico, 1982; Shumate, 1988). There are (1) the common emitter saturating switch; (2) the shunt driver; (3) the common collector (or emitter follower) driver; and (4) the emitter-coupled or current routing circuit. Circuits (1), (2), and (4) switch current through the device, while (3) supplies a voltage step. Circuits (2) and (3) are considered to be low impedance drivers. The first three circuits, (1), (2), and (3), have limitations that make them unsuitable for practical high-speed transmitter design.

The common emitter saturating switch, shown in Fig. 5, is not adequate for high-speed operation. In general, its modulation bandwidth has to be traded to obtain adequate current gain; the bandwidth is also limited by the delay time required to remove charge from the collector to base capacitance, as the transistor is brought out of saturation. Use of Schottky clamps between collector and base, although effective, can lead to high-frequency noise transients on the power supply rails (Shumate and DiDomenico, 1982).

The shunt driver (Jarrett, 1974; Ostoich et al., 1975), which operates by shunting current around the optical device, is capable of modulating lasers at 1 Gb/s (Abbott et al., 1978). However, this circuit is not suitable for practical transmitters when a temperature-dependent or feedback-controlled dc bias has to be added to the laser or LED transmitter.

The emitter follower driver (White and Burrus, 1973), which attains a high modulation rate by providing a low impedance voltage step, has some serious limitations for practical design. Not only is it prone to oscillate at a high data rate if the emitter load is not purely resistive (Chessman and Sokal, 1976), but temperature compensation of the driver is quite difficult (Shumate and DiDomenico, 1982).

The most versatile of the modulators is the *emitter-coupled current routing switch*, shown in Fig. 16. It is a variation of the common emitter driver, although it resembles a linear differential amplifier. However, it is operated in the switching mode and, despite being overdriven at the input, remains out of saturation. This results in very fast switching speeds. Using Si bipolar devices, a similar modulator has been demonstrated to switch 60 mA at speeds slightly in excess of 1 Gb/s (Gruber *et al.*, 1978; Bosch *et al.*, 1984). More recently, a similar design in fine-line NMOS (Swartz *et al.*, 1986; Yanushefski *et al.*, 1988) has shown 2 Gb/s performance. Chen and Bosch (1988) tested a GaAs MESFET current routing switch modulator up to 4 Gb/s. The Si bipolar and GaAs MESFET IC technologies continue to improve in performance. Si bipolar ICs should be capable of attaining 3 Gb/s operation, and GaAs MESFET ICs will be required for the very high bit rates above 3 Gb/s. This current routing circuit offers several distance advantages.

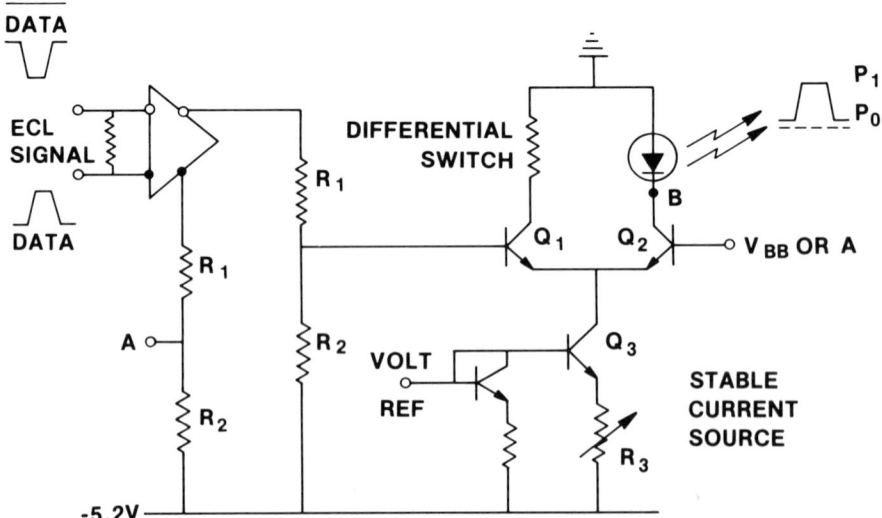

FIG. 16. A basic direct-current modulator, the emitter-coupled current routing switch.

1. Since it has constant current source, Q_3, no significant transients are placed on the power supplies during switching. The modulation current can be readily adjusted by R_3, or temperature compensated at the "volt ref" node without affecting the switching operation. The magnitude of the current available from Q_3 is limited by the 1.5 to 2.5 volt V drop required across the optical diode.

2. Direct current biasing of the optical diode can be readily added at the *n* contact of the optical diode, node B. Since Q_1 and Q_2 form a nonsaturating current switch, Q_2 is always in its active region, and its switching action is unaffected by the added dc current source. Such biasing circuits are discussed in detail in the third part of this section.

3. The differential switch can be used with an input logic gate (e.g., an ECL line receiver that is compatible with Si bipolar transistors). This is very convenient, as this gate can be used as a line receiver for the coded data stream and to compensate for any initial timing delay in either LED or laser transmitter circuits.

The circuit in Fig. 16 is shown with an ECL gate and level shifting resistors R_1 and R_2 (high-speed diodes could be used). This simultaneously satisfies the requirements for a "high" to "low" voltage swing of -0.8 V to -1.8 V for ECL logic, an approximately 2 V drop across the optical diode, and the grounding of its *p* contact. Such level shifting is not necessary if the *p* contact can be connected to $+V_{cc}$.

The transistor, Q_2, which has the optical diode in its collector circuit, can be switched against a fixed voltage, V_{BB}, set between the "high" and "low" logic levels. Alternately, it may be switched with a signal complementary to that imposed at the base of Q_1 (e.g., "node A"). This produces faster switching; however, care must then be taken to avoid "ringing" of the switched current pulse.

B. *Biasing Circuits for LEDs*

The biasing circuits for LEDs, added to the basic modulator, are relatively simple. Their functions are twofold: (1) to speed up the rise and fall times of the pulse response shown in Fig. 13, and (2) to compensate for the 2 to 4 dB decrease in optical power observed at constant modulation current, I_M, as the temperature increases from 0 to 60°C.

Some improvement in rise time can be obtained by supplying a small forward dc bias to keep the space charge capacitance of the diode, C_s, charged at all times (Lee and Dentai, 1978). This avoids a delay in the carrier injection into the active region, and hence a delay in the light output (Lee, 1975). The small current required, ~1 mA, has a negligible effect on the optical extinction ratio, ε.

An even greater improvement in the rise and fall times can be achieved by adding *current peaking* and *charge extraction* circuits to the basic modulator, as shown schematically in Fig. 17a. These circuit functions produce a *preshaped* driving pulse to the LED with the current waveform shown in Fig. 17b. It can be seen that the current peaking circuitry raises the current in the diode to a value, I_P, larger than the final desired level, I_M, for a short period of time, followed by a controlled decay back to the desired level.

A theoretical justification for such current peaking has been derived (Zucker and Lauer, 1978). They analyzed the transient response of an LED and were able to relate the rise time of the spontaneous emission to the device parameters and the characteristics of the current pulse. When a current step of amplitude I_P, is applied to the diode, it is assumed that it divides into a diffusion current and a capacitive current. This is represented by a parallel combination of a diffusion capacitance, C_d, and the space charge capacitance, C_S. In this treatment, C_S is considered to be constant. The calculated 10% to 90% rise time, t_r, is then given by

$$t_r = \left[t_c + \frac{2C_S}{BI_P} \right] \ln 9, \tag{7}$$

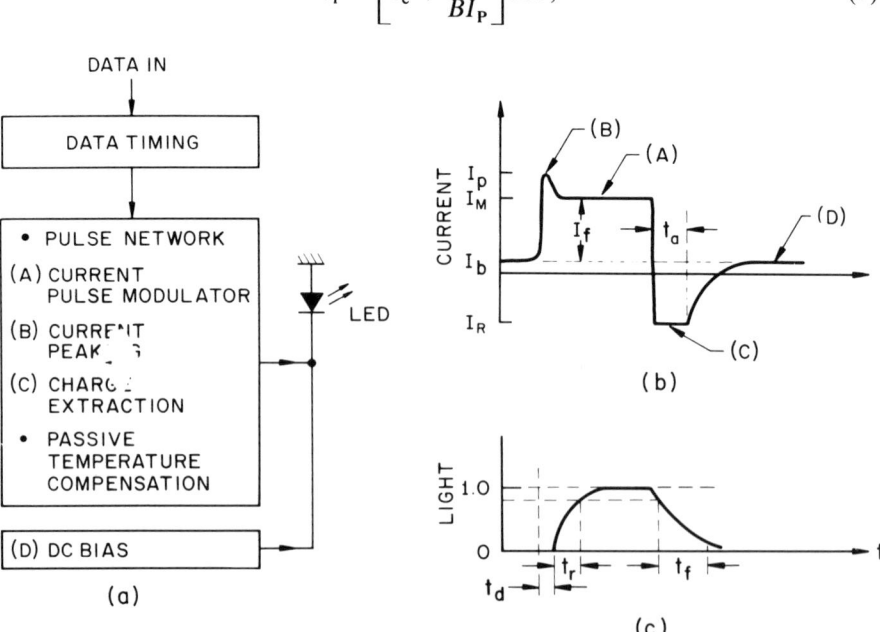

FIG. 17. An LED transmitter with current peaking and charge extraction circuits: (a) schematic; (b) pulse waveform of the diode current; (c) resultant waveform of the light output.

where t_c is the effective carrier lifetime, and $B = q/2kT$, where q is the electron charge, k is Boltzmann's constant, and T is temperature in Kelvin. This expression shows that the rise time decreases with increasing current step I_P and justifies the use of an external means, such as current peaking, to reduce the rise time.

Similar analyses of the optical fall time, t_f, of the LED (Uhle, 1976) show that it is of the form

$$t_f = t_c l_n \left[1 + \left(\frac{I_M}{I_R}\right)^2 \middle/ 1 + \frac{2I_M}{I_R} \right]. \tag{8}$$

Consequently, the fall time can be made considerably smaller than the effective carrier lifetime, t_c, by forcing a large reverse current, I_R, through the diode. This reverse current, applied for a short time, t_a, after the driving pulse, rapidly sweeps out the remaining injected carriers instead of allowing them to recombine more slowly by emitting spontaneous radiation.

As the observed rise time of a wideband Burrus LED is 1.5 to 2.5 times shorter than the fall time (Dawson, 1980), the maximum usable bit rate of an LED will depend primarily on the success of the techniques used to reduce the turnoff time to less than that controlled by the material recombination time. The application of a reverse bias at the on–off transitions has been used to increase the bit rate of an LED from 200 Mb/s to 300 Mb/s (Dawson, 1980). In this case, a reverse bias pulse of 3 V, and 3 ns duration, was delivered to the LED at turnoff. The initial LED rise and fall times were 1.3 ns and 3.5 ns, respectively, using normal drive. However, the fall time was improved to 1.5 ns by use of the reverse bias pulse. Similar techniques have been used to generate a 4 Gb/s pulse rate in a laser diode (Lau and Yariv, 1980).

Control of the average power with increasing temperature is possible by making the modulation current source of Fig. 16 increase by the amount necessary to overcome the decrease in slope efficiency of the diode. Such *predictive* compensation can be used because of the uniformity of the temperature dependence of the $L-I$ characteristic from device to device, and its invariance with time.

The results obtained with biasing circuits that accomplish these improvements in rise and fall times and temperature stability of the average optical power are described in Section V.A.

C. Biasing Circuits for Lasers

The biasing circuits for laser transmitters, added to the basic modulator of Fig. 16, are more complex than those used for LEDs. This arises because their objectives are greater and include attempts to reduce timing delay t_d, timing jitter t_j, and pattern dependence ΔP_P, as well as to control changes in the extinction ratio ε and average power P_{AV} with both temperature and time.

The circuits discussed in this section do not meet all these objectives simultaneously. We attempt to point out the limitations of each circuit strategy, without recommending any particular approach. The best choice of the biasing strategy to be adopted is complex. It depends to a great extent upon the data rate needed, the desired transmitter cost, and the type of packaged laser chosen and its initial and aged light-current characteristics. Some aspects of the ramifications of biasing strategies have been discussed by Garrett and Midwinter (1980), Swartz and Wooley (1983), and Dean and Dixon (1984).

Some early circuit designs, for data rates below 10 Mb/s (Salter *et al.*, 1977), used a fixed zero dc bias, $I_B = 0$. Feedback control was added to the modulation current, I_M, to account for changes in the threshold current, I_{th}, and differential quantum efficiency, η, with time and temperature. The limitations of such a bias strategy have been analyzed by Garrett and Midwinter (1980). Most importantly, timing delay, t_d, is approximately 1 ns for every 10 mA the bias point is below I_{th} (Salter *et al.*, 1977), making such a bias method inappropriate for data rates of interest to us, using lasers that typically have threshold temperature dependencies as large as 0.5 mA/°C, and aging rates of 1%/hr.

All the biasing circuits for use above 10 Mb/s, which we shall now describe, use a dc bias current, I_B, chosen to be close in value to I_{th} (see Fig. 18). This dc

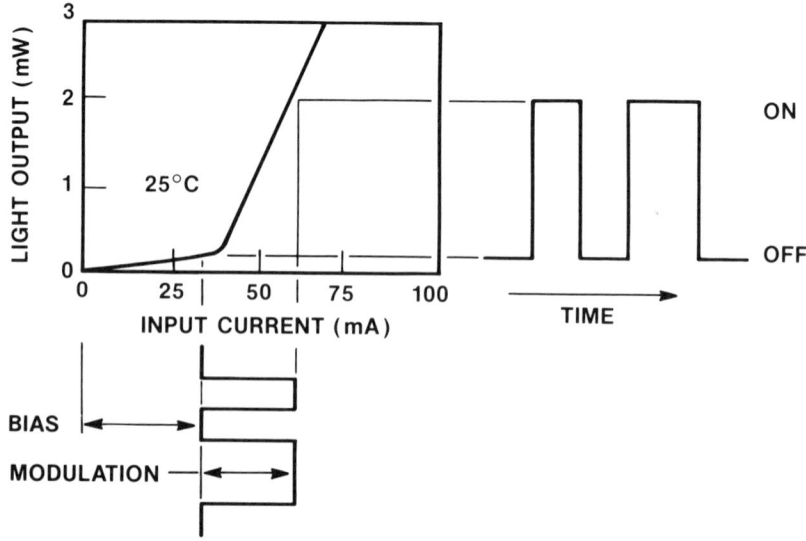

FIG. 18. Laser and light–current characteristics with bias (I_b) and drive current (I_{mod}) superimposed.

bias is always under some form of feedback control to ensure that the ratio of (I_B/I_{th}) does not deviate significantly from unity when either I_{th} or η change. The desired optical pulses can then be obtained by superimposing upon I_B a *fixed* pulsed current, I_M, from the modulator. In some circuits, *double-feedback* control is used by including feedback control to I_M, at the dc current source of the modulator ("volt. ref." in Fig. 16).

1. Biasing Conditions

Before discussing these single and double feedback control circuits, we must discuss the ramifications of the choice of the bias point, I_B/I_{th}. The choice of I_B/I_{th} is important because it involves a compromise of the ideal objectives, stated earlier, for the laser biasing circuits, and also because it partly dictates the feedback strategy to be used. There are three possible choices for I_B/I_{th}.

a. $0.80 < I_B/I_{th} < 1$ This produces the highest extinction ratio, $\varepsilon = P_1/P_0$, but at the expense of increased timing delay, t_d, and its attendant pattern dependence, ΔP_P (Gruber et al., 1978). It is usually the bias point of choice for single feedback loop strategies with only average power, P_{AV}, control. This is because it allows the greatest margin for changes in η with time, and hence ε when I_M is fixed.

b. $I_B/I_{th} = 1$ Reduced ε is obtained, especially in some gain-guided lasers with soft lasing transitions. This arises in these lasers from a large spontaneous contribution to the longitudinal modes and high lateral carrier diffusion (Petermann, 1981b). Timing delay and pattern dependence, which are important at high bit rate, are eliminated; but the maximum intrinsic laser noise is observed at this biasing point. This can be particularly severe in index-guided lasers with a high degree of coherency, which makes them particularly susceptible to optical feedback (Arnold, 1981). This is unfortunate, as this biasing condition is ideally suited for index-guided lasers with sharp lasing thresholds, because there are double feedback circuitry designs that can easily track the time and temperature change of I_{th} in this case (Smith and Hodgkinson, 1980).

c. $1 < I_B/I_{th} < 1.20$ Since lasers are power limited, this obviously produces the lowest extinction ratio. On the other hand, time delay is absent, and intrinsic laser noise is reduced. The modulation bandwidth is increased, but relaxation oscillations can become particularly severe, causing a new patterning effect, ΔP_P, especially in index-guided lasers with little spontaneous damping off the relaxation mechanism (Arnold et al., 1982). These relaxation oscillations can be the limiting factor in very-high-speed laser transmitters. Biasing above threshold makes the use of double feedback control of both I_B and I_M virtually a necessity to avoid decreases, in an already low ε, as η

decreases with time (Swartz and Wooley, 1983). As we shall see, this biasing condition is the simplest to implement when double feedback control circuits, which aim at controlling two of the three parameters P_0, P_1, and P_{AV}, are used.

The biasing condition also affects the degree of coherency of the light from pulsed lasers. This is the case for both gain- and index-guided structures (Petermann, 1981b). For $I_B/I_{th} < 1$, the spectral width is increased and the coherency is significantly reduced, which is helpful in reducing modal noise in multimode fiber systems (Epworth, 1978). On the other hand, this biasing condition can be a drawback in dispersion-limited fiber systems using GaAlAs lasers at 0.8 to 0.9 μm (Baack et al., 1978). However, the advent of very low dispersion monomode fibers at 1.3 μm has lessened the pressure to bias at $I_B/I_{th} > 1$ to reduce the spectral line width. The exception, of course, is for the very longest transmission distances, where fiber dispersion can still play a role. The ability to tolerate wider spectral widths at the 1.3 μm wavelengths means that biasing at $I_B/I_{th} < 1$ is almost always acceptable. This also has the added advantage of helping to reduce the reflection sensitivity of these narrow-linewidth index-guided devices.

2. Feedback Strategies

Four optical feedback concepts have been proposed. The simplest is (1) *average power*, P_{AV}, control, which has a *single* closed feedback loop on the bias current, I_B. The other three strategies have *double* feedback control. In addition to feedback control on the bias current, these designs have a second feedback loop on the pulse current, I_M, which allows control of (2) *extinction ratio* ε, (3) *timing delay* t_d, or (4) the *slope change at* I_{th}. The second loop function of strategies 3 and 4 is indirectly to control the "off" light level, P_0, and hence the extinction ratio.

All four strategies require an optical detector at the rear (or unused) facet of the laser (Shumate et al., 1978), or an optical detector in the ouput fiber cable (Karr et al., 1979). The bandwidths of the detector and the circuitry for designs 1 and 4 are not critical. However, for designs 2 and 3, the bandwidths for the second loop must be at least comparable to that of the transmission system in which the transmitter is used.

a. Average Power An average power feedback control circuit is shown schematically in Fig. 19 in conjunction with the emitter-coupled modulator circuit of Fig. 16. The function of the operational amplifier, A_1, is to compare the average current of the monitoring photodiode ("dc content") with a preset reference current I_{REF}. When the optical output changes, then A_1 changes the dc biasing current, I_B, from the current source, Q_4, until the output power level is re-established. In practical circuits, I_{REF} is derived, as shown, from a second emitter coupled switch or *data reference*. This ensures that the setting

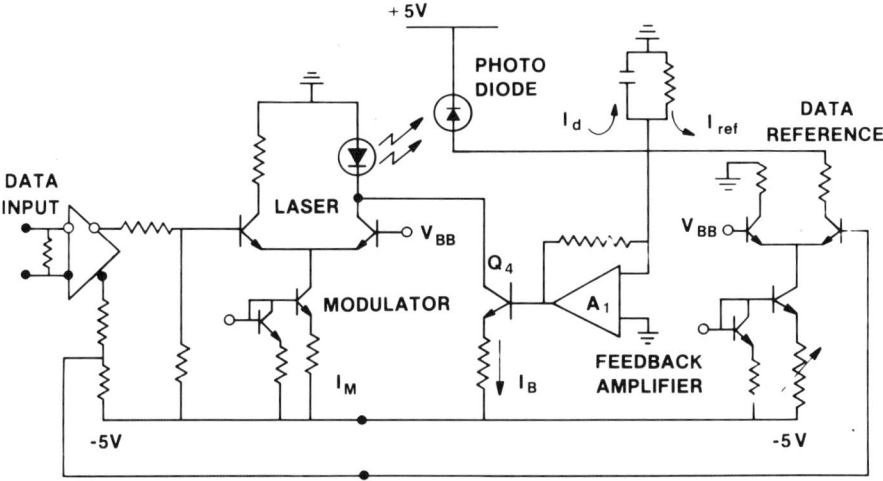

FIG. 19. Simplified circuit representation of a single-loop feedback transmitter for average optical power control.

of I_B is invariant to random fluctuations of the data stream or the duty cycle (Shumate et al., 1978).

This circuit, although very attractive, has some serious limitations that depend upon the individual temperature and aging characteristics of the laser (Dean and Dixon, 1984). Although the average power can be well controlled at all data rates, the extinction ratio is not. Changes in both I_{th} and η with temperature and time mean that the changing value of I_B produces a varying "off" light level, P_0, and with it a varying extinction ratio, $\varepsilon = P_1/P_0$. In general, ε decreases with both increasing temperature and time, and so to establish adequate system margin, the initial biasing condition must be $I_B/I_{th} < 1$. This circuit produces both timing delay, t_d, and rise-time, t_r, variations as I_B/I_{th} varies and approaches unity with time.

b. Extinction Ratio Several workers have proposed circuit designs to control the extinction ratio by using the "ac part" of the photodiode signal. Both P_1 and P_0 (Salter et al., 1977; Chown et al., 1980), $(P_1 - P_0)$ and P_0 (Gruber et al., 1978), or $(P_1 - P_0)$ and P_{AV} (Chen, 1980) have been extracted from the ac signal and used to control I_M and I_B, respectively, in double feedback loops.

A double feedback loop, similar to that described by Chen, that extracts $(P_1 - P_0)$ and P_{AV} is shown in Fig. 20. The monitoring photodiode is buffered by the transimpedance amplifier A_1. Its dc output controls I_B, and hence P_{AV}, via the difference amplified A_2 in the manner described for single-loop control.

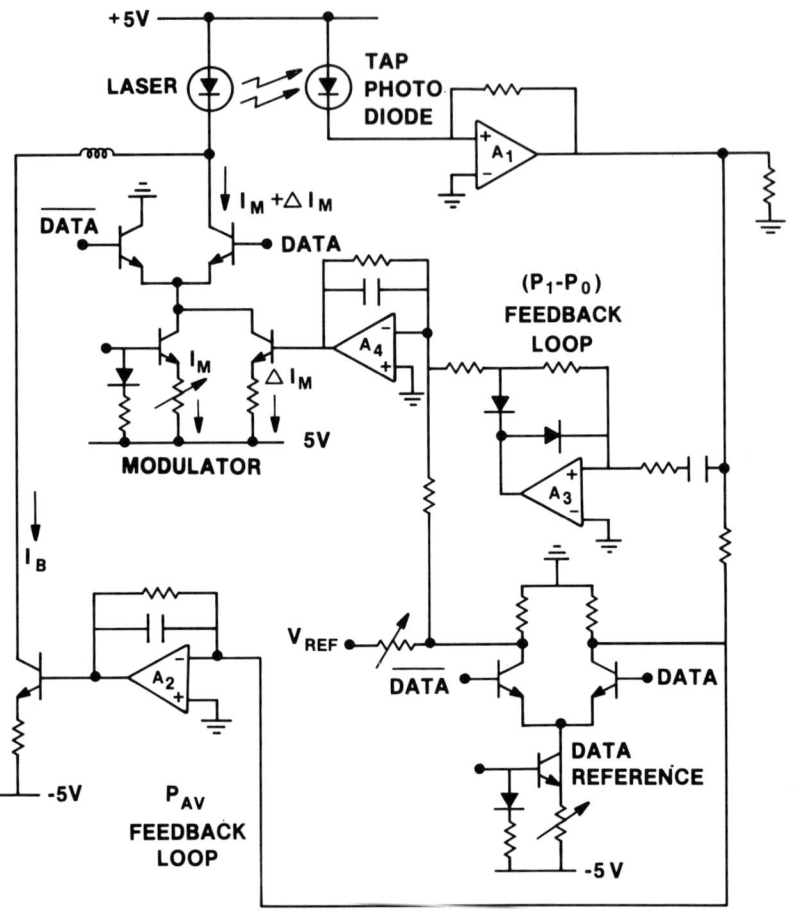

FIG. 20. Simplified circuit representation of a double-loop feedback transmitter for extinction ratio control from a front facet optical tap.

The quantity $(P_1 - P_0)$ is extracted from the output of A_1 by ac coupling into the operational amplifier A_3. This amplifier has high-speed Schottky diodes in its feedback loop that half-wave rectify and dc restore the "one" minus "zero" signal. This restored signal is then compared to a fixed "data" reference at the operational amplifier A_4. The output of A_4 drives a second modulation current source, ΔI_M, connected in parallel with a fixed modulation current source, I_M. Changes in ε with time and temperature are slow; thus the response time of A_4 is kept significantly longer than that of A_2.

The data reference in both feedback loops ensures that the circuit will operate under all duty cycle conditions, a disadvantage of some of the other double loop designs. However, in all these circuits, when no data is present (i.e., all zeros), the ac coupled loop is essentially "open." Care must be taken to bias A_4 such that ΔI_M decreases under such conditions. Otherwise, excessive current will be applied to the laser when the data is re-applied, with potentially disastrous results. Even large changes in I_{th} and η do not change ε significantly in this design because

$$P_{AV} = \sigma P_1 + (1 - \sigma)P_0 = \beta,$$

$$(P_1 - P_0) = \alpha, \tag{9}$$

$$\text{giving } \varepsilon = P_1/P_0 = \left[\frac{\beta + \alpha(1 - \sigma)}{\beta + \alpha\sigma}\right],$$

where σ is the duty cycle of the optical data stream, and α and β are constants.

The extinction ratio is not exactly constant with either temperature or time if $I_B/I_{th} < 1$, as a decrease in timing delay alters σ as I_B approaches I_{th} under the control of the average power loop. If $I_B/I_{th} \geq 1$, the stability of ε is exact, as σ is then a constant under all conditions (Swartz and Wooley, 1983). However, biasing at or above I_{th} can produce intolerably small ε values with some laser structures.

Although such circuits have been demonstrated to 280 Mb/s (Gruber et al., 1978), their major limitation is the need for wideband amplifiers. For example, amplifiers A_1 and A_3, in Fig. 20, must be broad-banded to retain the $(P_1 - P_0)$ information under all data-stream conditions. This increases circuit complexity, power supply drain, and transmitter cost, and potentially reduces reliability.

c. Timing Delay A double feedback strategy that is even more demanding of amplifier bandwidth is the timing delay technique (Salter et al., 1977). This circuit controls the P_0 value by developing a feedback signal proportional to the time difference between the electrical and optical signals—i.e., the timing delay, t_d. This is accomplished by using an "exclusive-OR" logic device, gated by the electrical "0" to "1" transitions and the corresponding time-delayed "0" and "1" optical transitions. The resultant voltage spikes are integrated and compared to a reference voltage obtained by integrating fixed-width reference pulses derived from the input data stream. The error signal produced is used to control I_B.

This technique thus provides an accurate means of tracking I_{th}, by keeping t_d, and hence $(I_{th} - I_B)$, constant. Average power and extinction ratio control are obtained by using the dc content of the photodiode current to control I_M. The technique has been demonstrated up to 140 Mb/s in both NRZ and RZ pulse format.

Again, the major drawback of this technique is the need for amplifiers and logic gates with very wide bandwidth and high power drain. The reference pulses, obviously, have to be significantly shorter than the data pulses, especially if I_B/I_{th} is set close to unity. Also, the strategy does not give good control of the value of ε, if the value of P_0 changes with time at a fixed $(I_{th} - I_B)$ because of softening of the lasing transition (Dixon and Dean, 1981).

d. *Slope Change* A much simpler double-loop method for tracking I_{th} with temperature and time uses low-frequency monitoring circuits (Smith, 1978). The basic circuit principle is shown in Fig. 21. It is essentially a *slope detector* that takes advantage of the high degree of nonlinearity of the electro-optical transfer function of lasers. A low-frequency (LF) signal (e.g., a 1 kHz square wave with 1% of the amplitude of I_M) is superimposed on the data signal. A Schottky diode gate, M, triggered by the data stream, ensures that this LF modulation is applied to the "0" level only. The resultant LF-modulated photodiode signal is detected with a phase-sensitive detector (PSD) to eliminate laser noise contributions. The output of the PSD is compared with a reference voltage and is used to set I_M. The second loop controls the average power by setting I_B in the usual way.

Several variations of this basic LF technique have been implemented using modulation of both "0" and "1" optical levels (Smith and Hodgkinson, 1980). These latter refinements allow the technique to be used with a wider range of laser devices such as the varying lasing transitions found with gain- and index-guided lasers. Circuits have been demonstrated to control the extinction ratio ($\varepsilon = 4$) to an accuracy of 10%, at data rates up to 160 Mb/s, without excessive power-supply drain.

A limitation of this strategy, in common with the timing delay technique, is there no control of ε when the lasing threshold region changes with temperature or time. Although ideally suited for index-guided lasers, with very sharp lasing transitions, the biasing point is always fixed at $I_B/I_{th} = 1$, where the intrinsic laser noise and reflection sensitivity are greatest. This inflexibility in choice of biasing point can be a drawback in very high bit-rate systems, for long transmission distances, where the desired spectral linewidth can only be attained at $I_B/I_{th} > 1$.

One strategy for tracking I_{th} that does not use optical feedback has also been proposed (Albanse, 1978). This "automatic bias control" circuit uses the saturation of the junction voltage of the laser diode at the onset of stimulated emission as the control parameter. This "pinning" of the bandgap voltage is sensed in a bridge circuit, and the out-of-balance condition acts as the error signal that controls I_B. Although this circuit requires no photodiode monitor, in a real-world system, feedback to I_M would also be required to compensate for changes in η, and hence ε and P_{AV}, with time and temperature. This would require, at least, a slow photodetector.

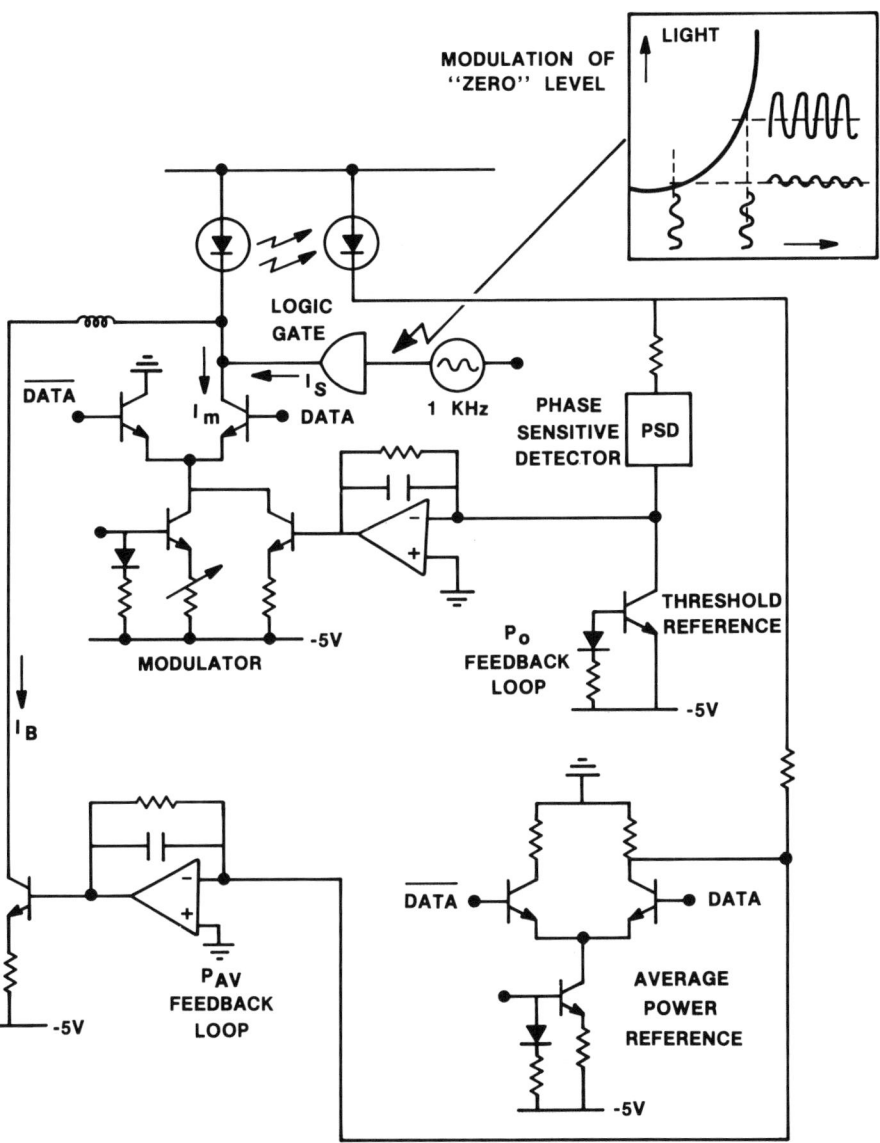

FIG. 21. Simplified circuit representation of a double-loop feedback transmitter with slope detector for tracking the threshold current.

D. Source Temperature Control

The characteristics of long-wavelength laser diodes have a strong dependence on temperature, including lasing threshold, lasing slope efficiency, aging rate, and center wavelength. For these reasons, some means of heating and cooling the laser diode to 20°C must be provided. A single stage thermoelectric heat pump, TEHP, can provide approximately 50°C of cooling and greater than 80°C of heating (Kraus and Bar-Cohen, 1983).

The TEHP is small, approximately 0.25 inches square by 0.08 inches high. The performance of an 18-couple single-stage TEHP is shown in Fig. 22. The laser temperature has been plotted against pump current for a series of heat loads. The hot side of the pump in this plot has been connected to an infinite heat sink at 80°C. The heat load on the TEHP is estimated to be 0.4 W (laser diode plus passive). At 1 A pump current, a temperature differential across the pump of 58°C can be achieved. The heating mode for the TEHP is shown in Fig. 23. In this mode, more than 80°C of heating can be provided.

A thermistor or a Si IC temperature sensor can be used to sense the laser temperature. A standard proportional control circuit is used to control the pump current. Typically the laser temperature can be held to within 2°C over

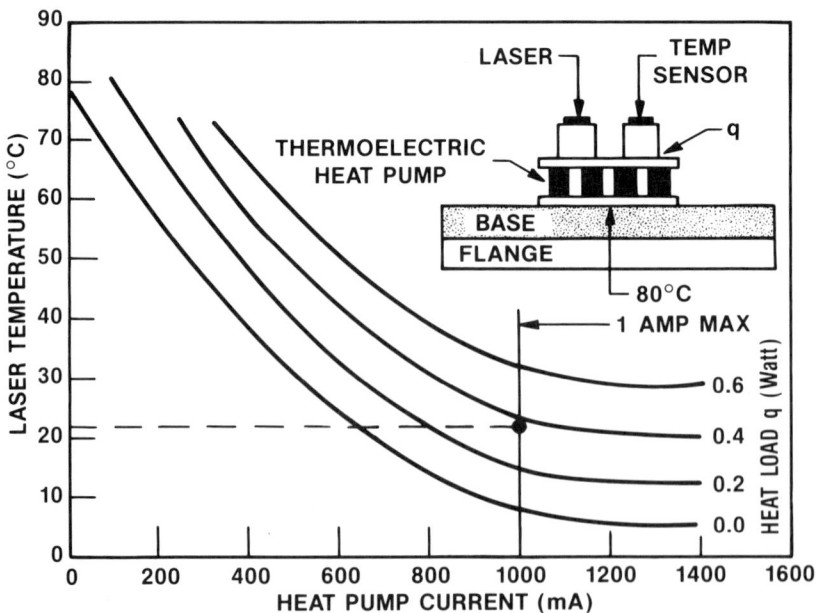

FIG. 22. Thermolectric heat pump ΔT-current characteristics in the cooling mode.

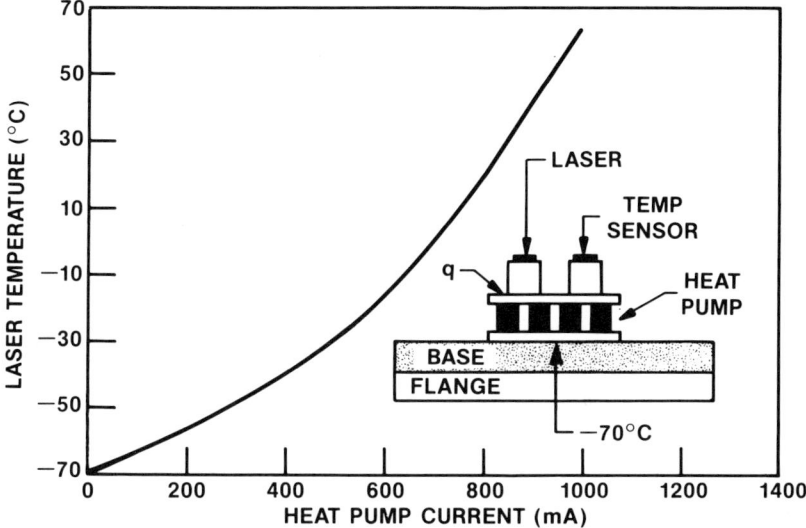

FIG. 23. Thermoelectric heat pump ΔT–current characteristics in the heating mode.

an ambient temperature range of $-40°C$ to $+65°C$. The TEHP is a high-current, high-power device, so care must be taken to heat-sink the hot side (case) of the pump adequately.

E. Equivalent Circuit Models

Many of the circuits just described were designed and evaluated using discrete components. As practical systems rapidly approach the Gb/s range, the demand for digital integrated circuits has intensified. Concomitantly, there is a rapidly increasing need for circuit models and nonlinear analysis programs to guide the development of such IC devices. In particular, there exists a need for accurate circuit models for different semiconductor laser structures and the influence of their packaging parasitics on transmitter performance.

Most derivations of equivalent circuit models of semiconductor lasers have, as a starting point, used the *single-mode* rate equations, with the additional approximations of homogeneous carrier inversion and linear gain. These two rate equations (Kressel and Butler, 1980) describe the dynamic characteristics of the electron inversion density, N,

$$\frac{dN}{dt} = \frac{I}{qv_a} - \frac{N}{t_{sp}} - GS, \tag{10}$$

and the photon density, S,

$$\frac{dS}{dt} = GS - \frac{S}{t_p} + \beta \frac{N}{t_{sp}}, \tag{11}$$

where I is the laser dc drive, q is the electron charge, v_a is the volume of the active region, t_{sp} is the spontaneous carrier lifetime, and t_p is the photon lifetime. The condition of linear gain is given by

$$G = \alpha(N - N_g), \tag{12}$$

where α is a constant of proportionality and N_g is the electron density at zero gain. The spontaneous emission factor β for a single mode is given by

$$\beta = \frac{\lambda^4}{2\pi^2 n^3} \frac{\gamma}{v_a \Delta \lambda}, \tag{13}$$

where λ is the operating wavelength, n is the refractive index, $\Delta\lambda$ is the bandwidth of the gain profile, and γ is the transverse optical confinement factor (Susmatsu et al., 1977). This term β acts as a damping term in the transient solutions of the rate equations. However, in general, β is used as an "effective parameter" to obtain better agreement with experimental results (Tucker and Pope, 1983). Greater values of β than those suggested by Eq. (13) are required because of the damping effects of multimode behavior (Danielsen and Mengel, 1978), lateral carrier diffusion (Furuga et al., 1978), gain saturation (Channin, 1979), and superluminescence (Lau and Yariv, 1980).

As the two coupled rate equations are nonlinear, approximate analytical solutions can only be obtained by linearizing them. In the *small-signal* limit, this can be accomplished, since N and S can be represented by dc values, N_0 and S_0, and small sinusoidal components, i.e., $N = N_0 - ne^{j\omega t}$ and $S = S_0 + se^{j\omega t}$ (Katz et al., 1981). This linear solution yields an intrinsic diode admittance above threshold, Y_d, which can be represented by the parallel combination of R, C, and L values shown in Fig. 24a. This admittance, given by

$$Y_d = \frac{1}{R_1} + j\omega C + \frac{1}{R_2 + j\omega L} = \frac{1}{Z}, \tag{14}$$

has components R_1, C, L, and R_2, which can be obtained from the intrinsic laser parameters t_p, t_{sp}, N_g, α, v_a, and β. Below threshold the laser term, $R_2 + j\omega L$, is large, and the admittance reduces to the shunt combination of R_1 and C used by previous authors in the spontaneous emission region (Lee, 1975; Dumant et al., 1980).

It has been shown (Tucker and Pope, 1983) that the small-signal light response, S, is merely proportional to the current through the complex admittance, Y_d. Thus the functional form of S with frequency can be derived

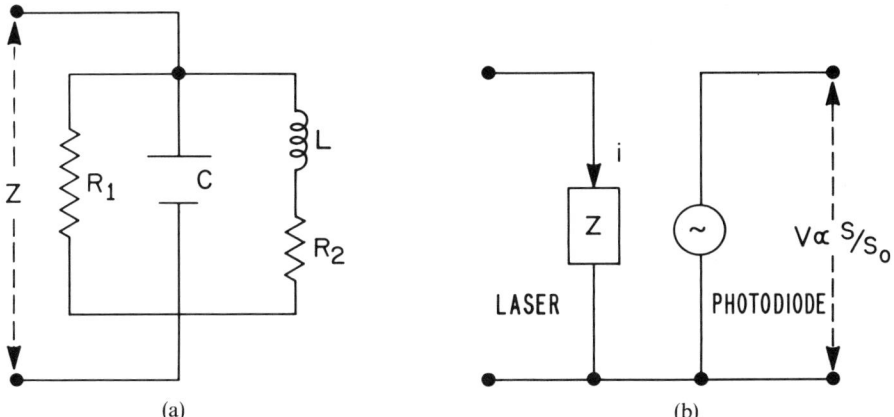

FIG. 24. (a) An equivalent circuit model for a laser (see Eq. 14). (b) Laser and photodiode combination represented as a current-controlled voltage source (Tucker and Pope, 1983).

simply as the voltage output of the current-controlled voltage source shown in Fig. 24b. The transimpedance of this source is $h = (qv_a G_0)^{-1}$, where G_0 is the dc component of the gain. The normalized light response (S/S_0) from this intrinsic model produces the flat response shown in Fig. 14a, with the peak at ω_r (given by Eq. 5) corresponding to the self-resonance of the complex admittance, Y_d (Katz et al., 1981).

Figueroa et al., (1982) has included the effects of package parasitics, and Tucker and Pope (1983), the effects of space charge capacitance, $I-V$ characteristics, and package parasitics. Both find that their modifications of the "intrinsic equivalent circuit" of Katz et al. (1981) more closely fit the empirical photon response functions of a number of different packaged lasers. Many of these devices show dips in (S/S_0) as large as 6 dB, at frequencies well below the resonance frequency, ω_r. Continued experimental and theoretical work in this area is necessary to identify these limiting contributions to the modulation bandwidth of high-bit-rate transmitters.

The only *large* signal model of a laser diode that has been published is that of Tucker (1981). It, too, is derived from the rate equations, (10) and (11), and yields a four-port model of the form shown in Fig. 24b. This, too, allows it to be used in transmitter circuit analysis as a device that simulates the input–output response of a laser–photodiode combination. However, the equivalent circuit elements of this large-signal model are nonlinear and require circuit analysis programs in which linearization is carried out numerically. Transient responses of the type shown in Fig. 14b are obtained, and the model does linearize to give the small-signal model shown in Fig. 24a. The availability of such a model allows a more accurate simulation of the closed-loop behavior of feedback strategies.

Tucker (1985) has augmented his earlier work to model the direct modulation performance of high-speed semiconductor lasers. A cascaded two-port model is used to characterize the laser. The model separates the electrical parasitics from the intrinsic laser and small-signal intensity modulation and frequency modulation response, and the large-signal switching transients and chirping are reviewed. Electrical parasitics associated with the laser package and chip can seriously degrade the modulation response at high frequencies. The bandwidth of the intrinsic laser is controlled by the relaxation oscillation resonance frequency and associated damping mechanisms. The dominant damping mechanism appears to be spectral hole burning. Development of low-parasitic chip and package design will lead to higher-speed performance.

V. Practical PCM Transmitters and Their Performance

In this section, we describe in more detail some of the problems experienced in designing transmitters for *real-world* systems. Of necessity, this section is not a survey but merely our collective design experiences. What is described is the evolution of transmitter design as the technology has shifted from the 0.83-μm wavelength window to that at 1.3 and 1.55 μm with improved fiber, detector, and source design and fabrication.

All the transmitter designs undertaken have had similar constraints, however. Available sources have been state-of-the-art, and their long-term packaged reliability under pulsed conditions is not well established. The desired data rates have also been at the limits of LED performance, although they have not come anywhere near the PCM capability of lasers. However, the optical power and pulse shape characteristics have been tightly specified to obtain the maximum transmission distance from the available fibers and receivers.

In two senses this section does have some generality. First, it draws from concepts and circuit design strategies discussed in the early sections of the chapter. Second, it illustrates the multiplicity of problems and trade-offs facing the optical transmitter designer each time a cost-effective, high-performance, high-reliability design is attempted in this evolving technology.

A. *LED Transmitters*

1. *0.87 μm—AlGaAs Sources*

In the 0.85–0.9 μm wavelength range, we previously indicated (Fig. 9) that a high fiber attenuation of about 6 dB/km, and a low chromatic dispersion limit of about 170 Mb/s × km, limit the usefulness of AlGaAs LEDs to short data links of a few kilometers and moderate data rates (≤ 50 Mb/s).

For PCM applications, these low to moderate bit-rate limitations can be conveniently handled with data derived from TTL logic inputs. Typical drive currents of 60 mA to 80 mA are required to produce the necessary peak power of approximately -13 dBm. Under constant current drive conditions, the output power decreases less than 0.012 dB/°C with increasing temperature, so compensation of the output power is *usually* not necessary. The 3 dB modulation bandwidths of these devices range from 70 to 80 MHz. This bandwidth corresponds to a rise time of approximately 4 ns (Gray and Meyer, 1977). This is more than adequate for bit rates below 50 Mb/s. Thus, any of the drive circuits discussed in an earlier review (Shumate and DiDomenico, 1982), particularly those that are TTL compatible, are adequate for these applications.

As we discussed in Section II.B, analog transmission requires large signal-to-noise ratios of >60 dB (Pan, 1981). In order to achieve this large signal-to-noise ratio, the $L-I$ characteristics of the LED have to be highly linear, with variations of $dL/dI < 3\%$ over the entire operating current range. The optical linearity with applied current is bounded at lower currents by turn-on effects in the device, and at high currents by the saturation effects in the diode. Over a limited current range, the $L-I$ characteristics of the AlGaAs LED are sufficiently linear for analog application. If we wish to operate the LED at higher currents, the LED output must be linearized with respect to drive current.

Several methods of linearization, including techniques based on complementary distortion (Asatanic and Kimura, 1978), feedback-selective harmonic compensation (Straus *et al.*, 1977), and optical feed-forward (Patterson *et al.*, 1979) have been used, but all these methods suffer from a generic problem. They require matching the LED characteristics either to a diode network or to another LED, which is extremely difficult as a function of both time and temperature. It remains to be seen if any of these approaches can be realized for system applications.

2. 1.3 μm InGaAsP Sources

In contrast to the short-wavelength region, LED systems in the 1.3-μm wavelength range have low fiber loss of less than 1 dB/km and a high chromatic dispersion limit of about 2.5 Gb/s × km. Repeater spacings of tens of kilometers and data rates of hundreds of megabits per second are feasible (Gloge *et al.*, 1980; Shikada *et al.*, 1982), which competes favorably with systems based on short-wavelength AlGaAs lasers.

In these InGaAsP LED systems, fiber loss limits the achievable repeater spacing more critically than fiber modulation bandwidth (Gloge *et al.*, 1980).

Thus, to maximize the span distance, at a given bit rate, the LED designer must optimize the coupling to the optical fiber to achieve the maximum LED output power. As we described in Section III.B.2, increasing the LED power does compromise the device modulation bandwidth, and thus a limiting power is always reached at a given data rate. The achievable LED output power, P, has been found to be related to the LED modulation bandwidth, B, by the relationship $P \propto B^{-v}$, where $v \cong \frac{2}{3}$ (Saul et al., 1985).

The typical maximum safe operating current for a 1.3-μm LED is about 150 mA. At this drive current, the fall times of wideband Burrus InGaAsP LEDs are usually two or three times longer than the rise time, making fall time the factor that limits the pulse response. Thus, modulation circuits with preshaped driving pulses are required to reduce the LED turn-off time to less than the material recombination time, as previously described in Section IV.B (Dawson, 1980). However, during the application of a reverse current pulse to the LED, care must be taken to avoid placing a reverse voltage bias on the diode, as they appear to be weak in reverse bias and tend to develop dark-spot defects that can seriously affect their reliability (Zipfel, 1985).

A practical circuit that encompasses all the transmitter features presented in Figs. 17a and 17b is shown in Fig. 25. The basic modulator is identical to that previously described in Section IV.A. This circuit is capable of producing

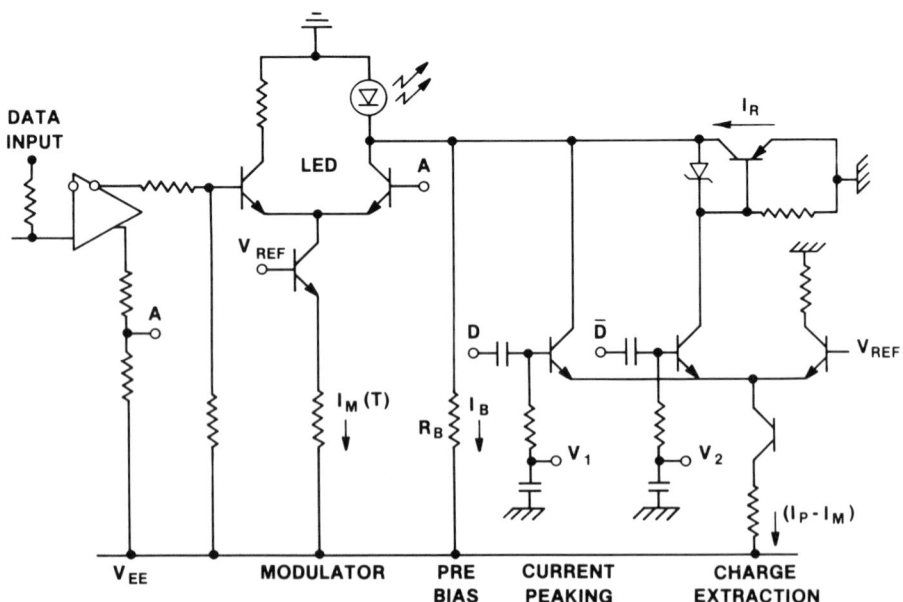

FIG. 25. A practical LED circuit, with both current peaking and charge extraction.

a 200-mA current pulse with less than 2 ns rise and fall times, when tested with a 10-ohm resistor load. A prebias current of a few milliamps is maintained on the LED at all times via the load resistor R_B.

Current peaking and charge extraction circuits, which speed up LED turn-on and turn-off, consist of short-duration (approximately 3 ns) current sources activated by differentiating the data signal. During LED turn-on, the short duration current pulse of amplitude $(I_P - I_M)$ is superimposed on the primary modulator current pulse, I_M, to provide current peaking. During LED turn-off, a similar short-duration current pulse (I_R) of opposite polarity supplies a reverse current to the LED to sweep out the injected carriers.

The performance of this circuit with and without such edge enhancement can be seen in Figs. 26a–d. In Figs. 26a and 26c, the light and current pulse traces have been superimposed. Without edge enhancement, turn-on of the light very nearly follows the current; but during turn-off, the fall time of the light is considerably slower than that of the current, as illustrated in Figs. 26a and 26b. The rise time improvement due to current peaking, and the more dramatic improvement of fall time due to charge extraction, can be seen in Figs. 26c and 26d. The effect of edge-enhancement circuitry on LED rise and fall time for LEDs with different bandwidths is summarized in Fig. 27.

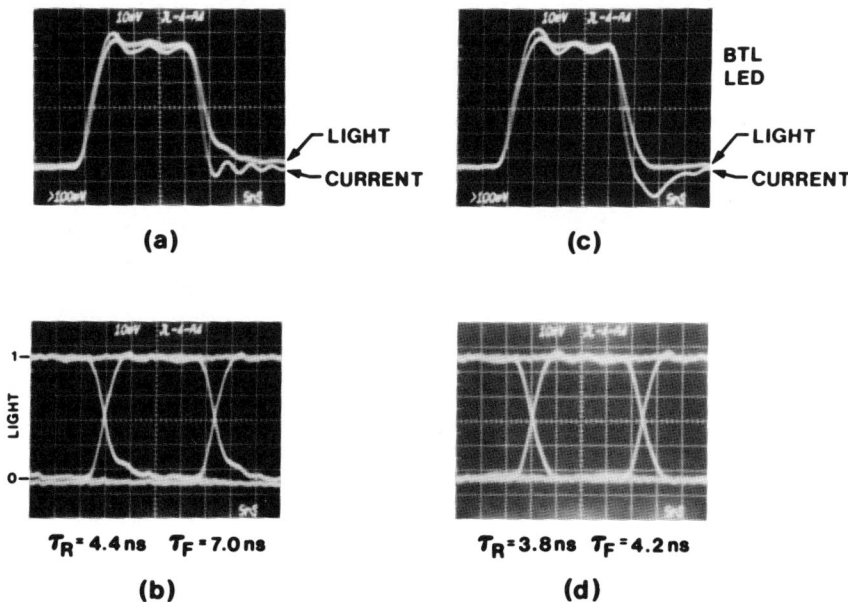

FIG. 26. Current and optical pulse waveform of the LED circuit shown in Figure 25: (a) and (b), with no edge enhancement; (c) and (d), with edge enhancement by the peaking and extraction circuits.

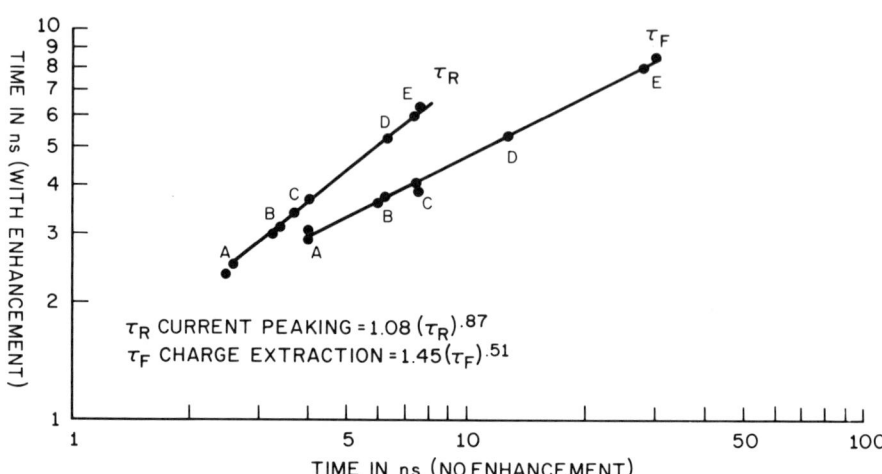

FIG. 27. The improvement of LED transmitter rise and fall times with edge enhancement plotted with reference to the unenhanced rise and fall times.

At a constant current, the temperature dependence of the output power of a 1.3-μm LED has a negative coefficient of approximately 0.025 dB/°C. This decrease in optical output power *could* be compensated by making the fixed modulation drive, I_M, proportional to temperature. Typically, about 1 mA/°C change in I_M would be required to compensate such a change in LED output with temperature. However, since 1.3-μm LED-based systems are usually loss limited, this strategy is not used, and the LED is generally operated at a fixed current that is the maximum safe operating current for all temperatures to be encountered. The resultant increase in power at lower temperatures is easily accommodated by the dynamic range of the receiver.

The above discussion has been concerned with surface-emitting LEDs. Other structures, such as edge-emitting LEDs and superluminescent LEDs, have higher potential for coupling their energy into a fiber than the surface-emitting LED. However, these devices have $L-I$ characteristics that are strongly temperature-dependent. This means that they may require feedback strategies similar to those necessary for lasers. Alternately, temperature control for output power stabilization could be used.

B. Laser Transmitters

1. AlGaAs Gain-Guided Lasers

PCM devices using these lasers have formed the first generation of optical transmitters. These have been used in both 45 and 90 Mb/s multimode fiber applications with unrepeatered spans of 7 km requiring mean optical powers lying between -2 and 0 dBm.

Our experience has been with transmitters using 10-μm and 5-μm stripe lasers formed by proton bombardment (Dixon, 1980). The design strategies and transmitter performance have been mainly dictated by the specific characteristics of these particular gain-guided devices. Most of our earliest transmitter design and evaluation involved the single-loop, average-power feedback loop described in Section IV.C.2.a, using back laser facet monitoring.

After significant functional life testing of both 10-μm and 5-μm lasers (Dixon and Dean, 1981; Dean and Dixon, 1982), we concluded that circuits using front facet monitoring and double feedback control of both the bias current, I_B, and modulation current, I_M, appeared desirable for these lasers. These latter circuits must handle significantly smaller monitoring signals from the front facet tap and also must be able to extract the ac information from this signal as described in Section IV.C.2.b.

We shall describe in the next few paragraphs the design considerations and optical performance realized with these transmitter designs.

a. Average Power Control with Back Facet Monitoring Early results with 10-μm stripe lasers formed by deep proton bombardment were dominated by three problems; laser self-pulsations, low-frequency noise associated with light jumps, and kinks at low power levels in the $L-I$ curves.

The laser self-pulsations introduced timing jitter, t_j, at the laser turn-off when the frequency of the self-pulsation was sufficiently low that its period became comparable to the pulse period. In Fig. 28 we show the experimental decrease in system margin as a function of pulsation frequency for a 45 Mb/s pseudo-random word sequence (Bosch, 1979). It can be seen empirically that significant degradation occurs when the pulsation frequency is <400 MHz (i.e., approximately nine times the bit rate). However, this measured result is over twice the frequency calculated for intersymbol interference by considering the indeterminancy of the turn-off (Duff, 1980). The source of the discrepancy between theory and experiment is unknown. It could be due to a low-frequency noise source associated with the light jump and subsequent self-pulsation. Alternatively, it could arise from a reflection generated at the end of a fiber pigtail, if the round-trip transit time equals the pulsation period and thereby magnifies the effect of the noise generated by the reflection.

Wide-stripe gain-guided lasers often exhibit regions of superlinearity in their $L-I$ characteristics (Paoli, 1979; Anthony et al., 1980). Associated with these "light jumps" is low-frequency noise that reaches a maximum at the center of the light jump (Dixon, 1984). The noise content, below 50 MHz, can be sufficiently large that it causes severe eye-degradation due to pulse-to-pulse pattern dependence, ΔP_p. It has been speculated that the noise generation and superlinearity are due to either deep traps (Copeland, 1978) or saturable

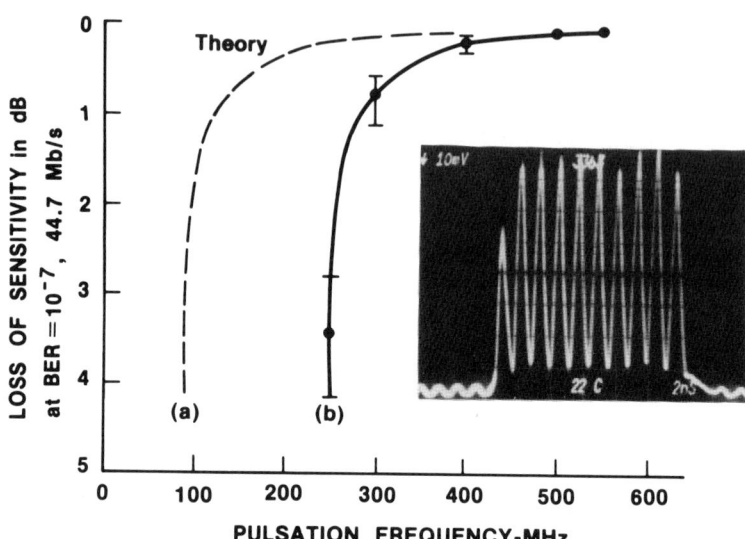

FIG. 28. The effect of self-pulsations of an AlGaAs laser on the eye-margin of a 45 Mb/s transmitter as measured by loss of receiver sensitivity, in dB at 1. E-7 BER: (a) theory (Duff, 1980); (b) experiment.

absorbers (Paoli, 1979). Figure 29a shows the variation of the noise spectrum of an AlGaAs laser (#467) as the cw optical power is increased through a light jump. Figure 29b shows the variation of the eye degradation of a 45 Mb/s transmitter containing laser 467, as the peak pulsed power is adjusted to different positions within a light jump. In the case shown, total eye closure occurred at the center of the pulse (point 2a) at an optical noise level at 10 MHz of −25 dBm. Above the center of the jump (point 3), the noise is no longer random, and the laser becomes a self-pulsator. The improvement in eye margin above the jump center is a result of the rapid increase in self-pulsation frequency with optical power (point 4). Although the laser population can be screened to eliminate lasers with these characteristics in their initial performance, an unknown factor is the onset of such phenomena with aging.

Improved stability against these effects, as well as an increase in the power level of kinks, came with narrower-stripe lasers with stripe widths from 3 to 8 μm created by using shallow proton bombardment (Dixon and Joyce, 1980). One phenomenon still affecting the initial performance of some of these narrower-stripe devices was severe pattern dependence, ΔP_p, when using a pseudo-random word sequence. A comparison of transmitters with strong and weak pattern dependence is shown in Fig. 30. The origin of ΔP_p arising from package parasitics was eliminated. Instead, the problem was traced to temporal instabilities in the gain-guiding of these lasers. This instability was

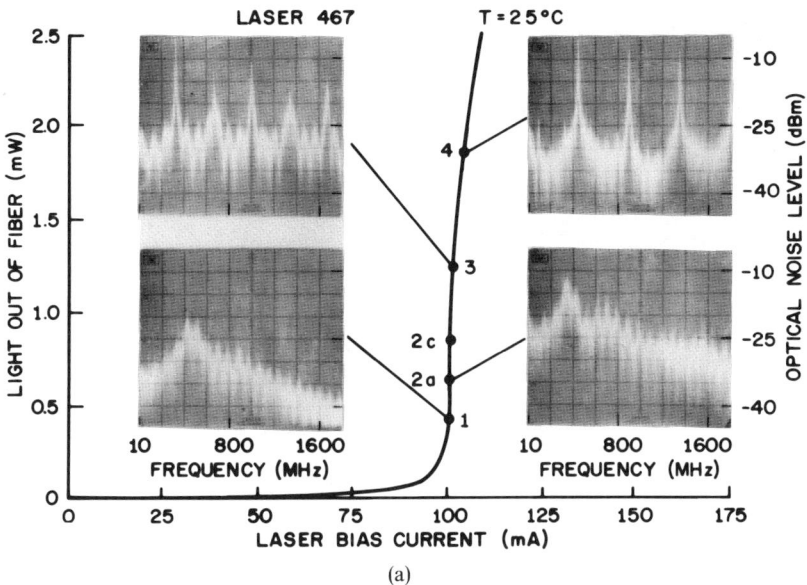

(a)

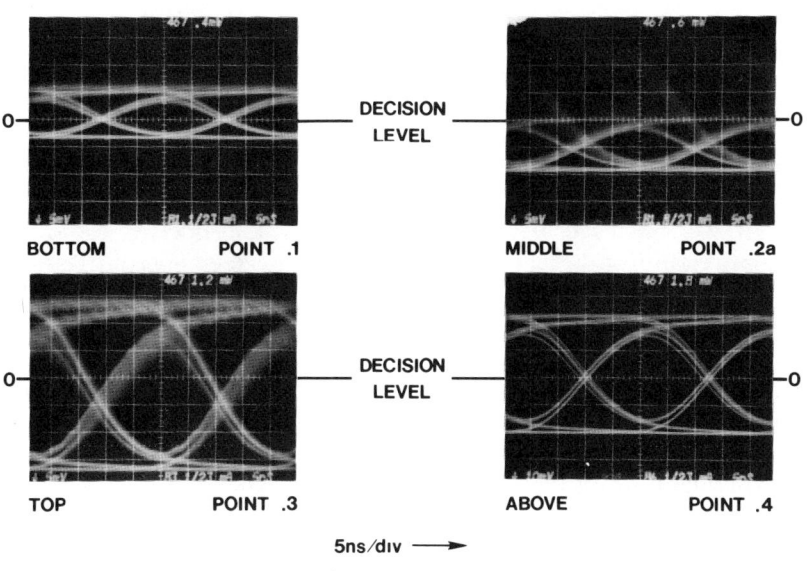

(b)

FIG. 29. (a) The variation of laser noise with optical power in a "light jump" of a proton-bombarded stripe geometry AlGaAs laser. (b) Optical "eye-diagrams" of a 45 Mb/s transmitter containing the laser shown in (a).

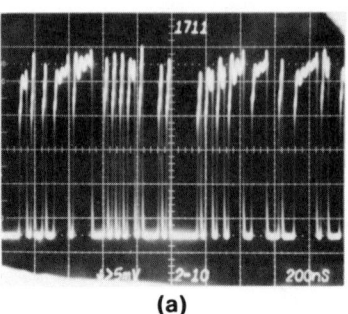

(a)

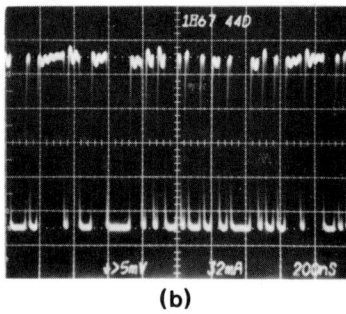

(b)

FIG. 30. Comparison of an optical pulse stream from a 45 Mb/s transmitter containing (a) an AlGaAs laser exhibiting strong pattern dependence; (b) an AlGaAs laser with weak pattern dependence.

characterized by a transient decrease in threshold, a slope efficiency increase, and a decrease in the full-width–half-maximum value of the far field pattern. The characteristic time for these changes is about 1 μs (Shen, 1982). These transient effects could not be explained by thermal focusing or free-carrier effects. It has been speculated that the pattern dependence was due to nonradiative absorption by proton damage at the edge of the stripe region even in shallow-bombarded lasers.

The major problem encountered with these narrow-stripe, shallow-bombarded lasers in transmitters with this circuit strategy has been long-term aging. Using lasers prescreened to eliminate devices with low-frequency noise, the dominant aging processes in the transmitter are power drift and extinction ratio degradation.

The mechanisms for these changes in gain-guided lasers, and specifically proton-bombarded devices, has been discussed extensively (Dixon and Dean, 1981; Dean and Dixon, 1982, 1985). The power drift has been shown to be associated with asymmetric aging of the two laser facets (Marcuse and Nash, 1982). The extinction ratio, degradation is mainly dominated by softening of the lasing transition, producing an increase in spontaneous light at the bias point, although changes in average power from asymmetric aging and decreases in radiative efficiency, η, are also contributors.

From the transmitter designer's viewpoint, a major question has been whether short- and long-term performance could be improved by a change in circuit strategy.

b. Average Power Control with Front Facet Monitoring The major problem associated with front facet monitoring is the magnitude of the control current available. For -2 to 0 dBm optical output power, laser packages with $\sim 50\%$ coupling efficiency to the front facet fiber, and good Si photodiodes, the control signal in transmitters with back facet monitoring is typically ~ 1 mA. However, for a front facet tap, minimizing the tap insertion loss is also a major objective. Control currents as low as 30 μA are obtained if taps with insertion loss below 1 dB are desired. Amplification is thus necessary before differencing with the data reference signal in Fig. 19. However, as only average power control is required, a low-bandwidth amplifier can be used, and excellent control is possible.

The results of using a transmitter with front facet monitoring using either a split fiber tap (Karr et al., 1979) or a GRIN lens tap (Dean and Dixon, 1985) reveal that a power drift of less than 0.3%/kh is attainable. This result is almost an order of magnitude better than observed with back facet control, when devices are used that have significant changes in coupling efficiency with time. In this work, the coupling into the front facet fiber changed because the beam wandered as the gain-guided lasers aged (Dean and Dixon, 1985). However, laser package relaxation with time can give similar coupling efficiency changes.

A further advantage of front facet monitoring is the small variation of control current from device to device. In practical transmitters with variations in laser coupling efficiency and far field pattern shapes, the variation in back facet diode current can range over 100% at constant front output power from the fiber. In contrast, a tap current variation is less than 10% at constant output power. This small variation reduces customization of circuit component values from transmitter to transmitter, which is desirable for manufacture.

Another advantage of front facet monitoring is better temperature control of optical power. Using a tap with a linear variation in tap ratio (i.e., tap signal to output signal) with temperature, an accurate electronic compensation of the temperature variation of the average optical power is possible. In sharp contrast, the large variable temperature coefficient of coupling efficiency of gain-guided lasers to the front facet fiber makes electronic temperature compensation difficult in back facet control circuits. In Figs. 31a and 31b, the relative variations in the temperature change of optical power,

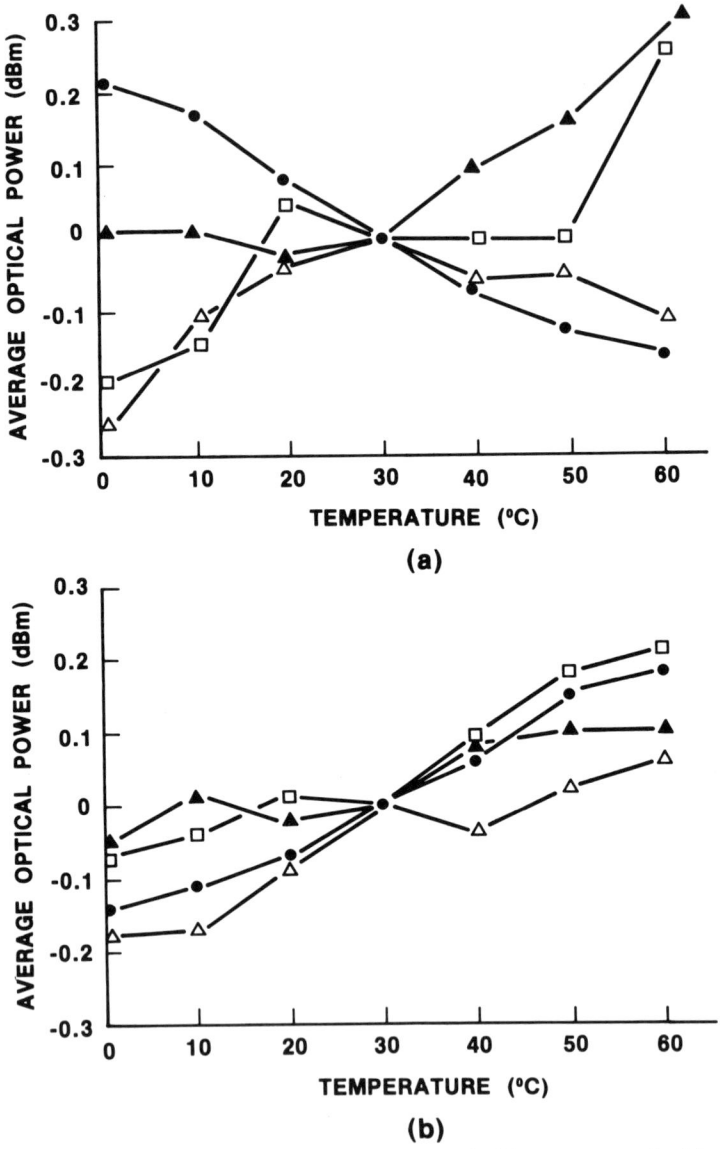

FIG. 31. The comparison of the optical power control with temperature for (a) a group of 90 Mb/s transmitters with back-facet photodiode control, and (b) the same group of laser packages under front facet tap control.

using both front-facet and back-facet control of packaged 5-μm shallow proton-bombarded lasers, are compared.

The outstanding question with this front facet strategy is whether double loop control can reduce extinction ratio degradation in such gain-guided devices.

c. Double Feedback Control Using Front Facet Monitoring Just as average power cannot be reliably monitored from the back facet because of different changes in front and back slope efficiencies, neither can the extinction ratio. This constraint means that P_1, P_{AV}, and P_0 have to be measured using the very small front tap signal. A further constraint exists: A second loop to control I_M by detecting the thresholds is not possible with these gain-guided lasers, as they age with a softening of the lasing transition.

The only circuit that can be used given both these constraints is one that measures P_{AV} and $(P_1 - P_0)$ directly. We have obtained control of extinction ratios as large as $\varepsilon = 10$, at data rates up to 90 Mb/s, with the circuit shown schematically in Fig. 20. This required a transimpedance amplifier to increase the tap current from 30 to 300 μA, and a high-speed operational amplifier with a unit gain frequency of ~ 600 MHz as the peak detector amplifier to obtain the $(P_1 - P_0)$ value. The high bandwidth requirements of the transimpedance amplifier and peak detector operational amplifier limit the use of this circuit to < 100 Mb/s. Also, as the transimpedance amplifier has an ac coupled output, the circuit has a strong tendency to oscillate if too much gain is used, or if extreme care is not taken to reduce phase shifts in the peak detector operational amplifier stage.

In Fig. 32, we show the control of the extinction ratio, at 45 Mb/s, as the temperature is changed. The laser in this work was deliberately chosen with large changes in the softness of the lasing transition with temperature. In Fig. 33 is shown the improvement in the system margin at an APD receiver by double-loop control of ε at 10, compared to the variable extinction ratio obtained with single-loop control.

Our conclusions, however, are that such a double-loop circuit is of dubious advantage for these gain-guided devices considering its cost, complexity, and limited bit-rate capability. Also, unless the extinction ratio is controlled with P_1 and P_0 above threshold (which gives barely useful extinction ratios of ≤ 5 in transmitters with less than -2 dBm output power), the softening lasing threshold causes $(I_{th} - I_B)$ to increase. Such a decrease of bias point to below I_{th} causes an increase in timing delay, t_d, and hence timing jitter.

A better solution appears to be to build in as large an initial extinction ratio as possible and live with the slowly degrading margin that results with extinction-ratio degradation of a tap-controlled single-loop transmitter. However, increasing the initial ε enormously does not appear possible in all

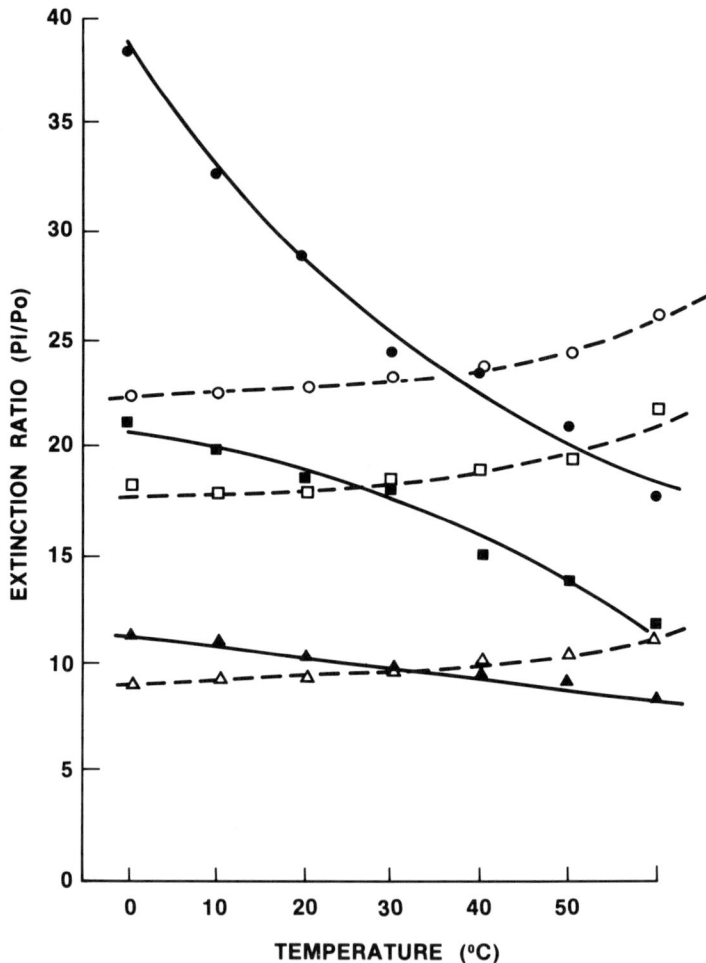

FIG. 32. The comparison of the variation of the extinction ratio with temperature of two 45 Mb/s transmitters with front facet optical taps and (solid symbols) single-loop feedback control with fixed modulation current; (open symbols) double-loop feedback control with variable modulation current, when using the same AlGaAs laser.

gain-guided devices, as there seems to be some evidence that the aging rate in proton-bombarded devices is a function of $(I_{th} - I_B)$ (Dixon and Dean, 1981; Dean and Dixon, 1985).

2. *1.3 μm InGaAsP Multifrequency Lasers*

The second generation of optical transmitters using 1.3-μm multifrequency lasers have completely replaced the short-wavelength AlGaAs/

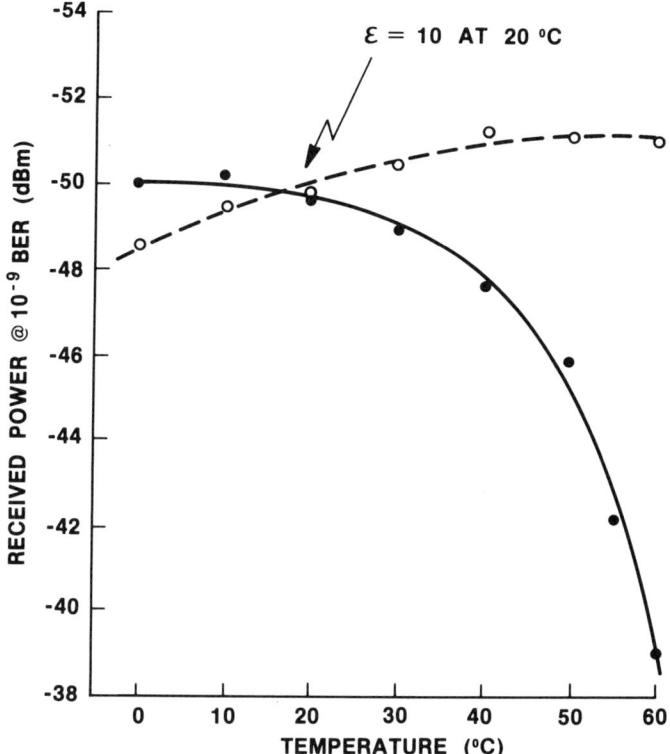

FIG. 33. The comparison of the system performance at 45 MB/s of a silicon APD receiver, at a 10^{-9} BER, when using the single-loop (solid circles) and double-loop (open circles) transmitters shown in Fig. 32. Both transmitters were tuned to give an extinction ratio of 10 at 20°C.

GaAs gain-guided lasers at 0.83 μm. Today, virtually all applications for this multifrequency source are for SM fiber transmission systems. However, for a short interim period these sources have found applications in MM fiber systems. In the United States' "Northeast Corridor" trunk line, wavelength division multiplexing was used to add a 1.3-μm laser system to an existing 0.83-μm system operating at 90 Mb/s (Jacobs, 1986). The 1.3-μm channel operates at 180 Mb/s and, in addition, bypasses every second repeater. However, the greater (span length) × (bit rate) product capability afforded by SM fiber soon led to the introduction of the AT&T FT Series G 417 Mb/s system (Gloge and Jacobs, 1988). Here nine DS3 channels at 44.7 Mb/s were electronically multiplexed to 417 Mb/s, and 40-km transmission spans were achieved. This system was recently upgraded to provide 1.7 Gb/s transmission capability (Fishman et al., 1986). In this system, 36 DS3 channels are electronically multiplexed, and 40-km transmission spans have been achieved. This same technology has been deployed in the recently installed AT&T SL TAT-8 trans-Atlantic subcable, where 60-km repeater spacings at 295 Mb/s operation have been achieved.

Our experience with long-wavelength transmitters has been primarily with narrow-spectrum, multifrequency, real-refractive-index guided lasers. We recently have become involved with single-frequency DFB lasers, which are discussed in the next section (VI). The multifrequency lasers combine a number of attractive properties. Their threshold currents at room temperature typically range from 8 to 30 mA, and their sharp laser thresholds means that biasing just a few milliamps below threshold can give large extinction ratios. The devices are kink-free up to at least 8 mW facet power, with slope efficiencies in excess of 0.1 mW/mA. The far field pattern is primarily lowest-order Gaussian, indicating fundamental stable lateral mode operation. Both the lateral and vertical beam divergence are less than 55 deg. FWHM, which permits 50–75% coupling into 50-μm core MM fiber and up to 60% coupling to SM fiber. The lasers are also capable of being modulated by direct carrier injection at frequencies of several GHz.

These specific characteristics of narrow-spectrum index-guided lasers have dictated both transmitter design strategy and the initial performance. In addition, the design is greatly influenced by the choice of fiber. In SM fiber applications the primary concern is the data rate, and in MM fiber applications it is modal noise.

In the next few paragraphs, we describe the design considerations and the optical performance realized with these real-index-guided lasers in both MM and SM fiber systems.

a. Average Power Control with Back Face Monitoring Preliminary functional life testing of index-guided InGaAsP lasers indicate that the single average-power feedback loop described in Section IV.C.2.a using back facet monitoring is adequate for these lasers. The dominant aging processes, power drift due to front-to-back facet mistracking and extinction ratio degradation, seen in the short-wavelength transmitters do not appear to be present in these devices. However, the laser-fiber coupling must be mechanically stable with respect to both temperature and time.

b. Temperature Control For InGaAsP DH lasers, T_0 is ~ 60–70 K. Along with this somewhat large temperature sensitivity, there is also a strong rollover degradation at elevated temperatures because of nonradiative recombination processes, as described in Section III.B.1. For these reasons, some means of cooling the laser to room temperature must be provided.

At present, the most convenient cooling of the laser uses a commercial single-stage thermoelectric heat pump. For a 200-mW thermal load, it is possible to cool the laser from a 65°C environment to room temperature, provided that the heat can be dissipated into a suitable heat sink. Usually these coolers require high current at low voltage (typically 1 A at 1.3 V). Unfortunately, this imposes an additional current load on the transmitter power supplies, which can be troublesome in some applications. In the case of a low-temperature environment, the cooler can also be used to heat the laser.

Thermoelectric heat pumps have demonstrated that they can be highly reliable and a convenient method of cooling and heating in a number of critical applications such as satellites and medical instrumentation; however, the reliability data for the laser transmitter environment is incomplete at this time.

c. Modal Noise in Multimoder Fiber Systems Modal noise produces unwanted amplitude fluctuations of the received signal in a PCM fiber system (Epworth, 1978). Its effect at a receiver, on a 90 Mb/s data signal, is shown in Fig. 34a where considerable "eye" closure, most notably "positive eye" closure, causes an increased bit error rate (BER).

Figure 34b shows the amplitude fluctuations observed during a long string of "ones," where it can be seen that the amplitude fluctuations occur on a time scale shorter than the bit period of approximately 10 ns. This means that these rapid fluctuations are too fast to be compensated for by the automatic gain control (AGC) of the receiver. Some eye closure of both the positive and

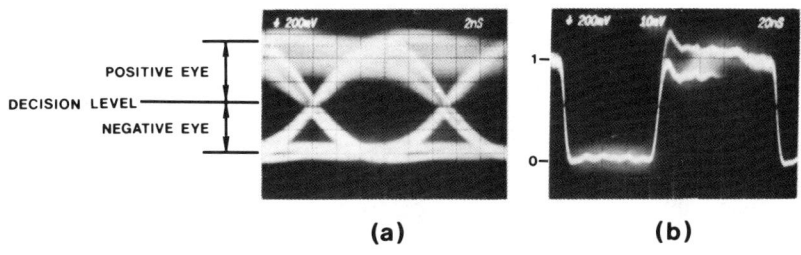

FIG. 34. Modal noise effects: (a) the severe positive eye closure of a received optical signal of −25 dBm due to modal noise; (b) the spurious amplitude modulation of the received optical signal during a string of eight consecutive "ones;" (c) positive and negative eye margin measurements, showing large degradation of the positive eye.

negative eye is permitted and must be allocated in the system margin. (The procedures for determining the amount of eye closure are discussed in Section V.3C.) Typically 10% to 15% of positive eye degradation (or closure) and 4% negative eye degradation are permitted. In Fig. 34c, it is shown that the amount of positive eye degradation (PED) for this particular etched mesa and buried heterostructure (EMBH) laser greatly exceeds the system requirements as a result of modal noise effects. The effect of modal noise on negative eye degradation (NED) appears to be minimal.

Three conditions are known to be necessary for the production of modal noise (Chown et al., 1980): (1) a narrow spectral width source, (2) modal or spatial filtering in the transmission medium, and (3) time variations of either (1) or (2).

Time variations of spatial mode filtering can occur because of mechanical perturbations of the fiber, but unless they are fast, they can be handled by the receiver AGC (Couch et al., 1983). A more dominant effect appears to be variations in the spectral content of narrow-wavelength sources—i.e., mode partition fluctuations (Peterman and Arnold, 1982; Epworth, 1981). Obviously, the coherence length of the source is a major factor defining the distance over which such modal interference effects can be generated (Throssell and Kanabar, 1983). Also, recent work has shown that the "speckle contrast," and hence the modal noise potential, is strongly dependent not only on the total spectral width (total number of longitudinal modes), but also on the individual width of the longitudinal modes (Danlicker et al., 1983).

Our own data support these observations that a dominant cause of modal noise is a result of mode partition effects in the source. This can be seen particularly clearly in the temperature-dependent variations of signal-to-noise ratio of a laser with a narrow spectral width launched into a multimode fiber (Nagai et al., 1983). In Fig. 35 are compared the variations in bit error rate (BER), at a received power of -25 dBm and a fixed decision voltage, of a surface-emitting LED and a channel substrate buried heterostructure (CSBH) laser as the temperature is changed. The BER for the LED is temperature-independent. In contrast, for the CSBH laser, the BER shows significant temperature variations in which the minima correspond to temperatures at which a single longitudinal mode is dominant. The maxima, on the other hand, correspond to temperatures where there are two competing longitudinal modes and hence a temperature of high mode partition noise (Joindot and Boisrobert, 1982). However, this temperature behavior can be readily modified in the presence of reflections that also increase mode partition fluctuations (Couch et al., 1983).

Twisting the fiber at several cycles/s does not significantly increase the BER fluctuations, indicating that these amplitude fluctuations are handled by the receiver AGC. These slow fluctuations can be converted to higher

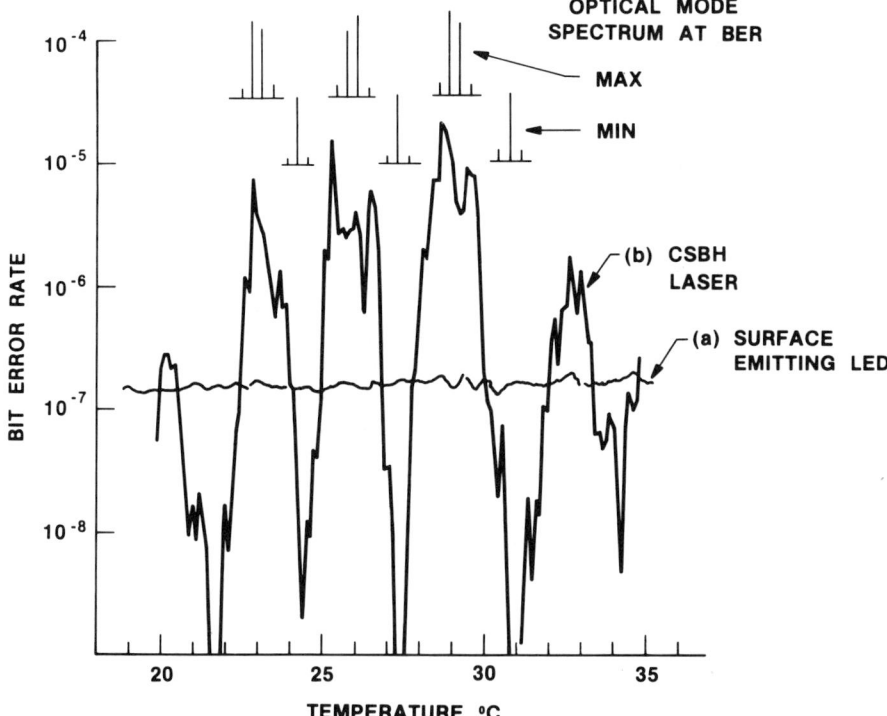

FIG. 35. The comparison of the temperature variation of bit-error rate, at a fixed decision level and a −25 dBm received power level, for (a) a surface-emitting LED and (b) a CSBH laser.

frequencies, however, if the twisting again causes reflection-induced laser mode partition fluctuation (Couch et al., 1983).

In multimode fiber systems, attempts have been made to circumvent modal noise by (1) using broad spectral width sources such as gain-guided lasers or LEDS; (2) by modulating lasers from below I_{th} with very short pulses, so that the wavelength spectrum of the laser is always in the transient state and hence has the widest spectral width (this technique has been accomplished in high-bit-rate systems by using RZ format [Epworth, 1981]); and (3) by reducing the source coherence by "microwave-dithering" the data signal at a frequency outside the receiver filter bandwidth (Bosch et al., 1980).

The effect of microwaves on modal noise suppression of a narrow-spectral-width source is shown in Figs. 36a–c. In Fig. 36a is shown the superposition of the microwave signal on the data pulses. In this experiment the laser was biased several milliamps below I_{th}, and it can be seen that the laser does not respond to the microwave signal when in the spontaneous

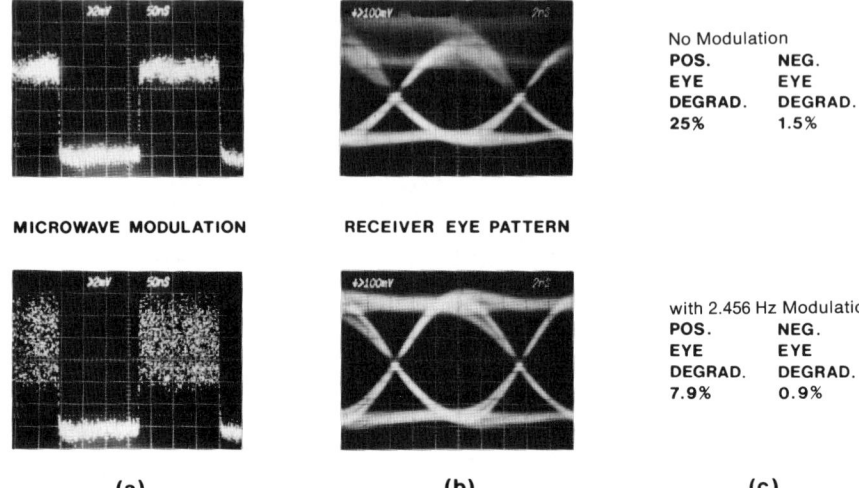

FIG. 36. The reduction of modal noise by external microwave superposition on the data signal. (a) Transmitted optical signals with and without 2.45 GHz microwave signal. (b) Received signals through optical fiber and connectors with and without 2.45 GHz microwave signal. (c) Positive and negative eye margins with and without 2.45 GHz microwave signal.

emission state. In Fig. 36b, the "clean" received eye indicates that the microwaves have significantly reduced the modal noise [compare Fig. 34a)]. In Fig. 36c, the measured effect of "microwave dither" on both the positive and the negative eye is given. The microwaves produce a substantial improvement $25\% - 8\% = 17\%$ in the positive eye margin (17%), but have a negligible effect on the negative eye margin. The influence of the microwaves is to broaden the individual longitudinal modes, as shown in Fig. 37, and hence to reduce the coherence length of the laser.

A similar reduction of modal noise can be observed with narrow-spectral-width lasers that develop self-pulsations. This can be seen in Fig. 38 for a CSBH laser that develops into a 4 GHz self-pulsation just below room temperature. The temperature dependence of the "positive eye" closure reveals that above 25°C eye closure is severe, but that below room temperature the modal noise is reduced and finally disappears as the self-pulsation deepens to 100% modulation. In this case, application of an external 2.45 GHz microwave modulation only has an effect above 25°C. However, it does not reduce the noise as effectively as the self-modulation, as it was impossible to create 100% modulation by external means.

d. Noise Effects in Monomode Fiber Applications Light reflected back into a laser from a fiber discontinuity, such as a connector, can enhance the

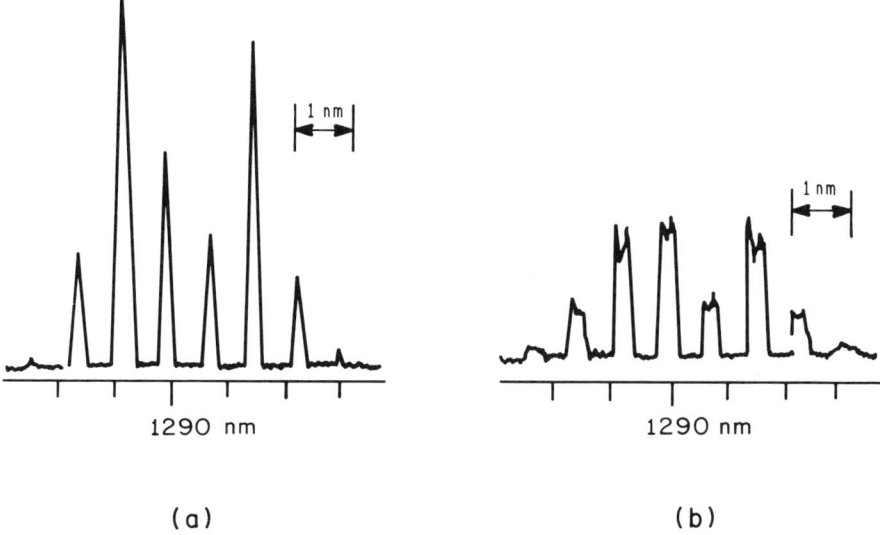

FIG. 37. The influence of a CW microwave signal on the time averaged optical spectrum of an EMBH laser: (a) no external microwave signal; (b) a 2.45 GHz microwave signal superimposed.

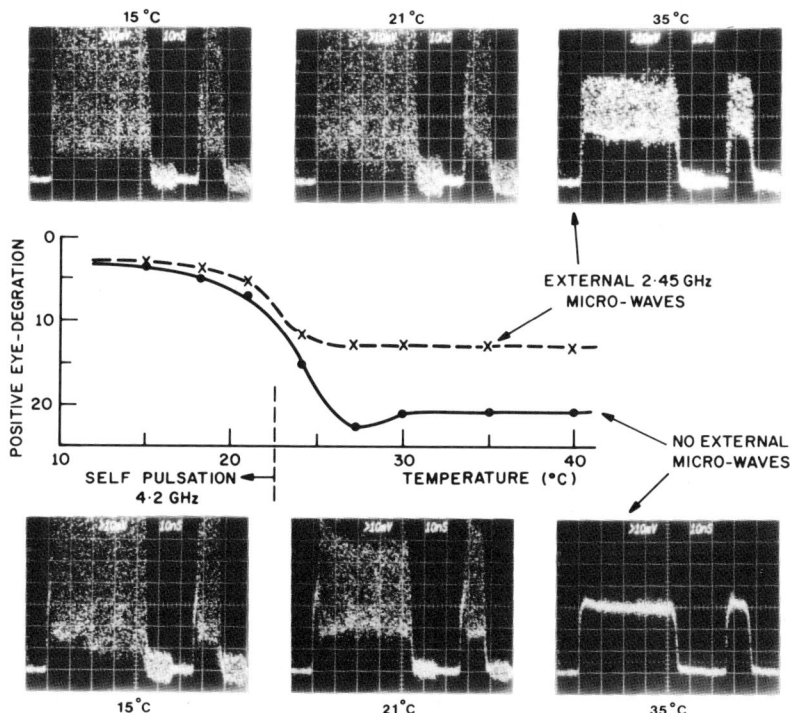

FIG. 38. The temperature variation of the positive eye-degradation due to modal noise for an EMBH laser that developed a 4.2 GHz self-pulsation below 22°C: (a) without an external microwave signal; (b) with a 2.45 GHz microwave signal externally imposed.

intensity noise of the laser and degrade its transmission performance (Peterman and Arnold, 1982; Lang and Kobayashi, 1980). However, from the transmitter designer's point of view, the critical question is whether this noise degradation is severe enough to warrant the use of an opto-isolator. The trade-off is whether the insertion loss penalty of the isolator is off set sufficiently by the decreased reflection-induced noise level.

At bit rates much below 2 Gb/s, it appears that the reflection penalty is small compared with the insertion loss of an isolator (Mazurczyk, 1981; Cheung, 1984). At higher bit rates or with DFB lasers reflections will become more significant, and an opto-isolator will probably be required (Farrington, 1981).

It is also possible that modal noise can result from the preferential selection, at a fiber discontinuity, of one of the two polarization states propagating in a single fiber (Heckman, 1981). More recently, modal noise has also been observed at single mode connectors placed close to the source where the fiber is overmoded and is operating close to fiber cutoff (Cheung, 1984).

C. Testing of Transmitters

In Section III, we defined the most desirable properties of optical pulses for optical communications purposes. Subsequently, in Section IV, we outlined the dc pulse modulation and biasing of lasers and LEDs that affect these pulse characteristics, and hence performance. We pointed out that achieving all these characteristics is not always possible. Thus, we are left with the technological questions of what are the trade-offs, and what is the best way to evaluate them if they vary from transmitter to transmitter.

During the development phase of a transmitter design, the optical pulse shape is the parameter most often studied. Several IC technologies (Si bipolar, Si fine-line NMOS, and GaAS MESFET) discussed in Section IV are capable of modulating a multifrequency laser well into the Gb/s range. In Fig. 39, we show the quality of an optical eye of the FT Series G transmitter operated at 1.7 Gb/s. The optical rise and fall time of the detected optical signal is 120 ps. The time-averaged optical spectrum of a modulated source as discussed in Section III is also carefully studied. Unfortunately, how this optical pulse traverses the fiber medium and is perceived by the receiver cannot be deduced with certainty from either pulse shape or optical spectrum information. The only way to evaluate the overall performance of a transmitter is by functional testing.

The experimental setup for functional testing is shown schematically in Fig. 40. The pulse pattern generator drives the transmitter. A calibrated regenerator is used to regenerate the digital data signal and to recover the clock signal. The clock and digitally regenerated data signal are connected to the error detector. The optical path contains the following

OPTICAL TRANSMITTER DESIGN

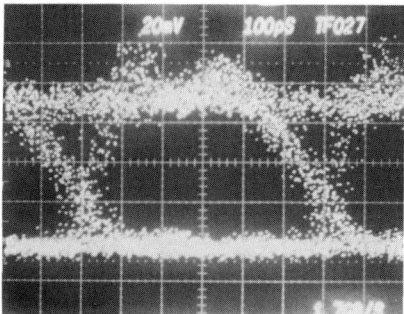

FIG. 39. Optical eye diagram of 1.3-μm CSBH multifrequency laser under 1.7 Gb/s, NRZ modulation $I_b = 0.81 I_{th}$.

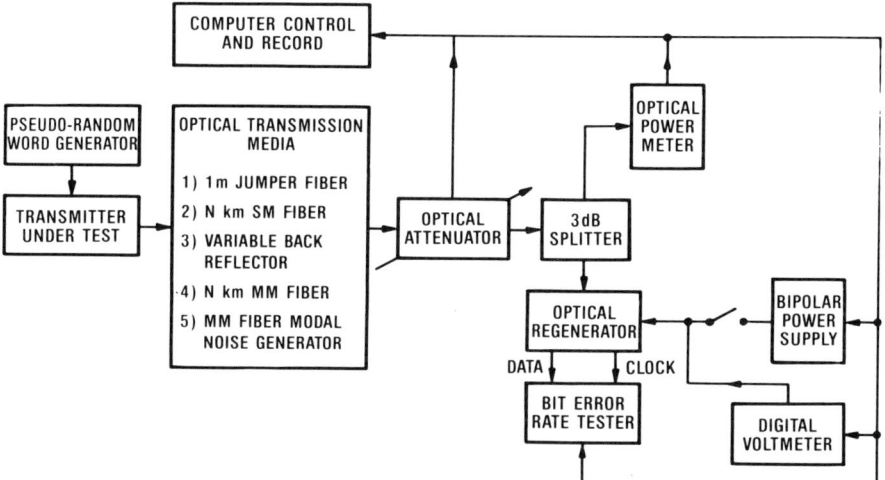

FIG. 40. Schematic representation of a functional test set for lightwave transmitter characterization.

elements: an optical medium, an optical attenuator, and a calibrated power splitter so that received optical power can be measured. The measurements consist of measuring the error rate as a function of received power. Typically, error rates down to 3 E-11 BER are measured. The measurements are first made with a short length of fiber as the transmission medium to establish a baseline. Then a given length of fiber with known dispersion characteristics is inserted, and the measurements are repeated. To test for reflection sensitivity, a variable back reflector is used as the transmission medium to impose deliberately a known reflection (typically 10%) on the laser, and the measurements are repeated again.

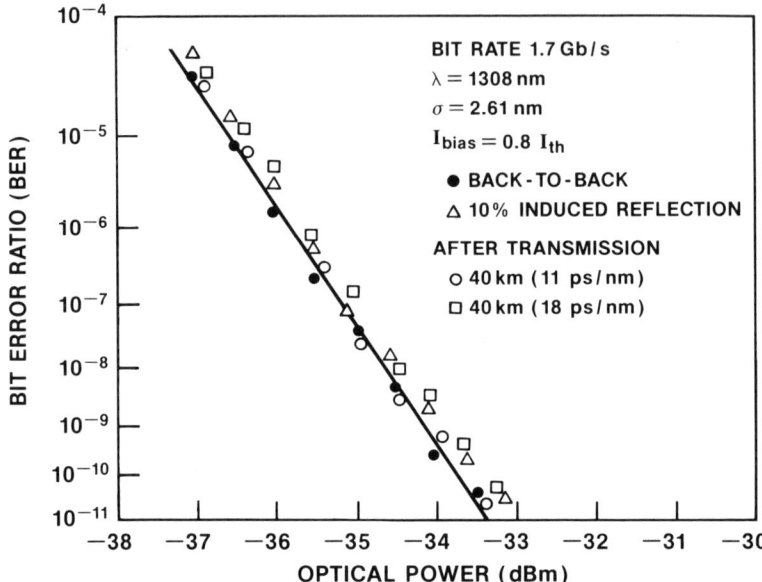

FIG. 41. Error rate curve for FT Series G transmitter with 1.3-μm multifrequency laser operating under 1.7 Gb/s, NRZ modulation. $I_b = 0.8 I_{th}$. Error rate curves of back-to-back (●) and penalties due to 10% deliberate reflection (△), and after 40-km transmission in 1.3-μm single-mode fiber with 11 ps/nm dispersion (○) and 40-km transmission in single-mode fiber with 18 ps/nm dispersion (□).

Figure 41 shows the results of a FT Series G transmitter operating at 1.7 Gb/s. The laser is biased at 0.8 I_{th}, and no optical isolator is used. The fiber medium is 40 km of standard SM fiber with zero chromatic dispersion at 1.3 μm. A baseline receiver sensitivity at 3 E-11 is −34.5 dBm. No dispersion penalty as a result of transmission through 40 km of fiber is observed. A deliberately induced 10% reflection toward the laser also did not result in a penalty. These functional tests allows an estimation of the performance of a particular transmitter, when it is built into an optical fiber system, to be made with a high confidence level.

VI. Multigigabit-per-Second Transmission Systems

A. System Overview

Long-haul trunking and submarine cable lightwave systems will need substantially increased transmission over long unrepeatered spans (> 100 km) because of the huge transmission capacity that will be provided for intracity communications with broadband ISDN, and the ever-expanding world communications needs. Present installed systems, described in the earlier sections, use 1.3-μm multifrequency Fabry–Perot lasers and now transmit at data rates up to 1.7 Gb/s over repeater spans of 40 km on "standard" SM fiber with zero chromatic dispersion at 1.3 μm and a fiber loss of 0.4 dB/km at this wavelength. Thus, future systems or upgrades of present systems will have to take into account the large quantities of standard SM fiber already installed. The 1.55 μm window with the low fiber loss of 0.25 dB/km is extremely attractive, but the large linear chromatic dispersion at 1.55 μm is approximately 15 ps/ks/nm, resulting in a most serious system impairment. Using a multifrequency Fabry–Perot laser at 1.55 μm with rms spectral width of ~1 nm would lead to a dispersion-limited bit rate–distance product of only 17 Gb/s × km, which is far worse than the 88 Gb/s × km demonstrated at 1.3 μm (Henry et al., 1988). There are two promising solutions to this problem: controlled-dispersion fiber and single-frequency DFB lasers. The third-generation lightwave systems will use both of these solutions.

1. Controlled-Dispersion Fiber

The zero chromatic dispersion of SM fibers can be shifted to the lowest-loss 1.55-μm region by compensating the material dispersion with increased waveguide dispersion (Croft et al., 1985). Typical values of the chromatic dispersion of DS fiber are shown in Fig. 3b. By using multilayer structures rather than simple step-index designs, it is even possible to make dispersion-flattened DF fibers in which the dispersion is <2ps/km/nm over the 1.3 to 1.6 μm wavelength range (Cohen et al., 1982). These fibers should be well suited to wavelength-division-multiplex (WDM) applications wherein lightwave channels at several different wavelengths are used simultaneously.

One concern related to all these fiber types is the power-handling capability of SM silica fibers. Laser output power continues to improve with sustained development. The maximum launch power of SM fibers is estimated to be around +6 to +10 dBm, limited by stimulated Brillouin scattering (Kimura, 1988).

2. Single-Frequency DFB Lasers

The other promising solution to the dispersive impairments at 1.55 μm is the use of an extremely narrow spectral-width single-frequency laser, which oscillates in a single longitudinal mode. The most successful approach today has been the distributed feedback (DFB) laser, which uses a periodically perturbated waveguide to provide the frequency-selective feedback (Suematsu et al., 1986). The ratio of the primary mode to the side modes, the side mode suppression ratio (SMSR), is typically >33 dB.

Injection lasers are inherently high-speed devices with modulation bandwidths bounded by the relaxation frequency. These devices are in commercial production, and typical small signal bandwidths (ft) are 7–14 GHz. Hence, they can be directly modulated at very high bit rates, and direct modulation is the easiest to implement. However, extreme care must be taken to minimize the laser package parasitics and to optimize the microwave interconnection path to the laser. Several IC technologies discussed in Sections IV and V can be used to modulate the laser with up to 100 mA well into the Gb/s range. These include Si bipolar ICs with ft = 8–20 GHz, Si fine-line NMOS with ft = 8–12 GHz, and GaAs MESFET ICs with ft = 10–50 GHz. Si bipolar and Si fine-line NMOS have demonstrated the ability to modulate lasers to 4 GB/s, and GaAs MESFETs will be required for the very high bit rates.

When a single-frequency laser diode is directly modulated, a number of unwanted optical effects occur. The first effect is a line broadening, which is frequency "chirp" and is associated with modulation-induced changes in the carrier density that lead to significant dispersion effects (Linke, 1985). Chirp can be minimized by biasing above threshold, but this can result in an extinction ratio penalty. The second effect is a type of mode partition noise generated between the primary mode and a residual side mode or submode. These effects in the presence of high dispersion can cause power penalties and can also produce error floors (Henmi *et al.*, 1988). Optical feedback from unwanted external reflections can also produce a number of deleterious effects that can impair BER system performance (Agrawal and Shen, 1986). In these cases an optical isolator is installed. These undesirable phenomena caused by the direct modulation of the laser have spurred many investigators to modulate the laser externally.

3. External Modulators

There are several reasons to operate the laser in a cw fashion and modulate its light using an external modulator. First is the relaxed requirements on the laser itself. Second, chirp is eliminated. Finally, higher modulation speed may be obtained.

There are several types of external modulators that have been successfully used to modulate a laser externally, including directional-coupler type, loss type, Mach–Zehnder, and total-internal-reflecting type. BWs of 5 GHz are attainable with drive voltages of about 8 V, which is consistent with the drive voltage available from a wide-band amplifier, making about 8 GB/s operation possible (Li and Linke, 1988). The improvements are dramatic. For example, at 4 Gb/s, the dispersion limit for a 1.55-μm DFB operating in a standard SM fiber (1.3-μm zero dispersion) system is increased from 7 km (direct modulation) to 110 km (external modulation) (Toughe et al., 1988).

B. System Performance Limits

Several investigators from the United States, Europe, and Japan have conducted long-distance experiments at 2 Gb/s, 4 Gb/s, and 8 Gb/s using 1.55-μm DFBs. Both direct modulation and external modulation have been studied. For a single channel at 1.55 μm being directly modulated, the highest bit rate reported was 16 Gb/s (Gnauck and Bowers, 1987). A summary of the long-distance transmission experiments that have been conducted in the last few years is shown in Fig. 42 (Li and Linke, 1988).

C. FT Series G 1.7 Gb/s 1.55-μm DFB WDM Channel

Recently, a second channel at 1.55 μm was added by WDM to the FT Series G 1.7 Gb/s system (40 km of standard single-mode fiber). The second

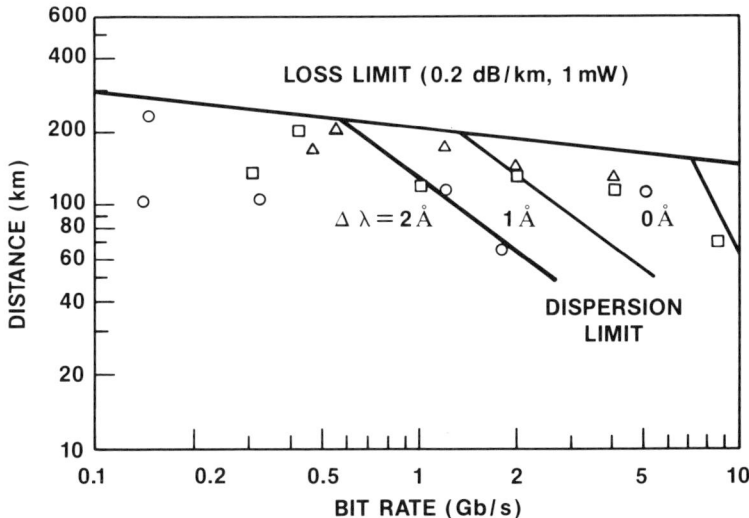

FIG. 42. Long distance transmission experiments conducted in laboratories in USA (□), Japan (△) and Europe (○) in the last few years. The solid lines represent theoretical limits imposed by fiber loss and dispersion, for systems employing conventional single-mode fiber and state-of-the-art devices, and for various laser spectral widths (Δλ).

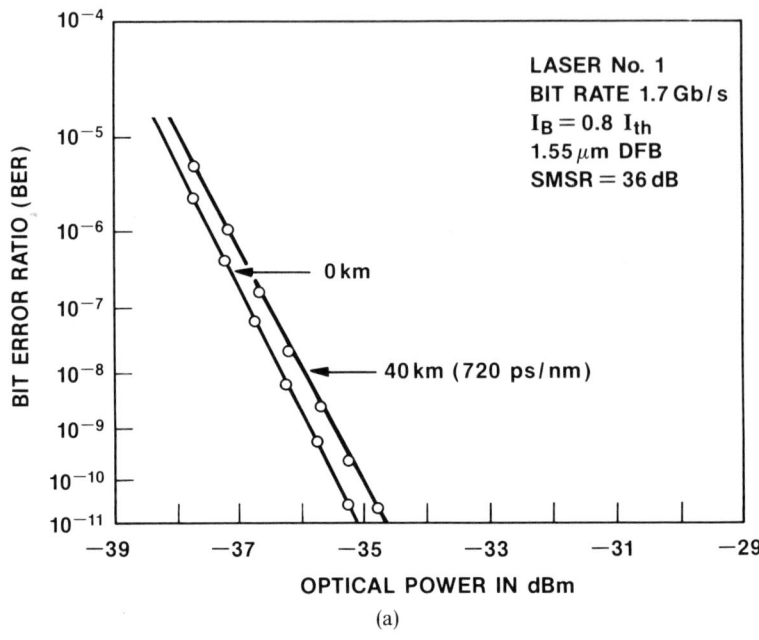
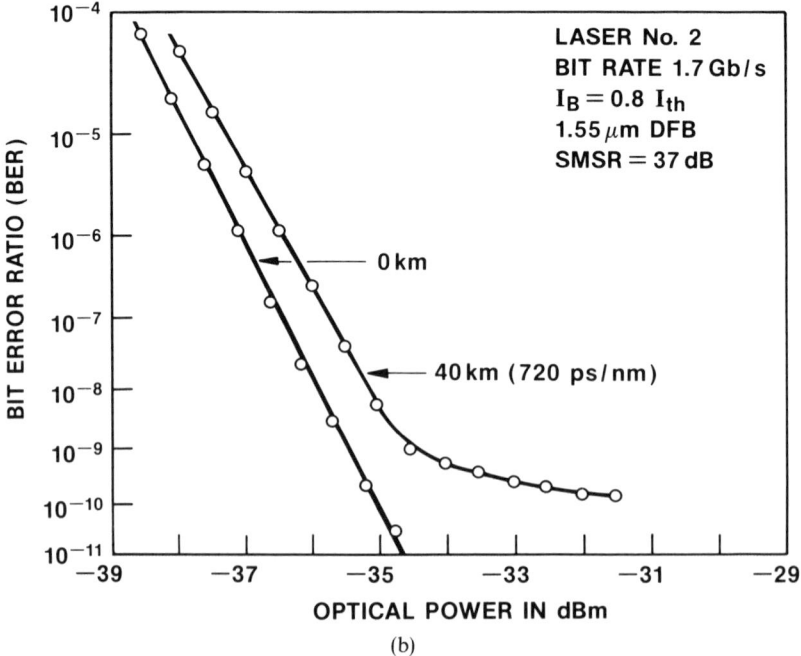

FIG. 43. Error rate curves for FT Series G transmitter with 1.55-μm DFB operating under 1.7 Gb/s operation, NRZ. $I_b = 0.8 I_{th}$; side mode suppression ratio (SMSR) $\geq$ 36 dB. Back-to-back baseline error curves compared with 40-km transmission through 1.3-μm single-mode fiber with 720 ps/nm dispersion. (a) Laser 1 exhibits normal behavior; (b) laser 2 exhibits an error floor.

channel uses a 1.55-μm DFB onto which an in-line optical isolator has been installed. Figure 43 illustrates some of the problems encountered. Shown are the measurement results of two directly modulated DFBs. Both lasers are biased at 0.8 I_{th}, and both have SMSR ratio > 33 dBm. The baseline BER measurements (0 km) are the same. Laser 1 shows normal performance with 40 km of fiber (720 ps/nm). Laser 2 shows an error floor. The causes of error floors are complex and are believed to be due to a combination of submode oscillations and dynamic chirping of the modulated DFB laser diode. Much more work is required to obtain robust direct-current modulation of DFB lasers at multigigabit-per-second data rates through highly dispersive fibers.

VII. Conclusion

A. Summary of Presenter Transmitter Design

First-generation fiber transmission systems, beginning in 1978, used MM fiber and operated at 0.83 μm using gain-guided AlGasAs lasers. They were available when systems to be installed demanded only 90 Mb/s and repeater spacings less than 7 km. These transmitters were prone to optical instabilities that compromise both performance and reliability. Below 50 Mb/s, they were complemented by 0.83-μm LED devices.

The second-generation transmitters have used 1.3-μm InGaAsP surface-emitting LEDs and multifrequency Fabry–Perot lasers. They took advantage of the lower loss and dispersion in the installed MM fibers. These multifrequency 1.3-μm lasers were first used to upgrade the 0.83-μm laser system by wavelength division multiplexing. SM fiber lightwave systems then began to replace MM fiber systems in 1984 for long-haul transmission. They then became the backbone for trunking applications with data rates starting at 180 Mb/s, quickly followed by systems operating at 417 Mb/s and 1.7 Gb/s. Repeater spans of 40 km are typical for these systems. These laser transmitters also have found application in transoceanic systems operating at 295 Mb/s with 60-km repeater spacings. More recently they found acceptance in the loop distribution system (<20 km) at 90 Mb/s, and experiments are being conducted to determine if the last few kilometers to the home might also be served. The 1.3-μm LEDs have experienced a tremendous growth in MM fiber data link applications up to 200 Mb/s.

The third-generation transmitters using the single-frequency lasers at 1.3 and 1.55 μm are now the subject of great research and development efforts to determine how best to take advantage of the narrow linewidth and large modulation bandwidth of these devices. Multigigabit-per-second operation over long transmission spans (100 km) is the goal of these efforts.

REFERENCES

Abbott, S. M., Muska, W. M., Lee, T. P., Dentai, A. G., and Burrus, C. A. (1978). *Electron. Lett.* **14**, 349.
Agarwal, G. P., and Shen, T. M. (1986). *IEEE J. Lightwave Technol.* **LT-4**, 58.
Agarwal, G. P., Anthony, P. J., and Shen, T. M. (1988). *IEEE J. Lightwave Technol.* **LT-6**, 620.
Aiki, K., Nakamura, M., Kuroda, T., Umeda, J., Ito, R., Chinore, N., and Maeda, M. (1978). *IEEE J. of Quantum Electron.* **QE14**, 89.
Albanase, A. (1978). *B. S. T. J.* **57**, 1533.
Anthony, P. J., Paoli, T. L., and Hartman, R. L. (1980). *IEEE J. of Quantum Electron.* **QE16**, 735.
Arnold, G. (1981). *Proc. 7th European Conference on Optical Commun.*, Copenhagen, p. 10.4.1.
Arnold, G., Russer, P., and Petermann, K. (1982), *In* "Semiconductor Devices for Optical Communications" (H. Kressel, ed.), Chap. 7. Springer-Verlag, New York.
Asatanic, K., and Kimura, T. (1978). *IEEE J. Solid State Circuits* **SC-13**, 133.
Baack, C., Elze, G., Enning, B., and Walf, G. (1978). *Frequenz*, **32**, 346.
Bosch, F. (1979). Unpublished data.
Bosch, F., Dybwad, G. L., and Swan, C. B. (1980): *CLEOS*, San Diego, Paper TuDD7.
Bosch, F., Palmer, G. M., Sallada, C. D., and Swan, C. B. (1984). *IEEE J. Lightwave Techol.* **LT-2**, 952.
Botez, D., and Ettenberg, M. (1979). *IEEE Trans. Electron. Dev.* **ED26**(8), 1230.
Bowers, J. E., and Pollack, M. A. (1988), *In* "Optical Fiber Telecommunications II" (S. E. Miller and I. P. Kaminow, eds.), Chapter 13. Academic Press, New York.
Burrus, C. A., Casey, M. C. Jr., and Li, T. (1979), *In* "Optical Fiber Communications" (S. E. Miller and A. G. Chynoweth, eds.), Chapter 16. Academic Press, New York.
Campbell, J. C., Tsang, W. T., Qua, G. J., Johnson, B. C., and Bowers, J. E. (1988). *OFC '88 Tech. Dig.*, Paper TUC3.
Carter, A. C., Goodfellow, R. C., and Griffith (1979). *Proc. of I.E.D.M.*, Washington, D.C., p. 118.
Catania, B. (1986). *4th RACE Seminar*, Bruxelles.
Channin, D. J. (1979). *J. Appl. Phys.* 50, p. 6168.
Chen, F. S. (1980). *Electron. Lett.* **16**(1), 7.
Chen, F. S., and Bosch, F. (1988). *IEEE J. Lightwave Technol.* **6**, 475.
Chessman, M., and Sokal, N. (1976). *Electron. Des.* **24**, 110.
Cheung, N. K. (1984). *SPIE Proceedings*, Vol. 479, Fiber Optic Couplers, Connectors & Slice Technology, p. 12.
Chown, M., Davis, A. W., Epworth, R. E., and Farrington, J. G. (1980), *In* "Optical Fibre Communication Systems" (C. P. Sandbank, ed.), Chap. 9. John Wiley and Sons, New York.
Chynoweth, A. G., and Miller, S. E. (1979), *In* "Optical Fiber Communications" (S. E. Miller and A. G. Chynoweth, ed.), Chapter 1, p. 7. Academic Press, New York.
Cohen, L. G., Mammel, W. L., and Jang, S. J. (1982). *Elect. Lett.* **18**, 1023.
Conradi, J., Kim, M. B., Duck, G., Maciejko, R., Straus, J., and Springthorpe, A. J. (1982). *Technical Digest*, OSA/IEEE 5th Topical Meeting on Optical Fiber Communications, Phoenix, p. 26.

Cook, J. S. (1979). *Proc. F.O.C. '79*, Chicago (M. A. O'Bryant and P. Pollshuk eds.), p. 1.
Copeland, J. A. (1978): *Electron Lett.* **14**, 809.
Couch, P. R., Epworth, R. E., and Rowe, J. M. T. (1983). *Proc. 9th European Conference on Optical Commun.*, Geneva, p. 139.
Croft, T. D., Ritter, J. E., and Bhagavatula, A. (1985). *Conf. Opt. Fiber Commun.*, Tech. Digest, Paper WD2.
Danielsen, M., and Mengel, F. (1978). *Elect. Letts.* **14**(16), 505.
Danlicker, R., Bertholds, A., and Maystre, F. (1983). *Proc. 9th European Conference on Optical Commun.*, Geneva, p. 251.
Dawson, R. W. (1980). *IEEE J. of Quant. Electron.* **QE16**(7), 697.
Dean, B. A., and Dixon, M. (1982). *SPIE Proceedings*, Vol. 328 (Laser and Laser Systems Reliability), p. 35.
Dean, B. A., and Dixon, M. (1985). In "Semiconductors and Semimetals," (W. T. Tsang, ed.) Vol. 22 Part C, Chapter 4, Academic Press, New York, p. 153.
Dentai, A. G., Lee, T. P., and Burris, C. A. (1977): *Electron. Lett.* **13**, 484.
Dixon, M. (1984). Unpublished.
Dixon, M., and Dean, B. A. (1981). *Digest of 3rd Int. Conference on Integrated Optics and Optical Commun.*, San Francisco, p. 36.
Dixon, R. W. (1980). *B.S.T.J.* **59**, 669.
Dixon, R. W., and Joyce, W. B. (1980). *Bell Syst. Tech. J.* **59**, 975.
Duff, D. (1980). *Prac. 13th Circuits and Systems Int. Symposium*, Houston, p. 947.
Dumant, J. M., Guillansseau, Y., and Monerie, M. (1980). *Opt. Commun.* **33**, 188.
Dutta, N. K., and Nelson, R. J. (1981). *IEEE J. of Quantum Electron.* **QE17**(5), 804.
Dyment, J. C. (1967). *Appl. Phys. Lett.* **10**, 84.
Dyment, J. C., D Asaro, L. A., North, J. C., Miller, B. I., and Ripper, J. E. (1972). *Proc. IEEE (Lett.)* **60**, 726.
Epworth, R. E. (1978): *Proc 4th European Conference on Optical Commun.*, Genoa, p. 492.
Epworth, R. E. (1981). *Laser Focus*, September, p. 109.
Ettenberg, M., and Kressel, H. (1980). *IEEE J. of Quantum Electron.*, **QE16**(2), 186.
Farrington, J. G. (1981). *Proc. 7th European Conference and Optical Commun.*, Copenhagen, p. 14.2.1.
Favre, F., Jeunhomme, L., Joindot, I., Monerie, M. and Simon, J. C. (1981). *IEEE of Quantum Electron.* **QE17**(6), 897.
Figueroa, L., Slayman, C. W., and Yen, H. W. (1982a). *IEEE J. Quantum Electron.* **QE18**(10), 1718.
Fishman, D. A., Lumish, S., Denkin, N., Schulz, R. R., Chai, S. Y., and Ogawa, K. (1986). *OFC '86 Tech. Dig.*, Paper PD 11.
Forrest, S. R. (1988), *In* "Optical Fiber Telecommunications II" (S. E. Miller and I. P. Kaminow, eds.), Chapter 16. Academic Press, New York.
Fujita, S., Hayashi, J., Uji, T., and Shikada, M. (1987). *Electron. Lett.* **23**, 636.
Furuga, K., Suematsu, Y., and Hong, T. (1978). *Appl. Optics* **17**, 1949.
Garrett, I., and Midwinter, J. E. (1980), *In* "Optical Fibre Communications" (M. J. Howes and D. V. Morgan, eds.), Chap. 6. John Wiley & Sons, London.
Gloge, D. C., and Jacobs, I. (1988), *In*, "Optical Fiber Telecommunications II" (S. E. Miller and I. P. Kaminow, eds.), Chap. 23. Academic Press, New York.
Gloge, D., Marcatili, E. A. J., Marcuse, D., and Personick, S. D. (1979), *In* "Optical Fiber Communications" (S. E. Miller and A. G. Chynoweth, eds.), Chap. 4. Academic Press, New York.

Gloge, D., Albanese, A., Burrus, C. A., Thinnock, E. L., Copeland, J. A., Dentai, A. G., Lee, T. P., Li, T., and Ogawa, K. (1980). *B.S.T.J.* **59**(8), 1365.
Gnauck, A. K., and Bowers, J. E. (1987). *Electron. Lett.* **23**, 801.
Goodfellow, R. C., Debney, B. T., Rees, G. J., and Buss, J. (1985). *IEEE J. Lightwave Technol.* **LT-3**, 1170.
Gray, P. R., and Meyer, R. G. (1977), *In* "Analysis and Design of Analog Integrated Circuits," Chap. 7. John Wiley and Sons, New York.
Gruber, J. Marten, Petschacher, R., and Russer, P. (1978). *IEEE Trans. on Commun.* **26**(7), 1088.
Hanson, D. (1982). *Technical Digest,* OSA/IEEE 5th Topical Meeting on Optical Fiber Communications, Phoenix, p. 6.
Heckman, S. (1981). *Optics Lett.* **6**(4), 261.
Henmi, N., Koizumi, Y., Yamaguchi, M., Shikada, M., and Mito, I. (1988). *IEEE J. Lightwave Technol.* **6**, 636.
Henry, P. S., Linke, R. A., and Gnauck, A. H. (1988), *In* "Optical Fiber Telecommunications II" (S. E. Miller and I. P. Kaminow, eds.), Chap. 21. Academic Press, New York.
Hirota, O., and Suematsu, Y. (1979). *IEEE J. of Quantum Electron.* **QE15**(6), 142.
Ito, T., Machida, S., Nawata, K., and Ikegami, T. (1977). *IEEE J. of Quantum Electron.* **QE13**(8), 574.
Iwahashi, E. (1981). *J. Quantum Electron.* **QE17**(6), 890.
Jacobs, I. (1980). *Bell Laboratories Record* **58**(1), 2.
Jacobs, I. (1986), *In* "Digital Communications" (T. C. Bartee, ed.), Chap. 1. Howard Sams & Co., Indianapolis.
Jarrett, B. (1974). *Electron. Des.* **22**, 96.
Joindot, I., and Boisrobert, C. (1982). *Elect. Letts.* **18**(17), 304.
Jones, J. R. (1982). *IEEE J. Quantum Electron.* **QE18**(10), 1524.
Kaiser, P., and Keck, D. B. (1988), *In* "Optical Fiber Telecommunications II" (S. E. Miller and I. P. Kaminow, eds.), Chap. 2. Academic Press, New York.
Kaminow, I. P., and Li, T. (1979), *In* "Optical Fiber Telecommunications" (S. E. Miller and A. G. Chynoweth, eds.), Chap. 17. Academic Press, New York.
Karr, M. A., Chen, F. S., and Shumate, P. W. (1979). *Appl. Optics*, **18**(4), 1262.
Katz, J., Margalit, S., Harder, C., Wilt, D., and Yariv, A. (1981). *IEEE J. of Quantum Electron.* **QE17**, 4.
Kimura, T. (1988). *IEEE J. Lightwave Technol.* **6**, 611.
Korotky, S. K., and Alferness, R. C. (1988), *In* "Optical Fiber Telecommunications II" (S. E. Miller and I. P. Kaminow, eds.), Chap. 11. Academic Press, New York.
Kraus, A. D., and Bar-Cohen, A. (1983), *In* "Thermal Analysis and Control of Electronic Equipment," Chap. 18. McGraw-Hill, New York.
Kressel, M., and Butler, J. K. (1980), *In* "Semiconductor Lasers and Heterostructure LEDS," Chap. 17. Academic Press, New York.
Kressel, H., and Ettenberg, M. (1975). *Proc. IEEE* **63**, 1360.
Kressel, H., and Ettenberg, M. (1982), *In* "Semiconductor Devices for Optical Communications" (H. Kressel, ed.), Chap. 10. Springer-Verlag, New York.
Kressel, H., Ettenberg, M., Wittke, J. P., and Ladany, I. (1982), *In* "Semiconductor Devices for Optical Communications" (H. Kresse, ed.), Chap. 2. Springer-Verlag, New York.
Kuwahara, T., Watanabe, M., Suzuki, S., and Sudo, S. (1981). *Technical Digest*, 7th European Conference of Optical Communications, Copenhagen, p. 2.2-1.
Lang, R., and Kobayashi, K. (1980). *IEEE J. of Quantum Electron.* **QE16**(3), 347.
Lau, K., and Yariv, A. (1980). *Optics Commun.* **35**(3), 337.
Lau, K. Y., Bar Chaim, N., and Ury, I. (1984). *7th Topical Meeting on Integrated and Guided Wave Optics*, Orlando, p. WB2-1.

Lee, T. P. (1975). *B.S.T.J.* **54**, 53.
Lee, T. P. (1982). *S.P.I.E.*, Vol. 340, Future Trends in Fiber Optic Communications, p. 22.
Lee, T. P., and Dentai, A. G. (1978). *IEEE J. of Quantum Electron.* **QE17**, 150.
Lee, T. P. and Li, T. (1979), *In* "Optical Fiber Telecommunications" (S. E. Miller and A. G. Chynoweth, eds.), Chap. 18. Academic Press, New York.
Lee, T. P., Burrus, C. A., and Saul, R. H. (1988), *In* "Optical Fiber Telecommunications II" (S. E. Miller and I. P. Kaminow, eds.), Chap. 12. Academic Press, New York.
Li, T., and Linke, R. A. (1988). *IEEE Commun. Mag.* **26**, 29.
Linke, R. A. (1985). *J. Quant. Electron.* **QE-21**, 593.
Linke, R. A., Kasper, B. L., Burrus, C. A., Kaminow, I. P., Ko, J.-S., and Lee, T. P. (1985). *IEEE J. Lightwave Technol.* **LT-3**, 706.
Marcuse, D., and Lee, T. P. (1983). *IEEE J. of Quantum Electron.* **QE19**(9), 1397.
Marcuse, D., and Nash, F. R. (1982). *IEEE J. Quantum Electron.* **QE18**, 30.
Mazurczyk, V. (1981). *Electron Lett.* **17**(3), 143.
Miller, S. E. (1979), *In* "Optical Fiber Communications" (S. E. Miller and A. G. Chynoweth, eds.), Chap. 20, p. 663. Academic Press, New York.
Mills, R. J. (1982). *Technical Digest*, OSA/IEEE 5th Topical Meeting on Optical Fiber Communications, Phoenix, p. 6.
Nagai, S., Nita, S., Yamashita, K., Sogo, T. and Takamiya, S. (1983). *Proc. 9th European Conference on Optical Commun.*, Geneva, p. 143.
Nagel, S. R., Walker, K. L., and MacChesney, J. B. (1982). *Technical Digest*, OSA/IEEE 5th Topical Meeting on Optical Fiber Communications, Phoenix, p. 8.
Nakagawa, K., Aoyama, K., Yamada, J., and Yoshikai, (1986). *Conf. Rec. GLOBECOM 86*, Vol. 2, p. 1205.
Nakamura, M. (1983): *Technical Digest*, OSA/IEEE 6th Topical Meeting on Optical Communications, New Orleans, p. 96.
Ogawa, K., and Vodhanel, R. S. (1982): *IEEE J. of Quantum Electron.* **QE18**(7), 1090.
Ohtsuka, T., Fujimoto, N., Yamaguchi, K., Taniguchi, A., Naitou, H., and Nabeshima, Y. (1987). *IEEE J. Lightwave Technol.* **LT-5**, 1534.
Oron, M., Tamari, N., and Ballman, A. A. (1983). *Technical Digest*, Topical Meeting on Optical Fiber Communications, New Orleans, p. 96.
Ostoich, V., Jeppeson, P., and Slaymaker, N. (1975). *Electron. Lett.* **11**, 515.
Pan, J. J. (1981). *Proc. N.T.C. '81*, New Orleans, P. C1.1.1.
Pan, J. J. (1983). *Technical Digest*, OSA/IEEE 6th Topical Meeting on Optical Fiber Communications, New Orleans, p. 74.
Paoli, R. L. (1977). *IEEE J. of Quantum Electron.* **QE13**, 351.
Paoli, T. L. (1976). *IEEE J. of Quantum Electron.* **QE12**, 770.
Paoli, T. L. (1979). *Appl. Phys. Lett.* **34**, 652.
Patterson, R. E., Straus. J., Blenman, G., and Witkowicz, T. (1979). *IEEE Trans. Commun. Com.* **27**, 582.
Personick, S. D. (1979), *In* "Optical Fiber Communications" (S. E. Miller and A. G. Chynoweth, eds.), Chap. 19. Academic Press, New York.
Peterman, K., and Arnold, G. (1982). *IEEE J. of Quantum Electron* **QE 18**(4), 543.
Petermann, K. (1981a). *J. of Opt. and Quant. Electron* **13**, 323.
Petermann, K. (1981b). *Proc. 7th European Conference on Optical Communications*, Copenhagen, p. 10.1–1.
Pinnow, D. A. (1983). *Technical Digest*, OSA/IEEE 6th Topical Meeting on Optical Fiber Communications, New Orleans, p. 85.
Rashleigh, S. C., and Marrone, M. J. (1983). *Technical Digest*, OSA/IEEE 6th Topical Meeting on Optical Fiber Communications, New Orleans, p. 34.

Refi, J. J. (1986). *IEEE J. Lightwave Technol.* **LT-4**, p. 265.
Runge, P. K., and Trischitta, P. R. (1984). *IEEE J. Lightwave Technol.* **LT-2**, 744.
Salter, S. R., Smith, D. R., White, B. R., and Webb, R. P. (1977). *Proc. 3rd European Conf. on Optical Commun.*, Munich, p. 208.
Saul, R. H., Lee, T. P., and Burrus, C. A. (1985), *In* "Semiconductors and Semimetals," (W. T. Tsang, ed.). Vol. 22 Part C, Chapter 5, Academic Press, New York, p. 193.
Sermage, B., Eichler, H. J., Heritage, J. P., Nelson, R. J., and Dutta, N. K. (1983). *Appl. Phys. Lett.*, **42**(3), 259.
Shen, T. M. (1982). Unpublished data.
Shikada, M. Nomura, H., Suzuki, A., Minemura, K., and Sugimoto, S. (1982). *Technical Digest, OSA/IEEE 5th Topical Meeting on Optical Fiber Communications*, Phoenix, Arizona p. 12.
Shumate, P. W. (1988), *In* "Optical Fiber Telecommunications II" (S. E. Miller and I. P. Kaminow, eds.), Chap. 19. Academic Press, New York.
Shumate, P. W., and DiDomenico, M., Jr. (1982), *In* "Semiconductor Devices for Optical Communications" (H. Kressel, ed.), Chap. 5. Springer-Verlag, New York.
Shumate, P. W., Chen, F. S. and Dorman, P. W. (1978). *B.S.T.J.* **57**, 1823.
Smith, D. R., Hooper, R. C., Smyth, P. O., and Rejman, M. A. Z. (1982). *Technical Digest, OSA/IEEE 5th Topical Meeting on Optical Communications*, New Orleans, p. 16.
Smith, D. W. (1978). *Electron. Lett.* **14**, 775.
Smith, D. W., and Hodgkinson, T. G. (1980). *Proc. 13th Circuits and Systems Int. Symposium*, Houston, p. 926.
Smith, R. G., and Personick, S. D. (1982), *In* "Semiconductor Devices for Optical Communications" (H. Kressel, ed.), Chap. 4. Springer-Verlag, New York.
Straus, J., Springthorpe, A. J., and Szentzsi, O. I. (1977). *Electr. Lett.* **13**, 149.
Streifer, W. Scifres, D. R., and Burnham, R. D. (1982). *Appl. Phys. Lett.* **40**(4), 305.
Suematsu, Y., Akika, S., and Hong, T. (1977). *IEEE J. Quantum Electron.* **QE-13**, 596.
Suematsu, Y., Arai, S., and Kishina, K. (1986). *IEEE J. Lightwave Technol.* **LT-1**, 161.
Swartz, R. G., and Wooley, B. A. (1983). *B.S.T.J.* **62**, 1923.
Swartz, R. G., Voshchenkov, A. M., Chin, A. M., Finegan, S. N., Lau, M. Y., Morris, M. D., Archer, V. D., and Ko, P. K. (1986). *Int. Solid State Conference Tech. Digest*, p. 64.
Thompson, G. H. B., and Henshall, G. D. (1980). *Electron. Lett.* **26**, 42.
Throssell, W. R., and Kanabar, Y. (1983). *Proc. 9th European Conference on Optical Commun.*, Geneva, p. 447.
Toughe, T., Nishimoto, H., Okiyama, T., and Hamano, H. (1980). *IEEE International Conference on Communications*, Philadelphia, Pennsylvania (Paper 10.1) p. 301.
Tsang, W. T., Dixon, M., and Dean, B. A. (1983). *J. Quantum Electronics*, **QE-19**(1), 59.
Tseng, C. W., and Chen, B. V. (1983). *Technical Digest*, OSA/IEEE 6th Topical Meeting, New Orleans, p. 88.
Tsukada, T. (1974). *J. Appl. Phys.* **45**, 4899.
Tucker, R. S. (1981). *IEEE Proceedings*, **128 Pt. I**(5), 180.
Tucker, R. S. (1985). *IEEE J. Lightwave Technol.* **LT-3**, 1180.
Tucker, R. S., and Pope, D. J. (1983). *IEEE Trans. on Microwave Theory and Techniques* **MTT-31**, 289.
Uhle, M. (1976). *IEEE Trans. on Electr. Devices* **ED-23**, 438.
White, G., and Burrus, C. A. (1973). *Inst. J. Electron.* **35**, 751.
Yanushefski, K. A., Yanushefski, M. J., Hokanson, J. L., Shastri, K. R., Smeltz, P. D., Wiand, G. T., and Runge, K. (1988). *ECOC '88 Tech. Digest*, PD 1.
Yasugi, T., Ueno, Y., Kajitani, M., and Matsuhashi, X. (1982). *Technical Digest*, OSA/IEEE 5th Topical Meeting on Optical Fiber Communications, Phoenix, p. 14.

Yuasa, T., Kaman, K., Shimazu, M., and Ishil, M. (1986). *10th IEEE International Semiconductor Laser Conference*, p. 2.

Yonezu, M., Sakuma, I., Kobayashi, K., Kamegima, T., Meno, M., and Nanniki, Y. (1973). *Japan J. Appl. Phys.* **12**, 1585.

Zipfel, C. L. (1985), *In* "Semiconductors and Semimetals," (W. T. Tsang, ed.), Vol. 22 Part C, Chapter 6, Academic Press, New York, p. 239.

Zucker, J., and Lauer, R. (1978). *IEEE Trans. on Electr. Devices* **ED25**, 193.

LIGHTWAVE RECEIVERS

GARETH F. WILLIAMS[*]

NYNEX SCIENCE AND TECHNOLOGY
WHITE PLAINS, NEW YORK

I. Introduction 79
II. Receiver and Device Requirements of Lightwave Systems 83
III. Receiver System and Noise Considerations 85
    A. Digital Receiver System Considerations 87
    B. Pin-Photodiode Receiver Noise and Sensitivity Calculations 91
    C. Avalanche Photodiode Receiver Noise and Sensitivity Calculations 107
IV. First- and Second-Generation Lightwave Receivers 112
    A. Voltage-Amplifier Optical Receivers 113
    B. Integrating Optical Receivers 114
    C. Transimpedance Optical Receivers. 117
    D. Integrating Transimpedance Optical Receivers 119
V. Active-Feedback Lightwave Receiver Circuits 121
    A. Introduction 121
    B. Micro-FET Feedback Receivers. 123
    C. Capacitive Feedback Receivers 127
    D. Dynamic Range Extenders 128
    E. Hybrid IC Active-Feedback Receivers 132
    F. IC Active-Feedback Receivers 135
    G. Design Scaling Laws for IC Receivers. 143
    H. Sensitivity Calculations for Present- and Future-Technology IC Receivers . . . 144
    References 148

I. INTRODUCTION

This chapter covers the development of optical receiver circuits as influenced by photodetector and transistor noise physics and by lightwave system requirements. These considerations led directly to the new high-sensitivity, wide-dynamic-range, active-feedback receiver ICs. This chapter also covers photodetector and transistor technology choices and some probably future device developments, in light of the receiver design advances,

[*] Chapter prepared at AT&T Bell Laboratories, Holmdel, New Jersey

the noise physics, and the system requirements. This chapter is written for both receiver designers and device physicists.

The earliest optical receiver circuits were simple voltage amplifiers and were very noisy. These receivers were followed by high-sensitivity integrating front ends (Personick, 1973a, 1973b; Goell, 1974). Because these receivers integrate the signal, they require subsequent equalization (differentiation) with its attendant problems and have limited dynamic range. Thus, although these early integrating receivers showed promise in laboratory experiments, neither they nor the voltage-amplifier receivers were ever used in a commercial system.

The transimpedance receiver was developed next and was the first commercially viable lightwave receiver. Its advantage is that it avoids signal integration; unfortunately, it is less sensitive than integrating receivers. The first commercial transimpedance receivers were developed for 0.8-μm wavelength, 45-Mb/s applications and used silicon avalanche photodetectors (APDs) to achieve high sensitivity (R. G. Smith et al., 1978). More recently, a silicon bipolar IC transimpedance receiver for the TAT-8 transatlantic cable was pioneered by Paski (1980) and perfected by Snodgrass and Klinman (1984); it operates at 296 MB/s and a 1.3-μm optical wavelength.

The superior sensitivity of integrating receivers has continued to attract interest; recently, integrating receivers have come back in long-wavelength (1.3–1.6 μm) receivers using low-capacitance conventional (nonmultiplying) pin photodiodes and fine-line microwave FETs (Hooper et al., 1980; D. R. Smith et al., 1980; Gloge et al., 1980; Ogawa et al., 1983). These improved integrating receivers offer sensitivities comparable to present 1.3- to 1.6-μm APDs below 100 Mb/s and good sensitivities to at least 1 Gb/s (Linke et al., 1983). However, they still require equalization and have limited dynamic range. As a result, these improved integrating receivers have primarily been used for laboratory demonstrations, where these difficulties do not matter, and in a very few leading-edge systems. Nearly all commercial systems have continued to use conventional non-integrating transimpedance receivers.

A new type of lightwave receiver, the active-feedback receiver, offers high sensitivity with a wide-band nonintegrating response, plus the widest dynamic ranges yet achieved (Williams, 1982, 1985; Fraser et al., 1983; Williams and LeBlanc, 1986). Active feedback receivers avoid the signal integration of previous high-sensitivity designs, yet do not compromise the sensitivity; in fact, these new receivers are somewhat more sensitive than corresponding integrating receivers. The dynamic-range-extender circuitry incorporated in these receivers provides the wide dynamic range; in fact, most implementations of these new receivers cannot be saturated by present lightwave transmitters. Active-feedback receivers can be realized inexpensively in IC form using any fine-line FET IC technology and are the first step towards the single-chip lightwave regenerators of the future. Although specific commercial designs are beyond the scope of this chapter, note that IC versions for 1- to 50-Mb/s datalinks and local-area networks are in production by AT&T Technologies (Morrison, 1984; Steininger and Swanson, 1986); high-speed versions for transmission systems operating at bit rates to 1.7 Gb/s have also been announced (Gloge and Ogawa, 1985; Dorman et al., 1987).

Because APDs were the only way to achieve high sensitivities with early receiver circuits, lightwave device physicists focused on APDs. With today's high-sensitivity receiver circuits, sensitivity improvements can come from advances in either photodetector or transistor technology. The lightwave device physicist must now be aware of both.

For example, present long-wavelength (1.3–1.6 μm) APDs offer little sensitivity advantage over nonmultiplying pin photodiodes below 100 Mb/s when used with either the integrating receiver circuit or the new high-sensitivity, active-feedback receiver circuits. Where this dividing line will be in the future depends on the rate of long-wavelength APD development versus that of the fine-line FET technology used in the receiver.

In addition, at bit rates up to a few hundred Mb/s, reducing the long-wavelength APD dark current to that of pin photodiodes is presently more important than achieving a higher ionization coefficient ratio, k, for lower excess multiplication noise; because receiver amplifiers are now so sensitive, the optimal APD gain is small and k is less critical. This makes high-sensitivity, long-wavelength APDs easier to develop.

These receiver advances and their device implications can also influence research into new types of devices. For example, these receiver advances and their implications led directly to the staircase APD proposal of Williams et al. (1982). Only one type of carrier should impact ionize in this structure, giving very low noise multiplication. However, as mentioned in the previous paragraph, a low leakage current is just as important as low-noise multiplication. The low operating voltage and large average band gap of the staircase APD should allow even smaller leakage currents than in long-wavelength nonmultiplying pin photodiodes, where the entire depletion region is narrow gap. Finally, the staircase APD might not have been proposed without the new receiver circuits, because the maximum multiplication of the detector may typically be only ~ 30; the high sensitivities of the receiver circuits mean that higher multiplications will not be needed in order to realize the full sensitivity of the detector.

The circuit and device developments to date have been driven by systems requirements, especially the move from the 0.8-μm to the 1.3-μm wavelength. The early first-generation 0.8-μm transmission systems used high-gain, low-noise silicon APDs, which solved both the sensitivity and dynamic range problems of the early receiver circuits. However, second-generation transmission systems operate in the 1.3- to 1.6-μm region, where the lower fiber loss allows wider repeater spacings. Unfortunately long-wavelength APDs comparable to 0.8-μm silicon APDs do not as yet exist. These receivers must presently use either pin photodiodes or APDs with lower gains and higher noise than the 0.8-μm silicon APDs. What is lost in the detector has had to be made up in the following preamplifier. In addition, APDs and their associated variable high-voltage supplies are presently too expensive for datalinks and local-area networks and not reliable enough for transoceanic submarine cables. In these systems, the low noise amplification and wide dynamic range must at present be achieved in the preamplifier rather than in the detector.

Section II briefly discusses lightwave systems requirements and assesses their impact on present and future optical receivers and devices. This area is sometimes ignored, but it will continue to determine which new device and receiver developments will be used.

Section III lays the foundation for the receiver circuit discussions of Sections IV and V. Almost all lightwave systems are digital; Section III begins by reviewing how optical-fiber digital regenerators are designed to achieve the best sensitivity of which the optical receiver is capable and reviews the derivation of the digital bit-error rate as a function of the receiver noise. It then reviews the classic Smith and Personick (1980) calculations of the receiver input device noise, presents FET and photodetector technology figures-of-merit, and discusses sensitivity calculations for the new active-feedback IC receivers of Section V. These calculations are then compared with sensitivity results reported in the literature. This section then covers technology choices between GaAs FETs and silicon FETs and between pin photodiodes, InGaAs/InP APD detectors, and Ge APD detectors.

Section IV reviews and discusses the first- and second-generation lightwave receiver circuit designs. The early first-generation commercial 0.8-μm transimpedance receiver amplifiers traded sensitivity for bandwidth and used a silicon APD to make up the sensitivity loss; in the design of R. G. Smith *et al.* (1978), the APD improved the sensitivity by 15 dB over the same receiver with a pin photodiode. The later silicon bipolar transimpedance receiver ICs (Paski, 1980; Snodgrass and Klinman, 1984) were designed for 1.3-μm applications and use nonmultiplying pin photodiodes. They have achieved good sensitivities at bit rates of several hundred megabits per second. These receivers are a genuine *tour de force* and probably represent the end of their line of development.

The second-generation 1.3- to 1.6-μm pin-FET receiver circuits use integrating or "high-impedance" receiver designs. Section IV discusses both the simple integrating receivers, which have been introduced commercially in a few leading-edge systems by the British Post Office (Hooper *et al.*, 1980) and the integrating transimpedance receivers favored by Gloge *et al.* (1980) and Ogawa *et al.* (1983). Both types achieve very high sensitivities; however, they integrate the signal and have little dynamic range.

Section V describes the new active-feedback receivers. This section first reviews the basic features of these designs, including both the novel feedback techniques and the dynamic-range-extender circuitry. Response, stability, and sensitivity considerations are covered in detail. This section then briefly discusses an illustrative design for a complete hybrid IC active-feedback receiver. This is followed by an extended discussion of active-feedback receiver ICs. Two generic IC designs are presented; these can be realized in any of the standard fine-line FET IC technologies, including NMOS, CMOS, and GaAs.

Section V also presents scaling laws, which are used to generate different versions of these IC receiver designs for different bit-rate applications, then presents calculations of IC receiver sensitivities for bit rates between 10 Mb/s and 4 Gb/s.

These high-sensitivity active-feedback IC receivers will achieve further sensitivity advances as the pin-photodiode and FET technologies develop. The development of submicron gate FETs is driven by very-high-speed GaAs and silicon digital IC programs; the new optical receiver ICs are compatible and will piggyback on these efforts. In addition, the pin-photodetector capacitance appears to be halving every two to three years. Section V concludes with theoretical calculations of the probable sensitivities to be expected from future IC receivers using these improved pin and FET technologies.

In long-haul terrestrial systems at several hundred megabits per second or above, these IC receivers often use InGaAs/InP APD detectors instead of pin photodiodes. Future IC receivers for these applications will require APDs with lower dark currents and lower junction capacitances than present devices. The importance of a small ionization coefficient ratio, k, will depend on whether the dark currents are decreased or the commonly used bit rates are increased faster than the FET technologists are able to achieve shorter channel lengths. Smaller APD junction capacitances would also reduce the importance of a small k ratio.

II. RECEIVER AND DEVICE REQUIREMENTS OF LIGHTWAVE SYSTEMS

This section first looks at the requirements of present and future systems and then at the likely impact of these requirements on present and future receiver and device technologies.

The important receiver design goals are high sensitivity, wide dynamic range, adaptability to high bit rates, and low cost. Most of the lightwave literature to date has focused on technology for long-haul transmission systems, which comprise the bulk of the present optical-fiber applications. The key goals in transmission systems are long repeater spacings, to minimize the number of repeaters, and high channel capacity. Accordingly, most of the receivers in the literature are optimized for high sensitivity and high bit rate; wide dynamic range and low cost are less important.

However, most future optical-fiber applications will be in the loop plant (between the central office and the subscribers), in local-area networks (e.g., on-premises office-of-the-future networks), in optical-fiber cable TV, and in data links (Miller, 1979). All require high sensitivity, wide dynamic range, and low cost; the bit rate varies with the application.

The new high-sensitivity, wide-dynamic-range active-feedback receivers discussed in Section V were originally invented for these applications (Williams, 1982, 1985). The IC versions can be realized economically in fineline digital VLSI technology and are adaptable to bit rates in excess of 1 Gb/s. They therefore are a superior alternative for high-bit-rate transmission systems as well.

These loop, local-area-network, cable-TV and datalink applications require high sensitivity because they all frequently include some type of very-high-loss optical data path. They all require wide dynamic range because the minimum loss path in each is a short, almost connectorless, fiber run with a loss of a few decibels or less; essentially the entire transmitter power can appear at a receiver. Field-installed optical attenuators are not a viable means to reduce the dynamic range requirement because of the expense of installing them and of changing them if the system is reconfigured, plus the need to keep a record of what value attenuator was installed where. Low cost is imperative because the total number of receivers in these applications is so large; each customer, work station, or computer may have its own receiver.

In loop feeder, the highest loss path is between the central office and the telephone-call concentrator of the furthest neighborhood. The higher the sensitivity, the wider the radius that a single central office can serve without repeaters; the number of customers increases as the square of the radius and every decibel is an additional 1–2 kilometers.

On-premises local-area networks also need high-sensitivity receivers, especially in star or tree configurations in which each transmitter's power can be divided among many receivers and several fiber branches. The same applies to fiber cable-TV systems. Typical local-area-network signal paths can also have many fiber connectors, each with a signal loss; in addition, a local-area network should be extendable to several buildings, e.g., on a campus or in an industrial park; link lengths can be up to a few kilometers.

The implications for circuit and transistor technology designers are clear. From now on, lightwave receivers will be IC receivers; the transistor technologies used will be IC technologies (with the possible exception of integrated pin-FETs). Most future applications are in loop, local-area networks, cable-TV, and datalink systems, which require relatively high performance and very low cost; those receivers absolutely must be realized as ICs. As mentioned in the introduction, GaAs and silicon FET IC technologies are presently preferred; any new transistor technology, e.g., high-electron-mobility transistors or heterojunction bipolar transistors, should also be realizable in IC form for these applications. In addition, the new nonintegrating high-sensitivity wide-dynamic-range receiver circuits developed for those applications are best realized in IC form (Section 5). Furthermore, the big development efforts that can be expected for loop-feeder receivers will ensure that IC receivers will always be at or near the sensitivity limit of the detector and transistor technologies, and hence superior for transmission systems as well.

New detectors will first be adopted in long-haul transmission systems, where sensitivity is most important and cost is secondary. They will be adopted in loop, LANs, cable-TV and datalink applications only if their cost comes down; these systems are cost critical, and the projected advances in pin-FET sensitivities should adequately meet the future needs of these systems. Most of the applications will run on 5 volts; few can justify the expense of a high-voltage supply for an APD. Therefore, low-voltage detectors, such as heterojunction phototransistors, integrated pin-FETs, or staircase APDs are preferred possibilities; high-voltage devices like conventional APDs will probably be restricted to transmission systems only.

III. Receiver System and Noise Considerations

This section covers the systems background and the device noise theory for the receiver amplifier circuit discussions of Sections IV and V. It begins with a tutorial on receiver systems, then reviews the classic Smith and Personick (1980) expression for the input device noise of pin-FET receivers. It then extends these expressions to include the noise of the rest of the receiver circuit and derives figure-of-merit expressions for pin, FET, and FET IC technologies. It then calculates theoretical sensitivities for FET IC receivers and compares present silicon and GaAs FET IC technologies on the basis of these noise calculations, experimental results from the literature, and circuit considerations. This section concludes by reviewing the APD receiver noise theory of Smith and Personick (1980) and calculating sensitivities for present-technology APD-FET lightwave receivers.

Lightwave receivers using bipolar transistors will not be considered because they are *presently* less sensitive than FET lightwave receivers at all bit rates. However, new bipolar technologies, such as heterojunction bipolar transistors (Kroemer, 1982, 1983) may someday change this.

This section focuses on digital receiver systems because most optical fiber systems are digital. Analog video is the only major exception, but will likely be replaced by digital video once economical video codec ICs become available. These codecs are realizable in the silicon or GaAs FET technologies favored for the receivers; ultimately, the receiver and codec will probably be on the same IC.

Section III.A is a brief tutorial for device physicists on the filtering and digital signal recovery circuits that extract the digital bit stream from the analog output of the optical receiver circuit and on how these circuits affect the digital-receiver sensitivity. This section contains just enough detail to place the optical receiver in the system and to motivate the noise treatment of Section III.B; circuit engineers may wish to consult Maione et al. (1978) to see a complete digital-signal recovery circuit.

Section III.B covers sensitivity and noise calculations for pin-FET receivers, including the active-feedback IC receivers of Section V. This section first reviews the connection between the signal-to-noise ratio at the output of the receiver and the digital bit-error rate at the output of the digital signal-recovery circuit, and the assumptions involved. Subsection III.B.1 then presents a semi-intuitive derivation of the Smith and Personick (1980) expression for the receiver front-end noise due to the input FET, the pin photodiode, and the input resistor. These devices account for the majority, but not all, of the receiver noise. Subsection III.B.2 extends this expression to include the noise contributions of the first-stage load device and of later stages; the resultant new expression can be used to calculate sensitivities for the complete IC receivers of Section V.F. Subsection III.B.2 also presents lightwave receiver figure-of-merit expressions for pin, FET, and FET IC technologies and then discusses the device implications. For example, microwave or VHSIC FETs are preferred even for 10- to 100-Mb/s lightwave receivers, but for noise, not speed, reasons. Subsection III.B.3 calculates theoretical sensitivities of present-technology silicon and GaAs FET IC receivers as a function of bit rate. Sensitivity results from the literature are also included. In theory, the two technologies should give essentially the same receiver sensitivities; GaAs FETs have a higher saturation velocity, but silicon FETs have a lower channel-noise factor.[1] In practice, GaAs is *presently* a few decibels superior at most bit rates, probably because fine-line GaAs FETs began as an analog microwave technology. Therefore, GaAs noise problems were important and were solved. Fine-line silicon FETs began as a digital IC technology in which analog noise was unimportant and was almost ignored.

---

[1] The extra channel noise in fine-line GaAs FETs is largely due to high-field scattering from the high-mobility central valley to low-mobility satellite valleys (Baechtold, 1972). This effect is used in Gunn diodes.

Section III.C discusses noise and sensitivity considerations for avalanche photodiode (APD) receivers. This section includes an APD receiver noise tutorial following the APD receiver noise theory of Smith and Personick (1980), which is based on the APD device noise theory of McIntyre (1966). It then shows optical receiver sensitivity calculations for present-technology APDs used with present-technology IC receivers. It considers only 1.3- to 1.6-μm wavelength APDs; 0.8-μm APD systems are obsolete because their repeater spacing is less than that of 1.3-μm pin-photodiode systems due to the much greater fiber loss at 0.8μm. These calculations indicate that receivers using present-technology low-leakage-current InGaAs/InP APDs are more sensitive at all bit rates than receivers using present germanium APDs and that at a typical maximum operating temperature of 85°C, present-technology pin-FET receivers can be more sensitive than present germanium APD receivers at bit rates below 200 Mb/s. In fact, these calculations indicate that even present InGaAs/InP APDs have little sensitivity advantage (at 85°C) over present-technology pin-FETs below 100 Mb/s and are less sensitive below $\sim$ 50 Mb/s due to their higher leakage currents.

A. Digital Receiver System Considerations

A typical digital receiver system block diagram is shown in Fig. 1a; associated waveforms are shown in Fig. 1b. For now, assume a non-return-to-zero (NRZ) signal format; if a bit is a "one," the signal is high during the entire bit interval; if a bit is zero, the signal is low during the entire bit interval. The receiver amplifier takes the photocurrent signal input, i_{ph}, and gives an output voltage, v_o, of the same waveshape; however, this output voltage also includes noise due to the detector and the receiver amplifier. The average noise shown on v_o is much less than the signal; however, occasionally a random fluctuation is bigger than the signal and can turn a 1 into a 0 or a 0 into a 1, causing an error. Typical systems require a bit-error rate (BER) of 10^{-9}, or one error per billion bits. The digital-receiver sensitivity is then the input optical signal power required for a signal-to-noise ratio high enough to get a BER of 10^{-9}.

The receiver output noise can be reduced by filtering; the receiver bandwidth is usually wider than necessary to pass the photocurrent signal; the extra bandwidth contains extra noise. This extra noise is removed by the channel filter, improving the signal-to-noise ratio at the input to the digital decision circuit. The total noise power on the channel filter output signal , v_f is the spectral noise power density $S(\omega)$ of the amplifier plus that of the photodetector integrated over the filter bandwidth; the lower the filter cutoff frequency, f_c, the lower the noise bandwidth and the less noise at the input to the decision circuit.

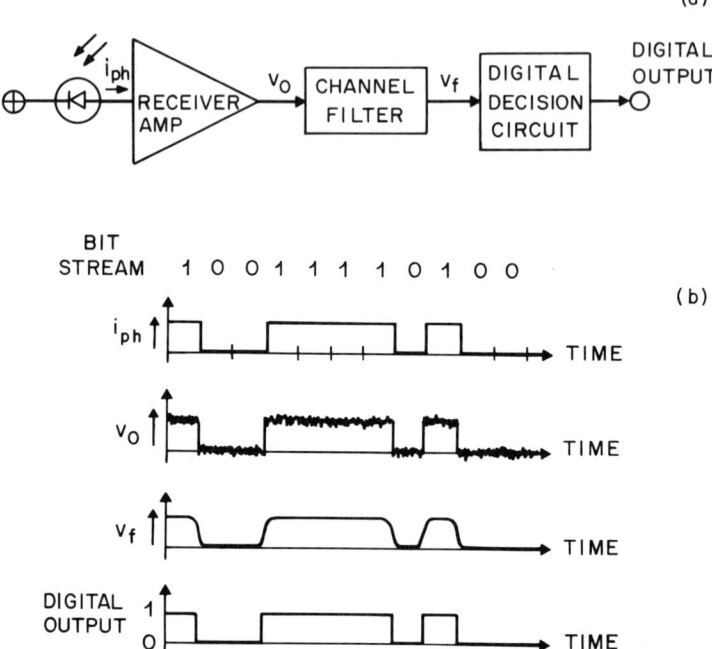

FIG. 1. Optical receiver system: (a) block diagram, (b) typical waveforms.

The digital decision circuit recovers the digital bit stream from v_f. There are two types of decision circuits. The asynchronous decision circuit is presently cheaper and has been used in datalinks. The synchronous decision circuit, in which a clock is recovered from the signal and is used to sample v_f at the center of each zero or one (Maione *et al.*, 1978), is more sensitive because it allows the use of a narrower channel-filter bandwidth, which removes more noise. In time, the synchronous circuit will be included in the receiver IC and will thus cost essentially nothing.

Note that the receiver amplifier typically includes automatic gain control (AGC) so that the peak-to-peak signal at the decision circuit input is held constant, independent of the input optical signal level. This ensures proper decision circuit operation. This AGC function has typically been implemented in a postamplifier (not shown) after the input receiver amplifier (Maione *et al.*, 1978); in the new high-sensitivity, wide-dynamic-range amplifiers of Section V, AGC is provided in the input amplifier as well.

The asynchronous decision circuit (Fig. 2) is just a discriminator. It gives a logic zero output when the filter output, v_f, is less than the decision threshold voltage, V_T; it gives a logic one output when v_f is greater than V_T. (Typically, V_T

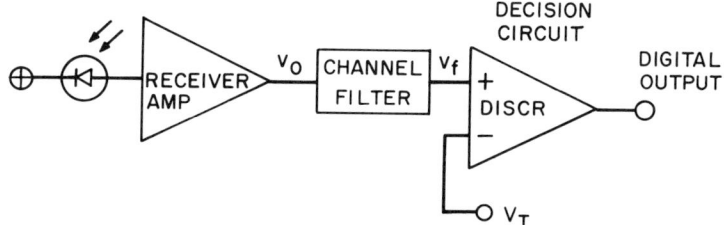

FIG. 2. Asynchronous decision circuit optical receiver system.

is midway between the "zero" and "one" analog signal levels.) Each digital transition (zero to one or one to zero) in the output bit stream occurs at the moment the signal v_f passes through V_T. Any noise as the signal transition passes through V_T will cause a transition timing error. These timing errors are minimized by maintaining fast rise and fall times on the signal v_f; this means a wide channel-filter bandwidth, hence high noise, thus reducing the achievable sensitivity.

The channel-filter bandwidth for an NRZ asynchronous receiver would typically be about twice the bit rare B, giving a risetime of about 0.17 times the bit interval. (The 10% to 90% risetime is $\tau_r \approx 0.35/BW$).

The synchronous decision circuit (Fig. 3a) samples the filtered signal, v_f, at the center of each bit interval, as indicated by the arrows in Fig. 3b, and decides whether it represents a zero ($v_f < V_T$) or a one ($v_f > V_T$). The circuit then puts out a standard length one or zero pulse exactly one bit-interval long. The sampling circuit is triggered at the center of every bit by the clock recovery circuit,[2] which reconstructs the transmitter's digital clock signal (timing) from the received signal. The decision circuit output bit intervals are each one cycle of the recovered clock. Since the signal is sampled only in the center of each bit interval, the rise and fall times can be very slow; this means a narrow channel-filter bandwidth and hence low noise. Thus, the synchronous decision circuit gives the best digital-receiver sensitivities.

The channel-filter bandwidth for a synchronous NRZ receiver is typically set at 0.56 times the bit rate (Maione et al., 1978; Smith and Personick, 1980), thus removing the high-frequency part of the NRZ signal spectrum. The resultant waveform, v_f, is shown in Fig. 3b; the risetime is now about 0.6 times the bit interval; the waveform is still satisfactory because the time between sampling points is one bit interval. The noise bandwidth is now only 28% of

---

[2] Typically, the clock recovery circuit is a phase-locked loop in which a voltage-controlled oscillator is synchronized on the bit transitions in the analog signal (Maione et al., 1978). This circuit effectively averages over many bit periods and many transitions; noise on any one transition is averaged out.

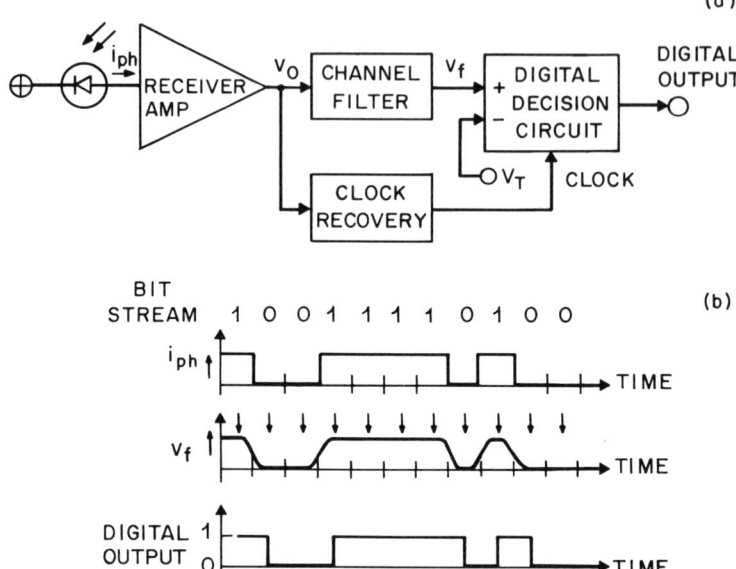

FIG. 3. Synchronous decision circuit optical receiver system: (a) block diagram, (b) typical waveforms.

that of the asynchronous receiver example; the receiver sensitivity is increased accordingly.

Typical optical receiver noise power spectra (Section III.B) contain a term independent of frequency, which dominates at low frequencies and a term proportional to the frequency squared, which dominates at high frequencies. The frequency-independent noise term is due to the device leakage currents and the circuit; the frequency-squared term is primarily due to the input FET. In high-sensitivity designs (Sections IV and V), the frequency-independent term can essentially be eliminated over most of the bandwidth. Therefore, the frequency-squared noise term is dominant; integrating this term over the signal bandwidth gives a total electrical noise power proportional to the receiver bandwidth cubed. Since the synchronous receiver example has only 28% of the noise bandwidth of the asynchronous receiver, for example, the electrical noise power is only $0.28^3 = 0.022$ times as much; the noise voltage is $\sqrt{0.022} = 0.15$ times as much. Since the signal voltage is proportional to the optical power, the minimum optical signal is 0.15 times smaller. This is an 8.3-dB optical sensitivity advantage for the synchronous detection example.

Until now, transmission systems have used synchronous decision circuits for maximum sensitivity; asynchronous circuits are used in some very-short-haul, low-sensitivity datalinks. However, since a linear channel and synchronous detector circuit can ultimately be very economically integrated on the same IC as the receiver amplifier, asynchronous systems will no longer be cheaper. This means that almost all transmission, loop, and local-area-network receivers plus many datalink receivers will use synchronous detection for better sensitivity. Therefore, the noise treatment of Section 3.2 assumes synchronous detection.

So far, this discussion has assumed an NRZ optical data transmission format. Other formats such as return-to-zero (RZ), biphase or Manchester, and block codes have been considered (Takasaki et al., 1976). Both RZ and biphase approximately double the receiver noise bandwidth, for an approximately 4.5-dB optical sensitivity penalty; they therefore are not preferred. Block-coding schemes (Rousseau, 1976; Brooks, 1980) in NRZ format have been used with the integrating receivers of Hooper et al. (1980) discussed in Section IV.B. These involve a smaller sensitivity penalty. The biphase and block-coding schemes reduce the low-frequency content of the signal, which can be helpful in ac-coupled systems or in integrating receivers; in addition, they provide more frequent signal transitions to help maintain synchronization of the clock recovery circuit. At present, the NRZ format is preferred for sensitivity; however, it may be used either with self-synchronizing scrambling or with block coding; both can be realized in IC form on the same chip.

B. pin-Photodiode Receiver Noise and Sensitivity Calculations

This section discusses noise and sensitivity calculations for lightwave receivers using (nonmultiplying) pin photodiodes and FET input stages, and lays the noise-theory foundation for the receiver circuit designs of Sections IV and V. Subsection III.B.1 is a tutorial review of the classic Smith and Personick (1980) expressions for the receiver front-end noise due to the input FET, the photodetector, and the input resistor. Subsection III.B.2 first extracts pin- and FET-technology figures of merit from the Smith-and-Personick front-end figure of merit, then discusses the device implications. It then extends the noise expressions to include noise from the first-stage load device and from later stages; the resultant expression gives accurate sensitivities for IC receivers. Subsection III.B.2 also defines a simple expression for an IC technology figure of merit. Note that the figures of merit help dictate device design; the noise expressions as a whole help dictate the receiver circuit designs of Section V. Subsection III.B.2 calculates theoretical sensitivities of present-technology GaAs and silicon FET ICs versus bit rate, and compares the two technologies on the basis of the calculations, some results from the literature, and circuit considerations.

The receiver-circuit noise expressions are for equivalent RMS input noise currents; the real noise sources in the amplifier and detector are replaced by a single equivalent noise current source at the input to a noise-free equivalent amplifier. This is convenient because the signal-to-noise ratio at the amplifier output is then just the photocurrent divided by the equivalent input noise current. In addition, these results are not affected by the feedback techniques used in the new receivers of Section V; the noise sources can be referred to the input before mentally closing the feedback loop; the loop is thus closed around the noiseless equivalent amplifier; by inspection, the equivalent input current noise source, which is outside the loop, is unchanged. Thus these open-loop noise calculations apply to all the circuits of Section V.

For digital receiver systems, the signal-to-noise ratio at the digital decision circuit input is given by the photocurrent signal divided by the receiver-amplifier-equivalent input-noise expression, provided that the noise bandwidth is taken as the channel-filter bandwidth. As mentioned in Section III.A, the channel-filter bandwidth for a synchronous NRZ receiver system is typically 0.56 times the bit rate B. The photocurrent signal can be had from the optical signal power by remembering that a 1-eV photon has a wavelength of 1.240 μm; at that wavelength, 1 watt of detected optical power gives 1 ampere of photocurrent. This then gives

$$\bar{I} = \eta \bar{P} \frac{\lambda}{1.240}, \tag{1}$$

where $\bar{I}$ is the average photocurrent in amperes, η is the photodiode quantum efficiency, $\bar{P}$ is the average optical power in watts, and λ is the optical wavelength in micrometers.

The problem now is to turn this signal-to-root-mean-square noise ratio at the decision circuit into a digital bit-error rate (BER). Consider a synchronous detection digital receiver system, as discussed in Section III.A. The digital decision circuit samples the signal once at the center of each bit interval; if the signal, s, is below the decision threshold D, the bit is read as a zero; if s is greater than D it is read as a one. The probability of error, i.e., the BER, is the probability that the noise at the sampling instant will bring the zero signal above D or the one signal below D.

By the central limit theorem, the noise amplitude probability distribution is Gaussian if the noise amplitude is the sum of many small independent physical process (Davenport and Root, 1958). The probability distribution for the zero-level signal is then, following Smith and Personick (1980).

$$P(s) = \frac{1}{\sqrt{2\pi}\,\sigma_0} e^{-[s-s(0)]^2/2\sigma_0^2}, \tag{2}$$

where $s(0)$ is the noise-free, zero-level signal, σ_0 is the signal variance, or root-mean-square zero-level analog noise. The probability E_{01} of mistakenly identifying a zero bit as a one is then (Fig. 4)

$$E_{01} = P(s > D) = \frac{1}{\sqrt{2\pi}\,\sigma_0} \int_0^\infty ds\, e^{-[s-s(0)]^2/2\sigma_0^2}.$$

The derivation of E_{10}, the probability of mistakenly identifying a one-bit as a zero, is similar. Changing variables, the general error probability is

$$P(E) = \frac{1}{\sqrt{2\pi}} \int_Q^\infty dx\, e^{-X^2/2} = \frac{1}{2}\mathrm{erfc}\left(\frac{Q}{\sqrt{2}}\right), \tag{3a}$$

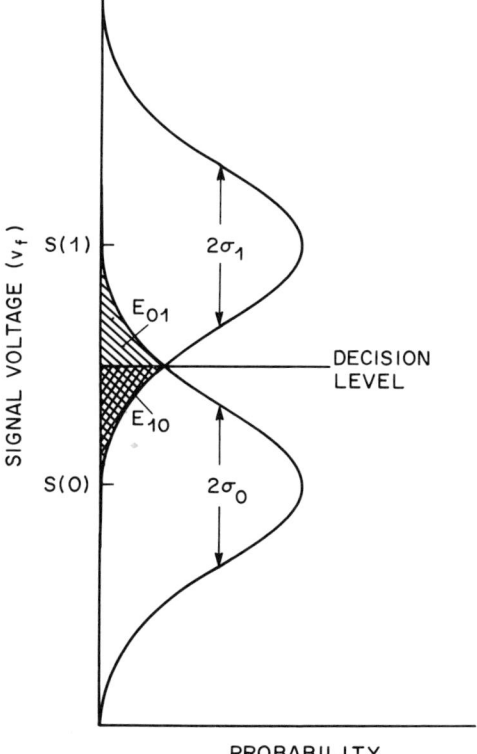

FIG. 4. Error probabilities for two-level digital system. E_{01} is the probability of mistakenly identifying a zero bit as a one bit; E_{10} is the probability of mistakenly identifying a one bit as a zero bit. [After Smith and Personick (1980).] Copyright © 1989, AT&T Bell Laboratories, reprinted by permission.

where

$$Q = \frac{(D - s(j))}{\sigma_j} \tag{3b}$$

and for $j = 0$, $P(E) = E_{01}$; for $j = 1$, $P(E) = E_{10}$.

For pin receivers, the zero-signal-level RMS noise, σ_0, and the one-signal-level RMS noise, σ_1, are essentially equal. Assuming a random bit stream (maximum information content), ones and zeros are equally frequent. The optimum decision level D is then midway between $s(0)$ and $s(1)$, and $E_{01} = E_{10}$. The bit-error rate (BER) is now

$$\text{BER} = \frac{1}{2}\text{erfc}\left(\frac{Q}{\sqrt{2}}\right), \tag{4}$$

where

$$Q = \frac{s(1) - s(0)}{2\langle s_n^2 \rangle^{1/2}},$$

where $\langle s_n^2 \rangle^{1/2} = \sigma$ is the RMS noise. Q is then just half the peak-to-peak signal to RMS noise ratio at the digital decider input. However, as mentioned, this SNR is just the photocurrent signal to equivalent input RMS noise current ratio. Assuming that the zero-level photocurrent is zero, the average photo current $\bar{I}$ is half the peak (one-level) photocurrent I_{pk} and

$$Q = \frac{\bar{I}}{\langle i_n^2 \rangle^{1/2}} = \frac{I_{pk}}{2\langle i_n^2 \rangle^{1/2}}, \tag{5}$$

where $\langle i_n^2 \rangle^{1/2}$ is the RMS equivalent input noise current.

Equations (4) and (5) give the pin-receiver bit-error rate (BER) in terms of the photocurrent signal to RMS equivalent input-noise current ratio. The BER as a function of Q is shown in Fig. 5. A typical system requirement is for a BER of 10^{-9}; this corresponds to $Q = 6$ or an average photocurrent of six times the RMS equivalent input-noise current. This corresponds to a peak-to-peak signal at the digital decider input that is 12 times the RMS noise. The optical power to get that photocurrent is then the digital optical receiver sensitivity.

The actual digital sensitivity of some early experimental receivers was less than that predicted from the measured RMS analog noise on the basis of the above Gaussian noise discussion (Williams 1982). The reason was non-Gaussian device noise due to surface leakage or bulk defects; later receives with better devices did not have this problem. The fundamental device noise sources (Sections III.B.1 and III.B.2) are almost always Gaussian because they are the sums of many small independent physical processes, as required by the central limit theorem (Davenport and Root, 1958). For example, a photodiode bulk-leakage current of 1 nA corresponds to 63 independent

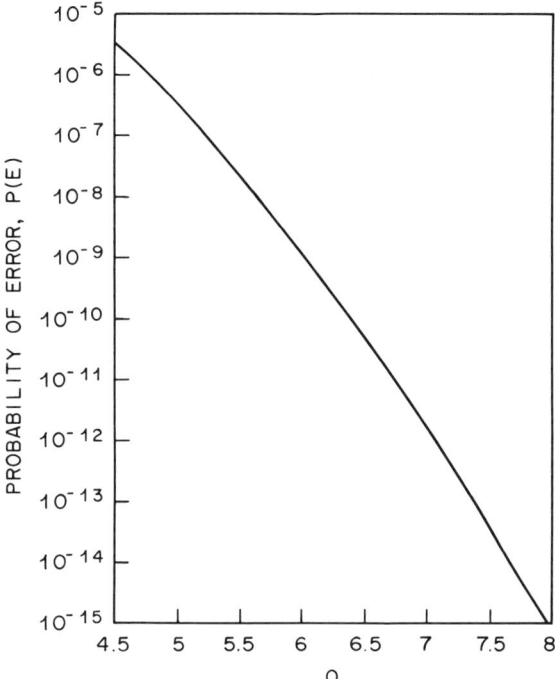

FIG. 5. Probability of error versus the average signal-to-RMS-noise ratio, Q. [After Smith and Personick (1980).] Copyright ©1989, AT&T Bell Laboratories, reprinted by permission.

carrier generations per bit at 100 Mb/s; similarly, the channel Johnson noise current of a typical input FET is produced by the Brownian motion of about 10^6 electrons. The result is the Gaussian noise amplitude probability distribution, which falls off exponentially in the square of the noise amplitude; this is why an average photocurrent that is only six times the RMS equivalent input-noise current can give a BER of 10^{-9}. On the other hand, device defeat or surface leakage noise can be the result of only a few physical processes at a few sites and therefore non-Gaussian; this means more bit errors for a given noise level.

Thus, both the RMS noise spectrum and the noise-amplitude distribution must be measured for both devices and receivers. In addition, early devices with defect and surface noise problems may have non-Gaussian noise, leading to lower digital receiver sensitivities. For example, even though the ideal shot noise of a 20-nA photodiode leakage current may be acceptable in theory, a real diode with 20-nA leakage is unacceptable if the leakage is a non-Gaussian process due to a few device defects or surface leakage paths. Thus, a high-leakage device may be acceptable in theory but unusable in practice if the leakage is a non-Gaussian process. However, a mature device technology, in which surface and defect noise sources are under control, usually has Gaussian noise.

1. Input Device Noise Theory for pin-FET Receivers

Figure 6 shows a generalized pin-FET receiver amplifier. This subsection presents a tutorial review of the classic Smith and Personick (1980) theory for the noise due to the input devices shown in Fig. 6, i.e., the channel noise of the input FET, the Johnson noise of the bias/feedback or input resistor, and the shot noise due to the input FET gate and pin-photodioide leakage currents. This theory gives much of the information about FET- and pin-technology requirements and amplifier design constraints. Subsection III.B.2 derives pin, FET, and FET IC technology figures of merit, and extends the Smith and Personick expression to include the noise of the input FET's load device and the noise of subsequent stages. This is useful for sensitivity calculations for the circuits of Sections IV and V.

As mentioned, the equivalent input noise current is the same whether or not the feedback techniques of Section V are used; for simplicity, the discussion below assumes no feedback.

The input bias or feedback resistor R_i contributes a mean Johnson thermal input noise current squared per frequency bandwidth df of

$$d\langle i_n^2 \rangle_R = \frac{4kT}{R_i} df. \tag{6}$$

The different frequency components of the noise are both independent and orthogonal. Therefore, the total mean-square input noise $\langle i_n^2 \rangle$ due to the input resistor is simply the mean-square spectral noise density of Eq. (6) integrated over the channel frequency bandwidth. If $|F(f)|$ is the magnitude of the normalized channel-filter frequency response, then

$$\langle i_n^2 \rangle_R = \int_0^\infty \frac{4kT}{R_i} |F(f)|^2 df; \tag{7}$$

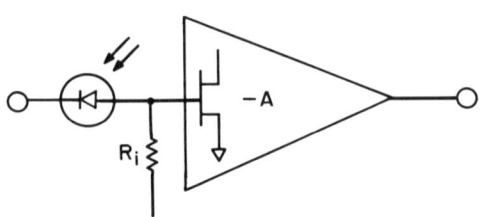

FIG. 6. General pin-FET receiver amplifier.

the independent noise currents squared are multiplied by the filter response magnitude squared, then integrated.

By inspection, this resistor-noise integral is proportional to the bandwidth or to the bit rate B times a numerical factor that depends on the shape of the filter frequency response function $F(f)$. (The same filter function is scaled up or down in frequency for different bit rates.) Changing variables from the frequency f to $y = f/B$ (the frequency normalized to the bit rate) gives

$$\langle i_n^2 \rangle_R = \frac{4kT}{R_i} I_2 B, \tag{8}$$

where

$$I_2 = \int_0^\infty dy |F^*(y)|^2 \tag{9}$$

where $F^*(y)$ is the filter frequency response shape; the numerical factor I_2 is called the second Personick integral.

The next question is how to normalize the channel frequency response $F(f)$. $F(f)$ must give the fraction of the input-noise-current frequency component that appears at the linear channel output to the digital decider. This is best done by first normalizing in the time domain so that a unit input photocurrent pulse maps to a unit output pulse to the decider. The corresponding filter frequency function normalization then applies to the noise as well.

The receiver input is a photocurrent; the input pulse shape $h_p(t)$ is normalized to correspond to a unit photocurrent over a bit interval $T = 1/B$:

$$\frac{1}{T} \int_{-\infty}^\infty h_p(t) \, dt = 1. \tag{10}$$

These receivers are used with synchronous decision circuits, which sample the waveform at the center of each bit interval when the pulse is a maximum (logic one bit) or a minimum (logic zero bit). Thus, a unit filtered pulse is one which is of unit amplitude at the sampling instant, t_o;

$$h_{out}(t_o) = 1. \tag{11}$$

Defining $H_p(f)$ and $H_{out}(f)$ as the fourier transforms of $h_p(t)$ and $h_{out}(t)$ then gives the normalized filter function:

$$F(f) = \frac{H_{out}(f)}{H_p(f)}. \tag{12}$$

Note that, for a given filter design, changing the input pulse shape changes the magnitude but not the frequency dependence of $F(f)$; $F(f)$ is normalized such that a unit average photocurrent pulse of whatever pulse shape chosen gives a unit amplitude input at the sampling instant to the decision circuit, but the shape of $F(f)$ depends only on the channel filter design.

The remaining question is how to arrive at the frequency dependence of the channel filter function $F(f)$; in other words, how to design the channel filter. By inspection, a narrow channel-filter bandwidth means smaller Personick integrals and less receiver noise; however, a narrow filter bandwidth also means slow rise and fall times, which cause the filtered bit-pulse $h_{out}(t)$ to spread into the neighboring bit intervals. Thus, choosing the filter function is a tradeoff between noise considerations in the frequency domain and pulse-shape considerations in the time domain. Since the decision circuit samples the waveform at the center of each bit interval, the bit-pulse $h_{out}(t)$ should peak at the center of its own bit interval and should be zero at the centers of the neighboring bit-intervals to avoid inter-symbol interference (ISI). The filter typically is designed by iterating between the time and frequency domains, alternately minimizing the ISI and the noise bandwidth, respectively. The result usually is a filter bandwidth of 0.5–0.6 times the bit rate for a NRZ signal. For such a typical filter design, I_2 worked out to be 0.56 (Maione et al., 1978).

The FET gate-leakage current and the pin-leakage current contribute an input shot noise current squared per unit frequency bandwidth of

$$d\langle i_n^2 \rangle_\ell = 2qI_\ell\, df, \tag{13}$$

where the total leakage current I_ℓ is the sum of the pin-photodiode leakage current $I_{\ell\text{pin}}$ plus the FET leakage current $I_{\ell\text{FET}}$, and q is the electron charge. Integrating over frequency, as before, gives the total input-leakage-current noise in the channel filter bandwidth:

$$\langle i_n^2 \rangle_\ell = \int_0^\infty 2qI_\ell |F(f)|^2\, df \tag{14}$$

or

$$\langle i_n^2 \rangle_\ell = 2qI_\ell I_2 B, \tag{15}$$

where I_2 is the second Personick integral as before.

The input FET channel noise is essentially the Johnson noise of the unpinched-off portion of the channel next to the source. This channel conductance is simply the transconductance of the FET; the drain current noise per unit bandwidth is then

$$d\langle i_n^2 \rangle_{\text{drain}} = 4kTg_m\Gamma\, df, \tag{16}$$

where Γ is a factor to account for high-field effects in short-channel transistors (Baechtold, 1972). For 1-μm gate length silicon MOSFETs, Γ is typically 1 to 1.2 (Ogawa et al., 1983); for 1-μm GaAs FETs, Γ is typically 1.4 to 1.8 (Ogawa, 1981).

This mean-square FET drain noise current must be turned into an equivalent input mean-square noise current. The drain noise current squared can be turned into an equivalent gate (input) noise voltage squared by dividing by the transconductance squared ($i_d^2 = g_m^2 e_g^2$):

$$d\langle e_n^2 \rangle_{\text{FET}} = \frac{4kT\Gamma}{g_m} df. \tag{17}$$

The corresponding mean-square input noise current is simply the mean-square gate noise voltage divided by the input impedance squared (without feedback.) For small input bias/feedback resistor values, that mean-square equivalent input noise current would be just $\langle i_n^2 \rangle_{\text{FET}} = \langle e_n^2 \rangle_{\text{FET}}/R_i^2$. However, this case is of no practical interest because the Johnson noise of that small-value resistor would swamp the FET noise and ruin the sensitivity. In high-sensitivity amplifiers, the input impedance (without feedback) is essentially the input FET gate capacitance, C_{FET}, in parallel with the junction capacitance of the pin photodiode, C_{pin}, plus any stray capacitance, C_s. The equivalent input noise current is then

$$d\langle i_n^2 \rangle_{\text{FET}} = \frac{4kT\Gamma(2\pi f C_T)^2}{g_m} df, \tag{18}$$

where the total input capacitance, C_T, is the sum of the FET input capacitance, C_{FET}, plus the photodiode capacitance, C_{pin}, plus C_s.

The physical assumption behind Eq. (18) is that in the absence of feedback, the photocurrent signal will be integrated by the input capacitance C_T ($V = 1/C_T \int i \, dt$); that R_i is so large (for low noise) that it can be neglected over most of the bandwidth. For example, in a 45-Mb/s receiver with $R_i = 1 \, M\Omega$, and $C_T = 1$ pF, C_T dominates above 80 kHz. This means that the input voltage produced by the photocurrent is inversely proportional to the frequency; since the input noise voltage is fixed, the signal-to-noise ratio is also inversely proportional to the frequency. When the signal is equalized (differentiated), the low-frequency signals (and noise) are attenuated; the high frequencies are boosted. Thus, with the signal now proportional to the input photocurrent, the noise is proportional to frequency, and the mean-square noise is proportional to the frequency squared, as in Eq. (18). Again, using feedback to avoid signal integration does not change the signal-to-noise ratio versus frequency.

The total equivalent mean-square input noise current of the FET is obtained by integrating over the channel-filter frequency response as before:

$$\langle i_n^2 \rangle_{\text{FET}} = 4kT\Gamma \frac{(2\pi C_T)^2}{g_m} \int_0^\infty f^2 |F(f)|^2 \, df, \tag{19}$$

or, changing variables from f to $y = f/B$ as before,

$$\langle i_n^2 \rangle_{\text{FET}} = 4kT\Gamma \frac{(2\pi C_T)^2}{g_m} I_3 B^3, \tag{20}$$

where

$$I_3 = \int_0^\infty |F(y)^*|^2 y^2 \, dy \tag{21}$$

is the third Personick integral.

The total mean-square equivalent input-noise current due to the input devices (pin photodiode plus FET plus bias resistor) is then (Smith and Personick, 1980)

$$\langle i_n^2 \rangle_T = \langle i_n^2 \rangle_{R_i} + \langle i_n^2 \rangle_\ell + \langle i_n^2 \rangle_{\text{FET}} \tag{22a}$$

or

$$\langle i_n^2 \rangle_T = \frac{4kT}{R_i} I_2 B + 2qI_\ell I_2 B + 4kT\Gamma \frac{(2\pi C_T)^2}{g_m} I_3 B^3, \tag{22b}$$

where $\langle i_n^2 \rangle_{R_i}, \langle i_n^2 \rangle_\ell$, and $\langle i_n^2 \rangle_{\text{FET}}$ are given by Eqs. (8), (15), and (20), respectively; the Personick integrals I_2 and I_3 are given by Eqs. (9) and (21), respectively; and $I_\ell = I_{\ell\text{FET}} + I_{\ell\text{pin}}$ and $C_T = C_{\text{pin}} + C_{\text{FET}} + C_s$.

Equation (22) for the total equivalent input noise due to the input devices contains much of the information about pin-photodiode and FET technology requirements and about amplifier design constraints. In high-sensitivity designs, the input resistor Johnson noise term $\langle i_n^2 \rangle_{R_i}$ is made almost negligible by making R_i large. The new receiver circuits, which can use large R_i values (e.g., $1 - 10$ MΩ at 45 Mb/s without sacrificing bandwidth, are described in Section V; earlier lower-sensitivity receiver circuits used low values of R_i to achieve the required bandwidth and were dominated by the resultant Johnson noise. (Typically R_i was from a few hundred ohms to a few thousand ohms at 45 Mb/s, giving $\sim 1000-10{,}000$ times more Johnson noise power.) The leakage current shot noise has been made negligible by reducing the pin photodiode and FET leakage currents. This leaves the input FET noise as the fundamental noise source; the receiver noise with low-leakage devices and state-of-the-art circuitry reduces to approximately.

$$\langle i_n^2 \rangle_T \cong 4kT\Gamma \frac{(2\pi C_T)^2}{g_m} I_3 B^3. \tag{23}$$

(This omits the noise of the rest of the receiver circuit, which will be discussed next in Section III.B.2.)

Thus, the circuit-equivalent input-noise power (or mean-square noise current) is approximately proportional to B^3, C_T^2/g_m and the channel noise factor Γ. One can write an input circuit figure of merit that is independent of bit rate (Smith and Personick, 1980):

$$M = \frac{g_m}{C_T^2 \Gamma}. \qquad (24)$$

The receiver optical sensitivity is inversely proportional to the root-mean-square noise current and therefore is approximately proportional to $\sqrt{M} B^{-3/2}$.

Note that if the mean-square noise current were proportional to B^2 rather than B^3, the photocurrent charge per bit would be constant. In fact, the charge per bit goes up approximately as the square root of the bit rate.

2. Complete Receiver Circuit Noise Expressions with Device Figures of Merit

This subsection first derives figure-of-merit expressions for pin-photodiode, FET, and FET IC technologies, then extends the Smith and Personick noise expression of Section 3.2.1 to include the noise from the input FET load device and the rest of the receiver circuit. The resultant expression is used to calculate the theoretical receiver sensitivities of Sections III.B.3 and V.H

The front-end figure of merit of Eq. (24) can be rewritten as the product of an FET technology figure-of-merit times a pin-photodiode figure-of-merit. In a typical high-performance FET technology, the source-to-drain spacing or channel length is fixed at the minimum reliable resolution of the lithography. The FET transconductance and input capacitance are both proportional to the channel width (typically a few tens to a few hundred micrometers); the g_m/C_{FET} ratio is set by the FET technology. Rewriting g_m as $(g_m/C)C_{FET}$, where C_{FET} determines the FET size, and taking $C_T = C_{FET} + C_{pin} + C_s$ gives

$$M = \left(\frac{g_m}{\Gamma C_{FET}}\right)\left(\frac{C_{FET}}{(C_{FET} + C_{pin} + C_s)^2}\right). \qquad (25)$$

Differentiating the second bracketed term with respect to C_{FET} says that the optimum-size FET in a given technology has a gate width such that

$$C_{FET} = C_{pin} + C_s = \tfrac{1}{2} C_T. \qquad (25a)$$

Assuming such an optimum-sized FET, one can now write

$$M = M_{FET} M_{pin}, \qquad (26)$$

where the FET-technology figure of merit is

$$M_{FET} = \frac{g_m}{\Gamma C_{FET}} = \frac{2\pi f_T}{\Gamma}, \qquad (27)$$

where $f_T = g_m/2\pi C_{FET}$ is the unity gain frequency of the FET technology.

M_{pin} the pin-photodiode (and stray capacitance) figure of merit is

$$M_{pin} = \frac{1}{4(C_{pin} + C_s)} \tag{28}$$

The optical receiver sensitivity is, again, proportional to the square root of the product of these figures of merit; these figures of merit will be used in Section V.H to discuss the sensitivity improvements calculated for future-technology receivers.

The FET figure of merit indicates that high-sensitivity receivers should be made with microwave or VHSIC technologies because such FETs have the highest f_Ts. The optical sensitivity of such receivers is approximately proportional to the square root of f_T; thus, these high-frequency FETs are preferred, even at low bit rates (e.g., 10 Mb/s), but for noise, not frequency-response, reasons.

The two presently preferred FET technologies are 1-μm gate-length GaAs MESFETs and 1-μm gate silicon MOSFETs. The two technologies offer roughly the same sensitivity in theory; the GaAs has higher g_m/C, but the silicon has a lower channel noise factor (Section III.B.3). In practice, GaAs is presently somewhat superior in performance, but silicon is much less costly.

The photodiode and stray-capacitance figure of merit of Eq. (28) indicates that the pin-photodiode capacitance C_{pin} and the stray capacitance C_s must be made as small as possible. Assuming an optimum-size amplifier input FET (gate width such that $C_{FET} = C_{pin} + C_s$), the optical receiver sensitivity is inversely proportional to the square root of $(C_{pin} + C_s)$. For small C_s, the optical sensitivity is inversely proportional to the square root of the pin-photodiode capacitance. Total front-end capacitances C_T are presently about a picofarad or less.

Thus, the pin-photodiode technology objectives are low capacitance, low leakage current, and high quantum efficiency. Low capacitance means either a low doping ($10^{14} - 10^{15}$/cc) for a wide depletion region, a small diameter (area), or both. For multimode systems, which are becoming obsolescent, the photodiode diameter is typically somewhat larger than the fiber core diameter for efficient optical coupling. Present photodiodes are typically 50–100 μm in diameter. However, the output of a single-mode fiber can be focused to a diffraction limited spot; this makes possible 5- to 25-μm diameter photodiodes.

How small a photodiode for a single-mode system can be made is primarily an optical coupling and device packaging problem; changing the mask to make a smaller diode would be easy. Note that the problem of efficiently coupling a single-mode fiber to a small-area optical device has already been solved for semiconductor laser transmitters. Such packages soon will be low cost as laser transmitters ride down the learning curve. A four-fold reduction in capacitance appears easily practical; this would correspond to a 3-dB improvement of the optical receiver sensitivity. However, for low-cost applications, the preferred alternative may be to leave the photodiode diameter large, use the simplest packaging, and concentrate on reducing the doping instead.

Thus, a high-sensitivity pin-FET receiver is a combined FET technology problem (high $g_m/C\Gamma$), pin-photodiode problem (low C_{pin}, low leakage current), circuit design problem (large R_i for low Johnson noise while preserving a wide bandwidth and dynamic range), and packaging problem (low stray capacitance C_s, small photodiode diameter if economic).

The pin-photodiode figure of merit can also be read as a total-input-capacitance figure of merit. For an optimized receiver, $M_{pin} = 1/(2C_T)$ by Eq. (25a); thus, for a given FET figure of merit, the sensitivity of an optimized receiver is inversely proportional to the square root of C_T.

Actual sensitivity calculations should also include corrections for the noise due to the input FET drain load device and for the noise of following stages. The Smith and Personick noise calculations are readily extendable to include these effects.

In the new IC optical receivers of Section V, the input FET's load device Q_1 is typically another FET. [In hybrid IC (HIC) optical receivers, the load device is typically a resistor but HIC receivers will become absolescent.] The IC load transistor Q_L adds an extra mean-square noise current at the drain of Q_1 of

$$d\langle i_n^2 \rangle_L = 4kT g_{mL} \Gamma df. \qquad (29)$$

where g_{mL} is the transconductance of Q_L. Equation (16) can then be rewritten as

$$d\langle i_n^2 \rangle_{drain} = 4kT(g_{m1} + g_{mL})\Gamma df, \qquad (30)$$

where g_{m1} is the transconductance of input FET Q_1. Retracing the derivation of Eqs. (17)–(20) then gives a total mean-square equivalent input noise current due to Q_1 (input FET) and Q_L (load FET) of

$$\langle i_n^2 \rangle_{Q_1+Q_L} = 4kT\Gamma(2\pi C_T)^2 \left(\frac{g_{m1} + g_{mL}}{g_{m1}^2}\right) I_3 B^3, \qquad (31)$$

where I_3, the third Personick integral, is given by Eq. (21).

The optimum ratio of Q_L size (gate width) to Q_1 size is a tradeoff. A large Q_L gives a higher drain current density in Q_1, and therefore a higher g_{m1}; a small Q_L gives less drain current noise due to Q_L. Assuming an optimum Q_L to Q_1 ratio for the particular technology, one can revise Eq. (27) to give a figure of merit for the FET IC technology:

$$M_{IC} = \frac{g_{m1}^2}{(g_{m1} + g_{mL})\Gamma C_{FET}}. \quad (32)$$

M_{IC} depends only on the IC technology, so long as the input FET Q_1 is scaled so that $C_{FET} = C_T/2$ and Q_L is scaled for minimum noise. The receiver figure of merit now is $M_{IC} \cdot M_{pin}$, where M_{pin} is given by Eq. (28) as before; the overall receiver sensitivity is approximately proportional to the square root of the receiver figure of merit. As will be shown below, the Q_L noise term typically is more important than the all following stage noise terms combined; thus, Eq. (32) is a good IC technology figure of merit and gives good approximate sensitivities.

Note that fine-line GaAs depletion-mode MESFETs typically require a higher drain current, hence a larger Q_L, than fine-line silicon enhancement-mode MOS technologies. This effect cancels part of the g_m/C advantage of GaAs, as does the higher channel noise factor Γ in GaAs.

The noise of the following stages of the receiver can be represented as an equivalent stage input mean-square noise voltage of $d\langle e_s^2 \rangle$ per unit bandwidth df. If a_{s-1} is the total voltage gain of the $s-1$ stages preceding stage s, the mean-square equivalent input noise voltage due to stage s is

$$d\langle e_s^2 \rangle_I = d\langle e_s^2 \rangle / a_{s-1}^2. \quad (33)$$

Assuming $\langle e_s^2 \rangle$ is constant in frequency, one can repeat the derivation of Eqs. (17)–(20), substituting $d\langle e_s^2 \rangle_I$ for $d\langle e_n^2 \rangle_{FET}$. This gives the mean-square equivalent input noise current due to stage s:

$$\langle i_n^2 \rangle_s = \frac{d\langle e_n^2 \rangle_s}{a_{s-1}^2}(2\pi C_T)^2 I_3 B^3. \quad (34)$$

The total mean-square equivalent input noise current of the receiver is then

$$\langle i_n^2 \rangle = \frac{4kT}{R_I} I_2 B + 2qI_\ell I_2 B + 4kT\Gamma \frac{(2\pi C_T)^2}{g_{m1}} I_3 B^3$$
$$+ 4kT\Gamma(2\pi C_T)^2 \frac{g_{mL}}{g_{m1}^2} I_3 B^3 + (2\pi C_T)^2 I_3 B^3 \sum_{s=2}^{N} \frac{d\langle e_n^2 \rangle_s}{a_{s-1}^2}, \quad (35)$$

where the first two terms are the input-resistor Johnson noise and the leakage-current shot noise, respectively, the third term is the input FET noise, the fourth term is the noise of the input FET load (Q_L), and the last term is the sum over the noise contributions of the following stages. Generally, the following-stage noise current is less important than the load device contribution; even the second-stage mean-square noise is divided by the first-stage voltage gain squared. Thus, Eq. (32) is a good figure of merit for comparing FET IC technologies.

3. Sensitivities of Present-Technology pin-FET Receivers

This subsection calculates theoretical sensitivities for present-technology silicon and GaAs FET IC receivers and compares the two technologies. It also includes sensitivity results from the literature. Section V.H will extend the calculations of this section to include sensitivity calculations for probable future-technology IC receivers.

For the numerical sensitivity calculations, g_m/C is taken as 70 mS/pF for 1-μm silicon MOSFETs and 90 mS/pF for 1-μm GaAs MESFETs; Γ is taken as 1.2 for the silicon and 1.5 for the GaAs. Since the FET technology figure of merit is $g_m/(C\Gamma)$ by Eq. (27), the silicon and GaAs sensitivities are essentially equal in theory; GaAs FETs have higher transconductances, but silicon FETs have lower channel noise. In practice, GaAs designs are *presently* a few decibels more sensitive.

The Personick integrals I_2 and I_3 can be roughly estimated by remembering that the channel-filter bandwidth is typically 0.56 times the bit rate B for synchronous detection NRZ receivers. Thus, taking $|F(y)| = 1$ for $y < 0.56$ and $F(y) = 0$ for $y > 0.56$ in Eq. (9) for I_2 gives $I_2 = 0.56$; using this approximation in Eq. (21) for I_3 gives $I_3 = 0.059$. In fact, I_2 is taken as 0.564 and I_3 as 0.0868 (Paski, 1980b). Values for I_2 and I_3 for different input and output (filtered) pulse shapes are found in Personick (1973) and in Smith and Personick (1980).

Figure 7 shows theoretical digital optical-receiver sensitivities versus bit rate for 1-μm gate length GaAs and silicon IC receivers using InGaAs pin photodiodes. The optical wavelength is 1.3 μm. The calculations assume the new high-sensitivity, micro-FET feedback IC receiver designs, as described in Sections V.A, V.B, and V.F, in which the input/feedback resistor noise is almost negligible; the sensitivities were calculated on paper designs using Eq. (35) for the total receiver noise (see Section V.H). Figure 7 assumes a 1-pF total front-end capacitance, e.g., $C_{pin} = 0.40$ pF, $C_s = 0.10$ pF, $C_{FET} = 0.5$ pF; the silicon input FET then has a transconductance of 35 mS; the GaAs input FET has a transconductance of 45 mS. The photodiode leakage current is taken as 1 nA at 20°C and 15 nA at a maximum operating temperature of 85°C (the 15x increase assumes that the leakage is a G-R current via midgap states.) The bit-error rate is 10^{-9}.

Since the theoretical sensitivities of the silicon and GaAs FET IC receivers are essentially the same, only one curve is plotted for both. Note also that the calculated sensitivities go approximately as $B^{-3/2}$; the first two terms in Eq. (35) for the input noise squared are negligible except at low bit rates; the

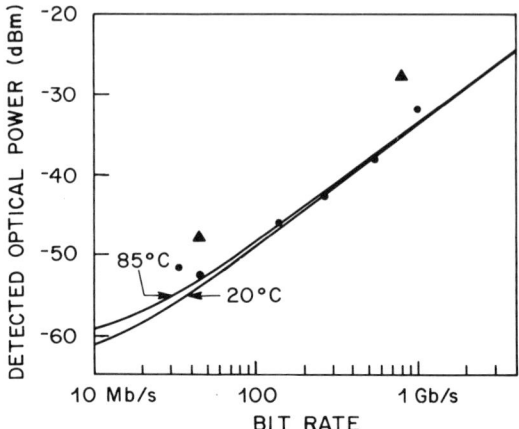

Fig. 7. Theoretical sensitivities of present-technology pin-FET lightwave receivers. Dots show GaAs FET receiver measurements from the literature; triangles show silicon FET receiver measurements.

others go as B^3 except for the (small) following-stage noise summation, which increases slightly faster than B^3 because the gain per stage is less for wider bandwidth stages. Finally, note that the calculated sensitivities are competitive with those calculated for present APD receivers to 100–200 Mb/s.

Section V.H presents further detail on these IC optical receiver sensitivity calculations and extends them to include sensitivity calculations for probable future-technology IC receivers. It also presents scaling laws for the basic IC design of Section V.F and uses the device figures of merit from the last section for an intuitive discussion for the calculated sensitivity improvements to be expected from the future-technology receiver designs.

Figure 7 also shows the best experimentally achieved pin-FET receiver sensitivities as of this writing. The GaAs FET receiver measurements at 34 Mb/s, 140 Mb/s, 280 Mb/s, and 565 Mb/s were by D. R. Smith *et al.* (1982); the 45-Mb/s measurement was by Williams and LeBlanc (1986); the 1-Gb/s measurement was by Linke *et al.* (1983). The 45-Mb/s silicon MOSFET receiver measurement was by Ogawa *et al.* (1982); the 800-Mb/s silicon result was by Abidi *et al.* (1984). All were measured at room temperature.

Unfortunately, the silicon MOSFET optical receivers are *presently* a few decibels less sensitive than either theory or RMS noise measurements would indicate. This may be because fine-line silicon FETs were developed as a digital IC technology in which noise was not important. The fine-line GaAs MESFET technology was developed as an analog technology for microwave receivers. Therefore, the GaAs noise problems were important and were solved.

At present, silicon FETs appear preferable for high-volume, low-cost designs in which a modest sensitivity penalty is acceptable; GaAs takes over for higher-sensitivity, higher-cost designs (e.g., long-haul transmission systems). Silicon may well take over the high-sensitivity applications when its noise problems are solved. On the other hand, the GaAs technology may become cheaper and its scale of integration may be increased. The two technologies are in a race with each other; the silicon technologists must improved their sensitivity; the GaAs technologists must reduce their costs.

GaAs FET IC technologies are also presently preferred over silicon for optical fiber receivers above 1 Gb/s, both because of the higher saturation velocity in GaAs and because GaAs FET ICs have lower parasitic capacitances. Short-channel GaAs MESFETs are fabricated on a semi-insulating substrate; the parasitic capacitances to the substrate are very small (Sze, 1981). Short-channel silicon MOSFETs are presently junction isolated and have the full junction capacitance to the substrate (Sze, 1981). In addition, the gates of silicon MOS transistors overlap both the source and the drain; GaAs MESFETs do not have this overlap capacitance; the Schottky gate metallization defines the channel and therefore cannot overlap the source and drain.

C. Avalanche Photodiode Receiver Noise and Sensitivity Calculations

In an avalanche photodiode (APD), the primary photocurrent is multiplied by impact ionization in the p-n junction, which is operated reverse-biased near breakdown. The multiplied photocurrent signal goes to the receiver circuit. Consider an APD in which the primary photocurrent is multiplied (on average) by a factor $\langle M \rangle$. If the multiplication process were noiseless and the APD leakage current negligible, the optical signal power required by the receiver would be decreased by a factor $\langle M \rangle$ and the optical receiver sensitivity increased by a factor $\langle M \rangle$.

In fact, the multiplication process is noisy because it is the result of random impact ionizations. The equivalent mean-square primary photocurrent noise $d\langle i_n^2 \rangle_D$ per bandwidth df due to the APD is the shot noise of the photocurrent I_{ph} plus leakage current I_ℓ times the McIntyre excess noise factor $F(\langle M \rangle)$ (McIntyre, 1966):

$$d\langle i_n^2 \rangle_D = 2q(I_{ph} + I_\ell)F(\langle M \rangle)df. \tag{36}$$

For noiseless multiplication, $F(<M>) = 1$, and the noise is just the photocurrent plus leakage-current shot noise; $F(\langle M \rangle)$ is the factor by which the real avalanche multiplication increases the noise over that of a noiseless multiplication.

If the avalanche is initiated by injection of photocarriers at one side of the avalanche region, $F(\langle M \rangle)$ is given by (McIntyre, 1966)

$$F(\langle M \rangle) = \langle M \rangle \left[1 - (1-k)\left(\frac{\langle M \rangle - 1}{\langle M \rangle}\right)^2 \right], \tag{37}$$

where k is the ratio of the electron and hole ionization coefficients. In silicon, electrons have the higher ionization coefficient; the 0.8-μm wavelength silicon APDs use photoelectron initiated multiplication, and k is the ratio of the hole ionization coefficient to the electron ionization coefficient. In InGaAs/InP 1.3- to 1.6-μm wavelength APDs, the avalanche is photo-hole initiated and k is the ratio of the electron ionization coefficient to that of holes. In silicon APDs, k is typically 0.02 (Melchior et al., 1978); in InGaAs/InP APDs, k is typically 0.4 at present.

The APD is not the "solid state equivalent of a photomultiplier" because both electrons and holes can impact ionize in an APD; in a photomultiplier, only electrons impact ionize (on the dynodes); there are no holes in a vacuum. In an electron-initiated APD, the electrons of the primary avalanche travel downstream, creating electron-hole pairs by impact ionization; the resultant holes travel upstream and can create more electrons, thus initiating secondary electron avalanches, etc. This hole feedback makes avalanche carrier multiplication much noisier than photomultiplier electron multiplication. When the electron-hole feedback loop-gain becomes unity, the avalanche is self-sustaining and the diode breaks down. The higher the multiplication, the closer to breakdown and the noisier the multiplication.

For $k = 0$ (best case), only one carrier ionizes and the APD multiplication process is similar to that of a photomultiplier. In this limit, $F(\langle M \rangle)$ becomes

$$F(\langle M \rangle) = 2 - \frac{1}{\langle M \rangle}. \tag{38}$$

For $k = 1$ (worst case), $F(\langle M \rangle)$ becomes

$$F(\langle M \rangle) = \langle M \rangle; \tag{39}$$

the avalanche multiplication is then more noisy due to the carrier feedback. In general, the lower the k, the less noisy the multiplication. Thus, the silicon APDs ($k \sim 0.02$) are much less noisy for a given multiplication than the InGaAs APDs ($k \sim 0.5$); unfortunately, silicon is transparent below 1.1 μm and cannot be used for 1.3- to 1.6-μm APDs.

Following Smith and Personick (1980), the total equivalent mean-square primary photocurrent noise $\langle i_n^2 \rangle_{\text{ph}}$ is the APD noise integrated over the receiver bandwidth plus the equivalent mean-square input noise current $\langle i_n^2 \rangle_T$

of the receiver amplifier, divided by $\langle M \rangle^2$:

$$\langle i_n^2 \rangle_{\text{ph}} = 2q(i_{\text{ph}} + I_\ell)F(\langle M \rangle)I_1 B + \frac{\langle i_n^2 \rangle_T}{\langle M \rangle^2}, \tag{40}$$

where the integral over frequency is $I_1 B$; I_1 is the first Personick integral. The receiver amplifier circuit noise $\langle i_n^2 \rangle_T$ is given by the expressions of Section III.B.

When a logic-zero bit is transmitted, the optical signal, hence the photocurrent i_{ph}, is ideally zero; the zero-level mean-square noise then is due only to the APD leakage current and to the amplifier noise. When a logic one bit is transmitted, the mean-square noise is increased by $2qI_{\text{ph}1}F(\langle M \rangle)I_1 B$, which is the noise due to the avalanche multiplication of the one-level photocurrent. Since an APD receiver's one-level noise is greater than its zero-level noise, the decision threshold D is typically set closer to the zero level than to the one level.

In practice, the optical transmitter is not completely turned off during zero bits, for reasons of transmitter response speed and (for laser transmitters) for reasons of optical frequency stability. Take the zero-bit optical signal level $P(0)$ as a fraction r of the one-bit optical signal level $P(1)$:

$$r = \frac{P(0)}{P(1)} = \frac{\langle i_{\text{ph}}(0) \rangle}{\langle i_{\text{ph}}(1) \rangle}, \tag{41}$$

where $\langle i_{\text{ph}}(0) \rangle$ is the expected photocurrent at the sampling instant for a zero bit; $\langle i_{\text{ph}}(1) \rangle$ is the photocurrent for one bit. r is called the transmitter optical extinction ratio.

Ideally, r should be zero; in practice, r is may be as high as 0.2. This means a smaller photocurrent signal component for a given average optical power. In addition, the zero-level photocurrent is the functional equivalent of a leakage current and adds a corresponding noise term to both the zero and one signals.

Taking the avalanche-noise amplitude distribution as approximately Gaussian, the error probability or bit-error rate (BER) is given by Eq. (3): BER $= \frac{1}{2}\text{erfc}(Q/\sqrt{2})$ where, for a zero bit,

$$Q_0 = (i_{\text{th}} - \langle i_{\text{ph}}(0) \rangle)/\langle i_{n0}^2 \rangle_{\text{ph}}, \tag{42}$$

and i_{th} is the decision threshold and $\langle i_{n0}^2 \rangle^{1/2}$ is the root-mean-square noise for a zero bit. Similarly, for a one bit,

$$Q_1 = \frac{(\langle i_{\text{ph}}(1) \rangle - i_{\text{th}})}{\langle i_{n1}^2 \rangle_{\text{ph}}^{1/2}}. \tag{43}$$

$Q_0 = Q_1 = 6$ gives a bit-error rate of 10^{-9} [Eq. (3), Fig. 4b].

Following Smith and Personick (1980), taking $Q_0 = Q_1$ equal to the Q required for the desired, BER, and using the APD receiver noise equation (40) to give $\langle i_{n0}^2 \rangle^{1/2}$ and $\langle i_{n1}^2 \rangle^{1/2}$ in Eqs. (42) and (43), yields the photocurrent signal required for the desired BER. Assuming that zero bits and one bits are equally probable, and converting the photocurrent to an input optical power, gives an equation for the average optical power required for a given BER:

$$\eta \bar{P} = \left(\frac{hc}{\lambda q}\right)\left(\frac{1+r}{1-r}\right)\left[(1+r)\frac{Q^2 q B I_1 F(\langle M \rangle)}{1-r}\right.$$
$$\left. + \left(\left(\frac{Q^2 q B I_1 F(\langle M \rangle)}{1-r}\right)^2 \cdot 4r + Q^2\left(\frac{\langle i_n^2 \rangle_T}{\langle M \rangle^2} + 2q I_{\ell m} F(\langle M \rangle) I_2 B\right)\right)^{1/2}\right], \quad (44)$$

where Q is the photocurrent signal-to-average-noise ratio for the given BER, multiplied by a prefactor that is unity for zero (ideal) extinction ratio r:

$$Q = \frac{\langle i_{ph}(1) \rangle - \langle i_{ph}(0) \rangle}{\langle i_{n0}^2 \rangle_{ph}^{1/2} + \langle i_{n1}^2 \rangle_{ph}^{1/2}} = \left(\frac{1-r}{1+r}\right)\frac{\bar{I}_{ph}}{\frac{1}{2}(\langle i_{n0}^2 \rangle^{1/2} + \langle i_{n1}^2 \rangle^{1/2})}, \quad (45)$$

$\langle i_n^2 \rangle_T$ is the equivalent mean-square amplifier noise, $I_{\ell m}$ is the leakage current of the APD that undergoes multiplication, and $F(\langle M \rangle)$ is the McIntyre excess noise factor of Eq. (37).

Note that a high multiplication $\langle M \rangle$ reduces the effect of the receiver amplifier noise but gives more avalanche multiplication noise and a higher $F(\langle M \rangle)$ in Eq. (40). A low $\langle M \rangle$ gives lower avalanche multiplication noise but increases the effect of amplifier noise. The optimum $\langle M \rangle$ is determined by this trade-off.

In theory, Eq. (44) can be differentiated to find the optimum gain; the result is an unmanageable expression of no particular physical interest. In practice, one is much better off using a minimum finder program to find the optimum gain $\langle M \rangle$ and the minimum $\eta \bar{P}$ numerically.

InGaAs/InP heterostructure APDs are presently preferred for 1.3- to 1.6-μm applications because they offer low leakage currents and a reasonable k-ratio. The light is absorbed in a low-field InGaAs ($E_g = 0.73$ eV) layer, and the avalanche multiplication takes place in a high-field InP layer ($E_g = 1.35$ eV). The use of a wide-gap avalanche region avoids the tunneling current problem of all-InGaAs APDs (Nishida et al., 1979; Kanbe et al., 1980; Susa et al., 1980). Until recently, these APDs were slow due to hole trapping by the InGaAs-to-InP valence-band step; this problem was solved by adding several intermediate band-gap layers between the InGaAs and the InP or by continuously grading the composition between InGaAs and InP (Matsushima et al., 1982; Campbell et al., 1983).

Figure 8 shows theoretical InGaAs/InP APD receiver sensitivities versus bit rate. The figure assumes that the APD is used with an optimized 1-μm gate

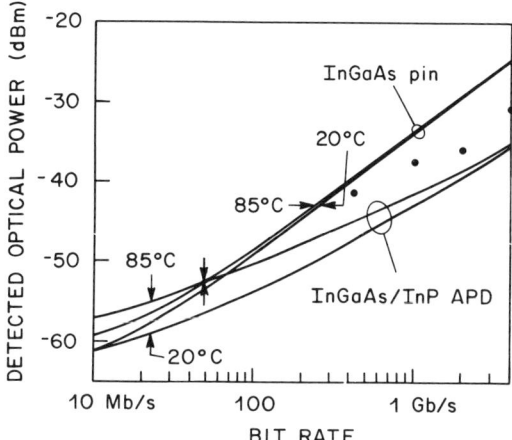

FIG. 8. Theoretical InGaAs/InP APD receiver sensitivities. $\lambda = 1.3$ μm, $I_L = 3$ nA at 20°C, 45 nA at 85°C, $C_T = 1$ pF, high-sensitivity receiver ICs. Dots show APD receiver measurements from the literature.

technology GaAs FET receiver amplifier, which is the most sensitive amplifier type at present. It also assumes a total receiver input capacitance of 1 pF (this includes the APD capacitance), a primary leakage current at 20°C of 3 nA, and an APD k-ratio of 0.4. Figure 8 shows the pin-FET receiver sensitivities from Fig. 7 for comparison, plus the 20°C and 85°C InGaAs/InP APD sensitivities. The 85°C leakage current is taken as 45 nA. This assumes that the APD leakage current is a generation-recombination current via midgap states in the InGaAs, which gives a 15x increase in leakage current from 20°C to 85°C.

Figure 8 also shows InGaAs/InP APD-FET receiver measurements from the literature. The measurements at 420 Mb/s and 1 Gb/s were by Campbell *et al.* (1983); those at 2 Gb/s and 4 Gb/s were by Kasper *et al.* (1985). All are the best values at those bit rates, as of this writing; all used the APD of Campbell *et al.*.

Note, however, that these measurements were made at room temperature; however, in field use, the maximum operating temperature is typically 85°C. Unless the receiver and APD temperature is controlled, the 85°C sensitivity is what matters in practice.

Figure 8 shows that present InGaAs/InP 1.3- to 1.6-μm APDs in theory offer little sensitivity advantage below 100 Mb/s at a maximum operating temperature of 85°C. This is due to the noise caused by the avalanche multiplication of the primary leakage current; an APD without leakage

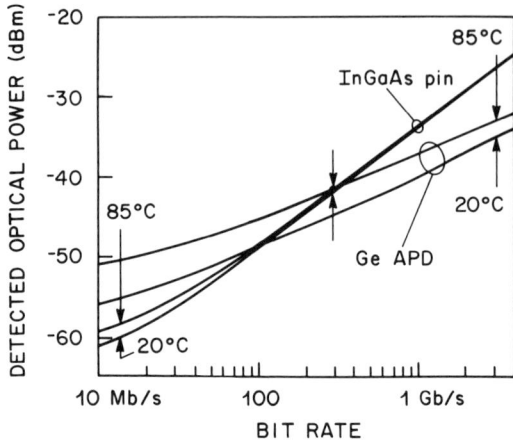

FIG. 9. Theoretical Ge APD receiver sensitivities. $\lambda = 1.3$ μm, $I_\ell = 100$ nA at 20°C, 1 μA at 85°C, $C_T = 1$ pF, high-sensitivity receiver ICs.

current would offer a sensitivity improvement at all bit rates. In addition, reducing the leakage current is presently more important than improving the k-ratio for bit rates less than ~ 500 Mb/s at 85°C.

Germanium APDs have also been used in 1.3- to 1.6-μm optical receivers (Mikawa *et al.*, 1981). Present Ge APDs have very high leakage currents but until recently were faster than InGaAs/InP APDs and therefore once were preferred for Gb/s applications. Figure 9 shows theoretical germanium APD sensitivities versus bit rate assuming $I_\ell = 100$ nA at 20°C, $I_\ell = 1$ μA at 85°C, $k = 0.5$, and a GaAs IC front end. The InGaAs/InP APD is presently superior to the germanium APD at every bit rate. In fact, at 85°C, InGaAs pin photodiodes are superior to germanium APDs for bit rates below 300 Mb/s.

IV. First- and Second-Generation Lightwave Receivers

This section traces the evolution of lightwave receivers from the early voltage-amplifier and transimpedance designs, in which the sensitivity was limited by circuit noise, up through the present versions of the integrating or "high impedance" receiver, in which the sensitivity is limited primarily by the fundamental device noise. Unfortunately, the high-sensitivity, integrating receivers require equalization, with its attendant problems, and have very limited dynamic ranges. Therefore, almost all commercial systems still use the transimpedance receiver, despite its lower sensitivity. In time, both transimpedance receivers and integrating receivers will be displaced by the high-sensitivity, nonintegrating active-feedback IC receivers of Section V.

Subsection IV.A describes the early voltage-amplifier optical receivers. These receivers had very low sensitivities but were the first lightwave receivers.

Subsection IV.B describes the simple integrating optical receivers (Personick, 1973; Goell, 1974), which can achieve very high sensitivities but are vulnerable to saturation on long strings of ones or zeroes in the data. This problem was solved by encoding the data stream (Brooks, 1980); upgraded versions of these receivers have been introduced in a few 1.3-μm transmission systems (Hooper et al., 1980; D. R. Smith et al., 1980).

Subsection IV.C describes conventional transimpedance receivers. These receivers do not integrate the signal but are less sensitive than integrating receivers. These were first used in 0.8-μm transmission systems; a silicon APD detector was used to make up for the lower sensitivity (R. G. Smith et al., 1978). A bipolar IC version for 1.3-μm systems has achieved good sensitivities at bit rates of several hundred Mb/s while using a nonmultiplying pin photodiode (Paski, 1980; Snodgrass and Klinman, 1984).

Subsection IV.D describes integrating transimpedance receivers (Gloge et al., 1980; Ogawa et al., 1983). These receivers are as sensitive as the simple integrating receivers of Subsection IV.B and do not require encoding of the data stream because they integrate the signal over less of the bandwidth. However, the integration pole frequency is temperature sensitive; these receivers are difficult to equalize reliably, e.g., if the temperature can vary over a commercial temperature range of $-25°C$ to $+85°C$.

A. Voltage-Amplifier Optical Receivers

Figure 10a shows a generalized schematic of a voltage-amplifier optical receiver. The photocurrent develops a voltage across a series load resistor R_L; this voltage is then amplified by a voltage amplifier of gain A. Ideally, the output voltage is $v_{out} = AR_L i_{ph}$. Note, however, that R_L must be small enough to keep the photocurrent signal from being integrated by the photodiode capacitance C_{pin} plus the capacitance C_X due to the input transistor of the amplifier, plus any stray input capacitance C_s. Explicitly, R_L must be less than the capacitive impedance $1/(\omega C_T)$ over the desired bandwidth, where $C_T = C_{pin} + C_X + C_s$ is the total capacitance seen by the photocurrent (Fig. 10b). For a bandwidth, or pole frequency, greater than the bit rate B,

$$R_L \leq 1/(2\pi C_T B). \tag{46}$$

Such voltage-amplifier receivers have low sensitivities because of the Johnson noise current of the small-value R_L needed to avoid signal integration

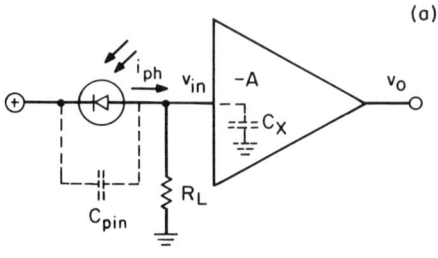

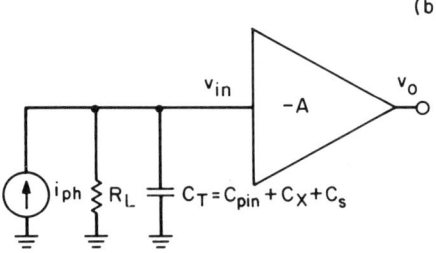

FIG. 10. Voltage amplifier receiver: (a) receiver circuit, (b) equivalent circuit.

by C_T (Personick, 1973b); by Eq. (8),

$$\langle i_n^2 \rangle_{R_L} = \frac{4kT}{R_L} I_2 B. \qquad (47)$$

Consider a 44.7-Mb/s design example; take $C_T = 4$ pF (typical for early receivers). By equation (46), $R'_L \le 890$ Ω. Assuming a noiseless voltage amplifier, so that the Johnson noise associated with R_L is the only noise source, the best possible sensitivity for the voltage-amplifier receiver at 1.3 μm [by Eqs. (1) and (8)] is only −35 dBm optical. The state-of-the-art receiver sensitivity at that bit rate is −51.7 dBm, an improvement of 16.7 dB optical or 33 dB electrical over the voltage-amplifier receiver circuit. If C_T for the voltage-amplifier receiver were reduced to the 1 pF of present state-of-the-art receivers, the sensitivity penalty still would be 13.7 dB optical. The simple voltage amplifier optical receiver now is not preferred for any application and has disappeared from the literature.

B. Integrating Optical Receivers

The integrating receiver or "high-impedance" receiver (Fig. 11a) can achieve excellent sensitivities because R_L is made large enough so that its Johnson noise is negligible, (Personick, 1973a, 1973b; Goell, 1974; Hooper *et al.*, 1980). However, the photocurrent signal is then integrated by C_T; the signal must be differentiated (equalized) later, as shown. Waveforms are shown in Fig. 11b. For 45-Mb/s receivers using state-of-the-art photodiodes and 1-μm gate-length GaAs FETs or silicon MOSFETs, R_L should be $\gtrsim 1$ MΩ for best sensitivity. For $R_L = 1$ MΩ and $C_T = 1$ pF, the resultant input current-to-voltage pole is 160 kHz (Eq. 46); above 160 kHz the photocurrent signal is integrated by C_T. Thus the output signal, v_0, is integrated over most of the signal bandwidth, even though the voltage gain, A, is flat. For the 45 Mb/s NRZ case, the Nyquist signal bandwidth $I_2 B$ is ~ 25 MHz; thus, the signal is integrated over seven octaves, from 160 kHz to 25 MHz. The equalizer must differentiate the signal over these same seven octaves; to do this, the equalizer zero frequency is set equal to the input pole frequency.

The equalization technique of Fig. 11a introduces a noise penalty. The reason is that the passive equalizer attenuates the signal; the peak attenuation is the equalization ratio $I_2 B/f_p = 25$ MHz/160 KHz $= 156X$ for the 45-Mb/s example. This signal attenuation enhances the effect of noise in stages following the equalizer, resulting in an equalizer noise penalty; here, if A is less than 156, the low-frequency signal voltage after the equalizer is smaller than the signal at the amplifier input! Active equalization, which would reduce this

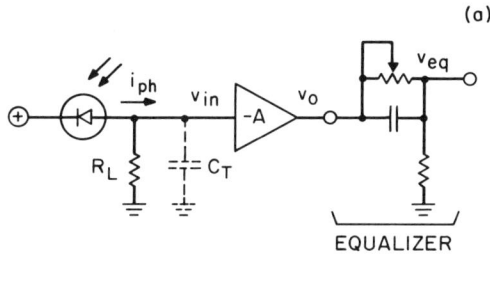

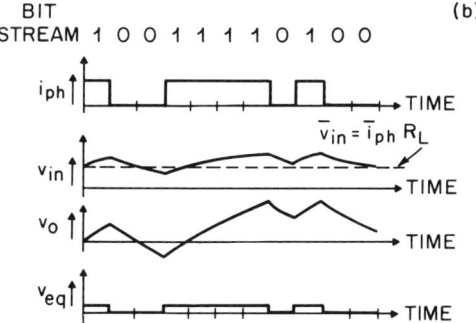

FIG. 11. Integrating receiver with equalizer: (a) circuit, (b) waveforms.

noise penalty, is difficult for these high frequencies and large equalization ratios, due to stability problems.

In theory, this equalizer noise penalty may be reduced by increasing the gain, A, before the equalizer. However, for a random bit stream (maximum information content), increasing A reduces the photocurrent for which saturation of the integrating amplifier on long strings of ones (or zeroes if ac coupled) causes an unacceptable bit-error rate. This reduces the dynamic range, resulting in a sensitivity versus dynamic-range tradeoff. Even with minimal dynamic range, A is still limited and the equalizer noise penalty is still appreciable at high bit rates.

A second sensitivity versus dynamic-range tradeoff is involved in choosing R_L. A high R_L improves sensitivity by reducing the input Johnson noise current but, for the maximum-information-content random-bit-stream, increases the probability of saturation on long strings of ones or zeroes because the integration pole frequency, f_p, is lower. A low R_L reduces sensitivity but improves dynamic range; both the Johnson noise and f_p are increased.

Both sensitivity versus dynamic-range tradeoffs may be reduced by encoding the data stream to limit the number of consecutive ones or zeroes; a 7B/8B encoding technique (Brooks, 1980) is used by the British Post Office (BPO) and others. In this technique, eight bits are transmitted for every seven bits of data; the redundancy allows the total disparity (difference between the number of ones and zeroes transmitted) to be kept close to zero. This removes the very-low-frequency signal components; large equalization ratios are not needed; the equalizer attenuation is reduced, and the allowable gain before the equalizer is increased; the equalizer noise penalty almost vanishes, and R_L can be made large.

These encoding techniques make the simple integrating receiver of Fig. 11 practical for transmission applications. However, there is a small sensitivity penalty because more bits must be transmitted to pass the same amount of information. If the receiver is FET noise limited, the mean square noise current goes as B^3 (Eq. 35) and the optical sensitivity goes as $B^{-3/2}$. A 7B/8B code requires an 8/7 higher bit rate, reducing the optical sensitivity by a factor $(7/8)^{3/2}$ or 0.9 dB optical. In addition, for encoding schemes in which any error in an eight-bit transmitted block causes all seven recovered data bits to be in error, the bit-error rate (BER) is eight times larger for a given signal-to-noise ratio. Using Eq. (3) or Fig. 5 for the BER versus signal-to-noise ratio says that this is equivalent to an additional encoding sensitivity penalty of 0.2 dB optical for a total encoding sensitivity penalty of ~ 1.1 dB. If a seven data bit plus one sign bit-encoding scheme is used, only a sign bit error will cause seven data bit errors; the BER is only doubled; the additional sensitivity penalty is then only 0.08 dB, for a total encoding sensitivity penalty of ~ 1 dB.

C. Transimpedance Optical Receivers

Transimpedance receivers were developed to increase the sensitivity achievable without signal integration; the ideal transimpedance circuit of Fig. 12a theoretically allows R_L to be increased to the same value as in an integrating receiver, but without integrating the signal. The reason is that R_L (now R_F) is connected around the voltage gain element; this negative feedback raises the input pole frequency. For the ideal circuit of Fig. 12a,

$$i_f = \frac{v_{in} - v_o}{R_f} = \frac{v_{in}(1 + A)}{R_F}. \tag{48}$$

This produces a virtual input resistance (Fig. 12b) of

$$r_e = \frac{v_{in}}{i_f} = \frac{R_F}{1 + A}. \tag{49}$$

r_e should be made small enough to keep the photocurrent signal from being integrated by the receiver input capacitance C_T; explicitly, r_e should be less than the capacitive impedance $1/(\omega C_T)$ over the signal bandwidth. Thus, the receiver bandwidth is simply the input pole frequency due to r_e in parallel with C_T:

$$f_p = \frac{A + 1}{2\pi R_F C_T}. \tag{50}$$

(a)

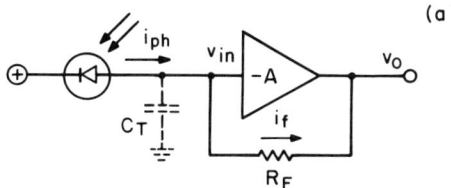

(b)

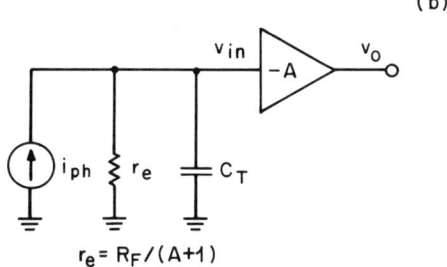

$r_e = R_F/(A+1)$

FIG. 12. Ideal transimpedance receiver: (a) circuit, (b) equivalent circuit.

Thus, the ideal transimpedance feedback configuration would theoretically increase the input current-to-voltage pole frequency or bandwidth by a factor of the gain, A, plus 1. Ideally this pole could be placed above the signal bandwidth by increasing A, eliminating integration of the signal. This would eliminate the need for equalization or coding and thus eliminate the associated noise penalty.

Unfortunately, real feedback resistors include a parasitic feedback capacitance C_R shunting the physical R_F (Fig. 13a). For frequencies greater than $f_{pR} = 1/(2\pi R_F C_R)$, the feedback resistor acts like a feedback capacitor and the signal is integrated even for large A. (The exact pole frequency is $f_p = 1/[2\pi R_F(C_R + C_T/(A + 1))]$ by inspection of Fig. 13b). In addition, even if an ideal feedback resistor had been available, the hybrid IC gain elements used in most conventional transimpedance receivers do not have enough voltage gain, A, to place the input pole frequency, f_p, above the passband. Adding stages would add extra phase shift inside the feedback loop and typically would cause instability problems.

Thus, the usual approach was to use a low-value R_F to avoid signal integration due to C_R and to reduce the gain required; the resultant Johnson noise penalty was accepted. The 0.8-μm wavelength, 45-Mb/s receivers of R. G. Smith et al. (1978) used $R_F = 4$ kΩ, and used a silicon APD to make up the sensitivity loss. The later 1.3-μm wavelength, 274-Mb/s design of Ogawa

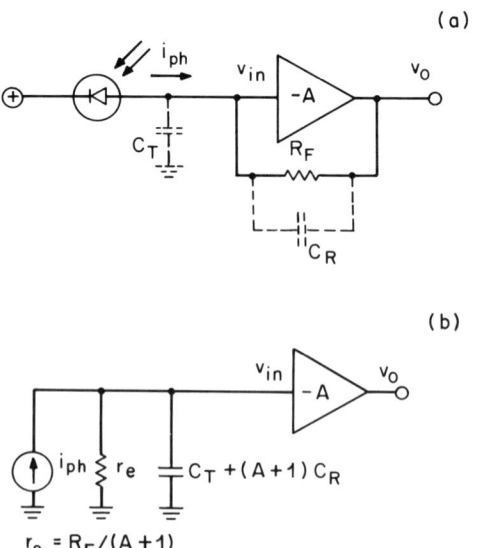

FIG. 13. Real transimpedance receiver: (a) circuit with parasitic capacitances, (b) equivalent circuit.

and Chinnock (1979) used $R_F = 5$ kΩ and accepted a modest signal integration in return for a somewhat smaller sensitivity penalty. Paski (1980) demonstrated both approaches in his experimental silicon bipolar receiver ICs; the commercial silicon bipolar receiver IC by Snodgrass and Klinman (1984) was designed to avoid signal integration.

D. Integrating Transimpedance Optical Receivers

The transimpedance amplifier circuit now is sometimes used as an integrating receiver (Gloge et al., 1980; Ogawa et al., 1983); this reduces the large equalization ratio often required with the simple integrating amplifier of Fig. 11. The best conventional high-value resistors have $C_R \sim 0.05$ pF. Thus, for the 45 Mb/s case discussed earlier, taking $R_F = 1$ MΩ, the amplifier must integrate the signal above $f_p = 3.2$ MHz. Equalization is still required, though the equalization ratio is typically 10–20 times less than for the simple integrating receiver example of Fig. 11. The equalizer noise penalty and the dynamic range/sensitivity tradeoffs for a nonencoded bit stream are correspondingly improved; at low bit rates ($B < 100$ Mb/s), further improvement might be obtained by active equalization. However, equalization noise is entirely eliminated for all bit rates by the nonintegrating receivers of Section V; the IC implementations are also much cheaper to manufacture than equalized receivers.

Both the simple integrating amplifiers of the BPO (Hooper et al., 1980; D. R. Smith et al., 1980) and the transimpedance integrating amplifiers of Ogawa et al. (1983) use a forward voltage amplifier with a FET input transistor followed by a bipolar junction transistor (BJT) cascode. The folded cascode of Ogawa et al. is illustrated in Fig. 14. The input transistor Q_1 is a 1-micron gate-length GaAs FET for low noise; this is followed by a level shifting PNP cascode stage Q_2 and an emitter follower output buffer Q_3. The input voltage, V_{in}, creates a signal current $V_{in} g_{m1}$ at the drain of Q_1. The emitter of Q_2 absorbs most of this current because it is a forward-biased diode and acts like a short. The Q_1 drain signal current passes through Q_2 to R_{e2}; the amplifier voltage gain is then

$$A = -g_{m1} r_{e2}, \tag{51}$$

where g_{m1} is the input FET transconductance. Typically these amplifiers are implemented in hybrid IC (HIC) form (~ 1 mil conductor and resistor patterns on a ceramic substrate with discrete semiconductor device chips). For the BPO-type receiver, the input resistor would be connected to a bias source; for a transimpedance design, the input resistor is connected to the output.

One problem with GaAs FET transimpedance integrating amplifiers using the voltage gain element of Fig. 14 is that their response pole frequency, f_p, is

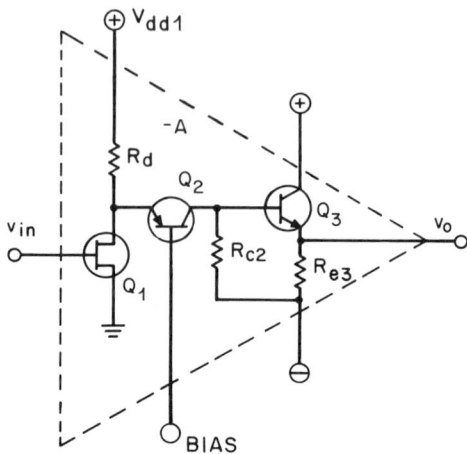

Fig. 14. FET-BJT cascode voltage amplifier.

approximately inversely proportional to the absolute temperature. For a $-25°C$ to $+85°C$ temperature range in the field, f_p would vary by 1.45:1. The equalizer zero frequency must track this variation in the receiver pole frequency for reliable operation; such tracking is difficult to achieve in practice.

This problem arises because f_p is approximately proportional to the voltage gain A (Eq. 50), A is proportional to the input FET transconductance g_{m1} (Eq. 51), and g_{m1} is proportional to the electron velocity in the channel. Now the electron velocity in GaAs near 300°K is approximately inversely proportional to the absolute temperature, due to phonon scattering (Ruch and Fawcett, 1970). This applies from low fields through to velocity saturation. Thus, f_p goes approximately as $1/T$ too.

The simple integrating receiver with restricted disparity encoding and limited equalization used by the BPO (Hooper et al. 1980; D. R. Smith et al., 1980) neatly side-steps the equalization tracking problem. The encoding removes the low frequencies from the signal; since no feedback is used, the pole frequency (above which the signal is integrated) is below the lowest frequency signal component. Exactly where the pole is does not matter; the equalizer can simply differentiate over the entire spectrum of the signal. The BPO design also allows ac coupling throughout the amplifier.

Of course, the nonintegrating, high-sensitivity active-feedback receivers of Section V avoid equalization problems entirely.

Finally, for both types of integrating receiver, the maximum photocurrent is still restricted by the dc drop across R_F, even if encoding is used. For the 45-Mb/s case cited, assuming 5 V supplies, this corresponds to a maximum photocurrent of $i_s \sim$ 2 to 4 μA, or a dynamic range of $\sim$22 dB. Thus, the best integrating receiver dynamic range still implies the need for field-installed attenuators in many applications.

V. Active-Feedback Lightwave Receiver Circuits

A. Introduction

The new active-feedback receivers of this section are the first lightwave receiver circuits to achieve high sensitivities without integrating the signal and the first to achieve wide dynamic ranges without compromising the sensitivity. In addition, the IC designs can be realized very inexpensively for commercial applications. The high sensitivity, plus the response, dynamic range, and cost advantages mean that these new lightwave receivers are suitable for local-area-network, datalink, and loop-plant applications, as well as for the long-haul transmission systems addressed by most previous receiver designs.

The previous lightwave receiver designs of Section IV offered a choice between high-sensitivity circuits that integrate the signal and lower-sensitivity transimpedance circuits that do not. The high-sensitivity circuits were eventually adopted for a few leading-edge transmission systems; however, because they integrate the signal, they have limited dynamic range and require subsequent equalization, with its attendant problems and sensitivity penalty. In addition, the data stream often must be encoded, introducing another modest sensitivity penalty. Finally, it would be almost impossible to extend the dynamic range of these integrating receivers; most possible techniques would vary the input pole frequency above which the signal is integrated, thus requiring a tracking equalizer.

The active-feedback receivers described in this section comprise a basic nonintegrating high-sensitivity receiver, plus a dynamic-range-extender/automatic-gain-control (AGC) circuit. The basic receiver can be realized using either the capacitive feedback circuit or the micro-FET feedback circuit; the former is preferred for hybrid IC designs, the latter is preferred for monolithic IC designs. Both basic receivers offer excellent sensitivity, but their dynamic range is little better than that of previous designs. Accordingly, the dynamic range extender circuits are necessary for most loop, local-area-network, and datalink applications and are useful for many transmission applications.

These active-feedback receivers are somewhat more sensitive than the integrating receivers of the prior art; the equalizer and its noise penalty are eliminated, encoding and its noise penalty are unnecessary, and the dynamic-range-extender circuits permit the basic receivers to be optimized solely for sensitivity, eliminating all sensitivity versus dynamic-range tradeoffs. The dynamic range is essentially unlimited (54 dB optical, 108 dB electrical demonstrated) and extends to higher optical powers than available from present transmitters.

A hybrid IC active-feedback receiver was demonstrated and monolithic IC active feedback receivers were proposed by Williams (1982, 1985); the first such IC was demonstrated by Fraser *et al.* (1983). IC versions are now in production by AT&T Technologies (Morrison, 1984; Steininger and Swanson, 1986), and a 1.7-Gb/s implementation has been announced. (Dorman *et al.*, 1987). However, these commercial designs are beyond the scope of this section.

In theory, the ideal transimpedance amplifier of Fig. 12a (Section IV.C) could have been used for the basic nonintegrating, high-sensitivity receiver. However, as was discussed in Section IV.C, previous attempts to realize this basic receiver encountered two problems, neither of which was solved in those designs. Both problems were caused by the need to use a large-value feedback resistor R_F for low Johnson noise.

The first problem was that the gain elements used did not have enough voltage gain to avoid signal integration by the receiver input capacitance if a large-value feedback resistor was used. Adding stages to these designs would have added phase shift within the feedback loop, which typically would cause ringing or instability. Solutions to this problem will be covered in the sections following.

The second, and more significant, problem was that conventional large-value feedback resistors act like feedback capacitors over most of the bandwidth, owing to their parasitic shunt capacitance C_R (Fig. 13a). This causes signal integration over most of the bandwidth; thus, using a large-value conventional feedback resistor for high sensitivity gives the integrating transimpedance receiver of Section IV.D. To avoid this problem in a typical 45-Mb/s design with $R_f = 2$ MΩ would require C_R to be less than 0.001 pF, which is impossible with conventional feedback resistors.

There are two ways to solve the problem of the parasitic feedback capacitance, C_R, and make a nonintegrating high-sensitivity receiver; C_R can either be eliminated by use of a nonconventional feedback resistor or C_R can be used as part of the feedback element. Section V.B discusses the micro-FET-feedback basic receiver, in which C_R is eliminated by using a special-design micro-FET as the feedback resistor; these designs are preferred for IC implementations. Section V.C discusses the capacitive feedback basic receiver, in which C_R is used as part of the feedback element; these designs are preferred for hybrid IC (HIC) implementations. Section V.D then describes the dynamic-range-extender/AGC circuits used with these two basic receiver circuits to form the complete high-sensitivity, wide-dynamic-range, nonintegrating receivers.

The remaining four sections discuss complete active-feedback receiver designs. Section V.E describes a hybrid IC design using the capacitive-feedback basic receiver circuit. Section V.F describes FET IC designs using the micro-FET-feedback basic receiver. The IC designs are preferred for commercial applications. Two principal IC circuit examples are presented, with a discussion of NMOS, CMOS, and GaAs implementations. Section V.G discusses how these IC receiver designs are scaled to different bit rates; Section V.H then presents calculations of IC receiver sensitivities for bit rates between 10 Mb/s and 4 Gb/s. These calculations include sensitivities of receivers using present pin-photodiode and FET IC technologies, plus sensitivities to be expected from probable future receivers, given expected advances in pin-photodiode and FET IC technologies. The sensitivity improvements expected in these future-technology receivers are then explained on the basis of the device figures-of-merit of Section III.B.2.

Finally, note that although the new active-feedback receiver designs behave much like the ideal transimpedance receivers of theory, the circuits are quite different. Both active-feedback designs use unconventional feedback elements never contemplated in the simple ideal-transimpedance receiver; for example, both include an active component (e.g., a transistor) in the feedback circuit. A second difference is that both active-feedback receivers usually include one of the new dynamic-range-extender/AGC circuits, which gives these receivers the dynamic range needed for most loop, local-area-network, and datalink applications. The new AGC circuits also eliminate undesirable tradeoffs in designing the new feedback circuits; this is particularly important for the micro-FET feedback IC receivers, which are preferred for commercial applications. Third, the new feedback and AGC techniques mean that the response and stability considerations differ from those of previous lightwave receivers. The fourth and most important difference is that ideal transimpedance receivers with high enough feedback resistors to get the highest sensitivities were never realized; in contrast, both of the active-feedback receiver designs have been demonstrated, and IC versions have entered production.

B. Micro-FET Feedback Receivers

The micro-FET feedback receiver of Fig. 15a uses a special-design micro-FET as a feedback resistor to form a current-to-voltage amplifier (Williams 1982, 1985; Fraser *et al.* 1983; Williams and LeBlanc, 1986); an equivalent circuit representation is shown in Fig. 15b. These receivers offer excellent

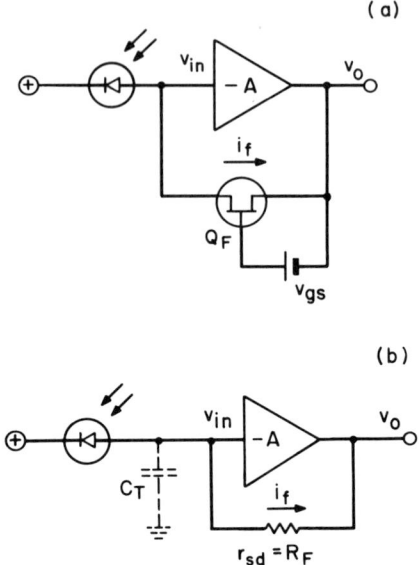

FIG. 15. Micro-FET feedback nonintegrating receiver: (a) circuit, (b) simplified equivalent circuit. [After Williams, 1982, 1985.]

performance, yet can be realized very inexpensively in IC form. Thus, these designs are preferred for commercial applications; typically, one of the dynamic-range-extender circuits of Section V.D is then integrated on the same chip as the basic receiver.

The three design goals for these new receivers are a high sensitivity, a wide-band nonintegrating response, and stability against ringing or oscillation. The noise analysis of Section III.B says that a high-sensitivity receiver must use a high-resistance feedback element for low Johnson noise, must have the minimum possible input capacitance C_T (typically $\lesssim 1$ pF), and, at present, should be realized in either fine-line GaAs FET or fine-line silicon MOSFET IC technologies. Conventional large-value feedback resistors cannot be realized in either IC technology; the feedback micro-FET used in the new receivers is easy to make in both. For a wide-band nonintegrating response, the current feedback must be independent of frequency; as was mentioned in Section V.A, even a small parasitic shunt feedback capacitance associated with a high-resistance feedback element will cause signal integration. In the example of a 45-Mb/s receiver with a feedback resistance of 2 MΩ, the parasitic shunt capacitance had to be less than 0.001 pF, which was impossible with conventional feedback resistors. This parasitic shunt capacitance is essentially eliminated in a properly designed feedback micro-FET. Finally, the current feedback through Q_F must be large enough to avoid signal integration by the receiver input capacitance C_T. Since the feedback resistance $R_F = r_{sd}$ must be large, the forward voltage gain A must be large. The problem is how to achieve this large voltage gain without inducing instability or oscillations; fortunately, the stability requirements for the micro-FET receiver are very lenient.

This section first discusses the micro-FET feedback element used in these receivers, then covers the forward voltage gain and stability requirements. The gain and stability requirements indicate that the forward voltage amplifier used in these receivers can be a simple multistage design, with a bandwidth only slightly greater than the photocurrent signal bandwidth. Such voltage amplifiers are readily realized in any of the fine-line FET IC technologies.

The special-design feedback micro-FET (Q_F) used in these receivers (Fig. 15a) acts as a feedback resistor (Fig. 15b) because it is operated near zero source-to-drain voltage, in the linear drain-current versus drain-voltage region. The channel is then an undepleted bar of semiconductor with the number of carriers, hence resistance, determined by the gate bias. The parasitic feedback capacitances are tiny ($C_R < 0.001$ pF) because Q_F is tiny, typically 1–2 μm wide by 1–5 μm from source to drain, with a total area of only a few square micrometers. Such transistors are readily fabricated in the high f_T/low noise 1-μm fine-line silicon and GaAs FET technologies presently preferred for IC receivers. Feedback resistances greater than 1 MΩ are readily achieved by biasing the gate of Q_F barely above the turn-on threshold. Thus, Q_F functions as an ideal feedback resistor with a negligible parasitic feedback capacitance and makes possible high-sensitivity, nonintegrating IC receivers.

Unfortunately, Q_F only acts as a resistor for voltages less than the gate voltage above the turn-on threshold. The channel of Q_F pinches off whenever the output voltage swing is greater than the gate voltage above threshold; Q_F then acts as a constant current source. Thus, the maximum photocurrent through Q_F is approximately the gate voltage above threshold divided by the feedback resistance; accordingly, the dynamic range of the basic IC receiver of Fig. 15a is somewhat limited.

In fact, the resistive micro-FET feedback receiver would be commercially impractical for some applications if used without a dynamic-range-extender circuit; fortunately, the dynamic-range extenders of Section V.D are readily integrated on the same IC as the basic receiver at essentially no additional cost. Furthermore, most applications (e.g., datalinks, local-area networks or loop feeders) would require a dynamic-range extender in any case. The complete IC receiver designs of Section V.F include both the basic micro-FET feedback receiver and a dynamic-range extender on the same IC.

The forward voltage gain, A, must be large enough so that the current feedback through Q_F is sufficient to keep the photocurrent signal from being integrated by the receiver input capacitance C_T. (C_T is the sum of the photodiode capacitance C_{pin}, plus C_{FET} due to the receiver input FET, plus the input stray capacitance C_s.) Since the simplified equivalent circuit of Fig. 15b is identical to the ideal transimpedance amplifier of Fig. 12a (Section IV.C), the receiver bandwidth is given by Eq. (50), taking $R_F = r_{sd}$ of Q_F. For a non-return-to-zero bit stream, the signal bandwidth is 0.56 times the bit rate B (Section III.B.1). If the receiver bandwidth is set equal to the bit rate B, the minimum forward gain is then

$$A = 2\pi R_F C_T B - 1. \tag{52}$$

For high sensitivity, C_T is made as small as possible and R_F is made large enough so that its Johnson noise current is minor compared to the noise of the receiver input FET. For the 45-Mb/s receiver example, in which C_T was 1 pF and R_F was 2 MΩ, Eq. (52) gives a minimum voltage gain, A, of 570.

The principal stability requirement is that the voltage-amplifier pole frequencies be above the receiver bandwidth so that the phase shift in the voltage gain A is less than 45° over the receiver bandwidth. Now the receiver photocurrent response was determined by the current feedback through Q_F, which therefore had to be frequency independent; the stability is determined by the voltage feedback, which is not frequency independent. The voltage feedback is generated by the frequency-independent current feedback driving the input capacitance C_T (Fig. 15b). Thus, the voltage feedback is inversely proportional to frequency over most of the loop bandwidth because the feedback current is integrated by C_T for frequencies above the $R_F C_T$ voltage-feedback pole frequency. (For the 45-Mb/s receiver example with $R_F = 1$ MΩ and $C_T = 1$ pF, the voltage-feedback pole frequency is 159 kHz). Thus, the voltage-feedback pole must be used as the dominant voltage-loop pole; since the voltage feedback gives a 90° phase shift over most of the loop bandwidth, 90° more due to poles in A would give positive feedback and cause oscillation. In practice, the phase shift in A should remain less than 45° as the frequency is increased, until the $R_F C_T$ feedback integration carries the loop voltage gain below unity. Mathematically, this unity loop gain frequency works out to be equal to the photocurrent-to-output-voltage response-pole frequency, or receiver bandwidth; thus, the stability requirement is that the phase shift in the forward voltage gain A be less than 45° over the receiver bandwidth.

Thus, the voltage amplifier used in these IC receivers can be a simple, multistage video amplifier, with a bandwidth a little greater than the receiver bandwidth so that the phase shift in A is less than 45° over the receiver bandwidth. The out-of-band rolloff is not important because the out-of-band voltage-loop gain is less than unity. These lenient stability requirements allow these receiver designs to be scaled to bit rates well in excess of 1 Gb/s.

C. Capacitive Feedback Receivers

Hybrid IC active feedback receivers use the capacitive-feedback, nonintegrating current-amplifier circuit of Fig. 16 (Williams, 1982, 1985; Williams and LeBlanc, 1986); as mentioned, any conventional large-value hybrid-technology feedback resistor is in fact a resistance in parallel with a parasitic capacitance, C_R, and acts as a feedback capacitor over most of the bandwidth. The feedback current, i_f, is (approximately) the total feedback capacitance $C_F = C'_F + C_R$ times the time derivative of the applied voltage, v_f; v_f is the integral of the output voltage, v_o. Since i_f is the derivative of the integral of v_o, i_f is proportional to v_o; this capacitive feedback element driven by an integrator acts like an ideal feedback resistor. Thus, the capacitive feedback amplifier functions as an ideal nonintegrating transimpedance current amplifier.

As mentioned in Section (V.A), the usable hybrid IC voltage gain A is low because of the extra gain rolloff and phase shift caused by the relatively large hybrid circuit capacitances. In the capacitive feedback current amplifier, the missing gain is supplied by the feedback integrator, which is typically an ordinary amplifier stage with an extra collector capacitance to make it integrate. In addition, the photodiode capacitance C_{pin} is almost always used

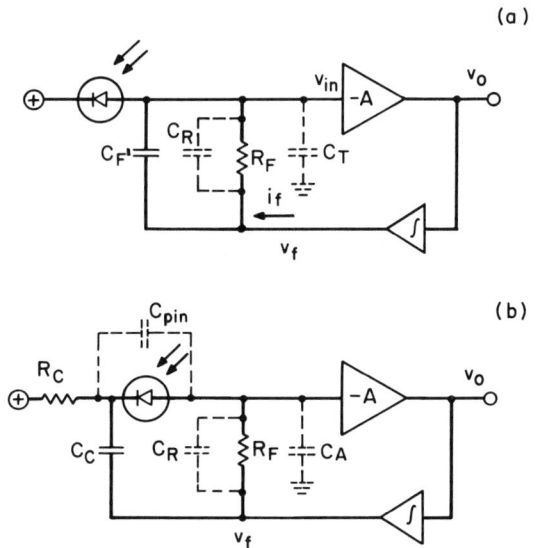

FIG. 16. Capacitive feedback nonintegrating receiver: (a) basic circuit (C'_F is optional), (b) photodiode capacitance feedback receiver (C_D acts as C'_F). [After Williams, 1982, 1985.]

as part of the feedback element by ac coupling v_f to the bias side of the photodiode, as shown in Fig. 16b; the high-frequency feedback impedance is then $\sim 10^2$ times smaller than that of R_F alone. The input-voltage-to-v_f high-frequency gain needed is reduced by the same factor (Section V.E). Note that using the photodiode capacitance as C'_F adds no extra capacitance to the front-end capacitance C_T and therefore does not reduce the receiver sensitivity.

A complete hybrid IC capacitive feedback receiver using capacitive feedback through the photodiode will be discussed in Section V.E; the frequency-response and loop-stability considerations will be presented with that circuit.

D. Dynamic Range Extenders

Both the resistive FET feedback IC basic receiver of Fig. 15 and the capacitance feedback hybrid IC basic receiver of Fig. 16 offer excellent sensitivity, but their dynamic range is typically little better than that of previous designs and is not enough for many commercial applications Therefore, the preferred receiver designs also include dynamic-range extender/AGC circuits.

Both the IC and the hybrid nonintegrating basic receivers can be represented by the equivalent circuit of Fig. 15b because both act as ideal transimpedance amplifers (Fig. 12a). Therefore, the same AGC circuit works for both. To the AGC circuit, both basic receivers look like a voltage amplifier, of gain $-A$, with the input shunted to ground by an almost noiseless virtual input resistor, r_e, produced by the feedback, as was shown in Fig. 12b. The input is also shunted by the total capacitance, C_T; however, to avoid signal integration, the design is such that r_e is low enough to be the dominant input admittance over the signal bandwidth.

The preferred dynamic-range extension/AGC technique (Williams, 1982) is to add a variable resistance input shunt device R_S to divert the excess photocurrent from the input of the high-sensitivity basic receiver (Fig. 17a); R_S and r_e divide the photocurrent. (Note that in Fig. 17a, the basic receiver is represented by the equivalent circuit of Fig. 12b.)

Usually (Fig. 17b) the variable input shunt device is a FET (Q_S) operated in the linear drain-current-versus-drain-voltage region; the shunt resistance is the source-to-drain resistance of Q_S. The value of R_S is controlled by the gate voltage of Q_S. The AGC servo circuit varies the value of R_S to maintain the amplifier output signal voltage, v_o, constant over the input shunt AGC range. The basic receiver shown in Fig. 17b is the resistive FET feedback IC design of Fig. 15a.

The input shunt cannot be used for AGC at low photocurrents without ruining the sensitivity. The reason is that R_S must be comparable to or less

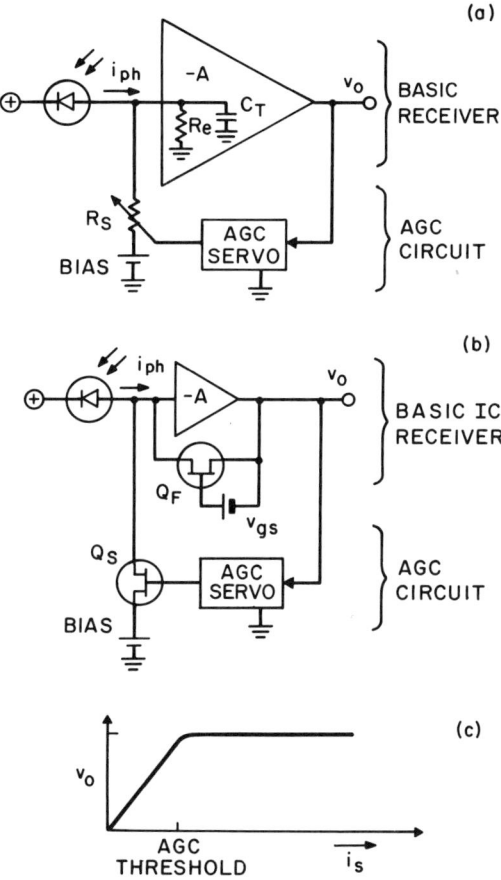

FIG. 17. Receivers with input shunt AGC: (a) general circuit, (b) FET input shunt AGC on basic IC receiver of Fig. 15a, (c) AGC characteristic. [After Williams (1982).] (© 1982 IEEE.)

than r_e to divert any appreciable photocurrent, and R_S, a real resistance, has Johnson noise; although r_e, a virtual resistance, does not. At moderate bit rates (10–100 Mb/s), r_e is typically 100–1000 times smaller than the feedback resistance R_F; since $R_S \lesssim r_e$, the Johnson noise is typically increased 10–30 times ($\langle i_n^2 \rangle^{1/2} \propto 1/R$) when the shunt is used. Therefore, the shunt is turned on only when the photocurrent is large enough that the extra Johnson noise does not matter. Above this AGC threshold, R_S is servoed to maintain the peak-to-peak v_o signal constant; below this threshold R_S or Q_S is OFF and any AGC at low photocurrents must be provided by a post-amplifier stage. The resultant receiver AGC characteristic is shown in Fig. 17c.

This AGC technique is implemented in the receiver of Fig. 17b by having the AGC servo circuit vary the gate voltage of input shunt Q_S to limit the output signal voltage, v_o, to a value equal to the desired AGC threshold photocurrent times the feedback resistance R_F of Q_F. For photocurrents less than the AGC threshold, v_o is less than the limiting value, the gate of Q_S is biased below the turn-on voltage, and Q_S is OFF. For photocurrents greater than the AGC threshold, the gate voltage is servoed to vary the source-to-drain resistance of Q_S so that the amplitude of the v_o signal is held constant. The AGC servo itself is typically just a peak detector followed by a slow integrator that drives the gate of Q_S with the integral of the difference between the peak detector output and a reference voltage.

Note that shunt AGC does not change the signal-frequency response of the receiver; the input pole frequency is already above the signal passband due to the virtual input resistance, r_e, caused by the feedback; R_S just moves the pole frequency even higher. Note also that there are no stability problems; the shunt AGC decreases the voltage feedback ratio, which actually improves the stability. In contrast, transimpedance AGC schemes, which use a variable-resistance feedback element, increase the voltage feedback ratio as AGC is applied. If this variable feedback were applied around the whole amplifier, the AGC range before the onset of instability would be limited unless the forward gain A were decreased as R_F were decreased. Therefore, shunt AGC is usually preferred at present, though transimpedance AGC applied around the first stage only may be used in IC versions for increased dynamic range (see Section V.F).

The variable input shunt R_S must usually be connected to a dc bias source so that the amplifier input bias does not change when the shunt is turned on; any change is multiplied by the voltage gain A. In high-bit-rate IC implementations the shunt bias source is typically an on-chip fixed-voltage source accurately matched to the input stage. (The matching is least critical in high-bit-rate designs because the forward voltage gain is low, due to the design scaling laws of Section V.G). In low-bit-rate designs, a slow shunt-bias feedback integrator or a digital equivalent is typically used; these also can readily be integrated on a receiver IC.

This shunt AGC circuit is adequate for many applications; however, for further dynamic-range extension, the forward voltage gain A can be decreased above a second AGC threshold after the input shunt FET (Q_S) has been servoed to its minimum resistance. Any voltage-gain reduction technique that preserves the bandwidth of the voltage gain and has the dynamic range needed can be used. 50:1 decreases in A, with corresponding increases in the maximum photocurrent, are readily achieved.

Note that although decreasing A decreases the current feedback, thus increasing r_e, the minimum R_S is much less than r_e and prevents signal integration by C_T.

For full AGC, this results in a three-stage AGC system (Fig. 18): post-amplifier AGC for the lowest photocurrents (AGC-1), followed by input shunt AGC (AGC-2), followed by reduction of the forward voltage gain, A, of the receiver (AGC-3). Both the receiver and post amplifier plus the control circuitry can readily be realized on one IC. A hybrid IC version (Williams, 1982), was tested at 45 Mb/s, using a modified AT&T Technologies FT3 regenerator board (of T.L. Maione *et al.*, 1978); a dynamic range of 54 dB optical (108 dB electrical) was achieved (Williams and LeBlanc, 1986) and the component count was less than that of the unmodified FT3 board.

Another way to increase the AGC dynamic range would simply be to increase the physical size of the input shunt FET so that voltage gain reduction would be unnecessary. At moderate bit rates, r_e is typically a few thousand ohms (for $C_T \sim 1$ pF); the dynamic range without AGC is typically ~ 20 or 25 dB. For a 50-dB plus dynamic range, the maximum photocurrent must be increased by three orders of magnitude. Thus, the minimum R_S would be only a few ohms; Q_S would then be larger than the amplifier input transistor. The extra input capacitance due to Q_S decreases the sensitivity; the dominant mean-square photocurrent noise terms go as C_T^2 (Section III.B).

The reason for this problem is that the maximum input-voltage swing without voltage-gain reduction is typically only a few millivolts in receivers designed for moderate bit rates; this is why Q_S would have to be a power FET with a saturation resistance of only a few ohms to handle a maximum

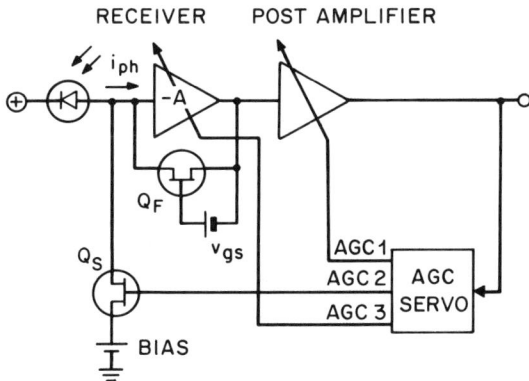

FIG. 18. Basic IC receiver with three-stage AGC.

photocurrent of only a few milliamperes. The voltage-gain reduction technique increases the input voltage swing; this drives the same photocurrent through a small, higher-resistance Q_S. The capacitance of this small Q_S is negligible, eliminating any sensitivity penalty.

Thus, the input-shunt/voltage-gain-reduction AGC is preferred for dynamic ranges of 50 dB or more. Such dynamic ranges are presently found only in low to moderate bit-rate (less than ~200 Mb/s) systems using semiconductor laser transmitters. These systems will become more common as semiconductor lasers become cheaper. Shunt AGC only is preferred for dynamic ranges of 40 dB or less, which is adequate for present LED-based systems, even at low bit rates, and for present laser-based systems at high bit rates.(High-bit-rate receivers are less sensitive and require less dynamic range for a given transmitter power.) Future lasers, however, may be more powerful and would require more dynamic range, even at high bit rates. A transimpedance AGC circuit for IC receivers is discussed in Section V.F.

E. Hybrid IC Active-Feedback Receivers

Figure 19 shows a capacitive-feedback, high-sensitivity receiver circuit with input shunt AGC, as developed for hybrid IC (HIC) optical receivers (Williams, 1982, 1985). A 45-Mb/s version achieved a sensitivity of -51.7 dBm at a 1.3-μm wavelength and 10^{-9} BER, with a dynamic range of 54 dB optical (Williams and LeBlanc, 1986); the sensitivity was within 1.5 dB of the best 1.3-μm APD result at the bit rate (Forrest et al., 1981), the dynamic range extended to an average optical power of ~1.8 mW. However, the IC versions of Section V.F will soon exceed this performance and are cheaper to manufacture. On the other hand, the HIC designs of this section are much easier to demonstrate in a laboratory and are particularly useful for early tests and models of receivers using new transistor technologies that are not yet available in IC form.

At low photocurrents, the input shunt FET Q_S is turned off by the AGC servo (Section V.D), and the circuit reduces to the photodiode capacitance feedback receiver of Fig. 16b. The forward voltage amplifier is a FET-BJT folded cascode with an emitter-follower output buffer (Fig. 14). This type of voltage amplifier was first disclosed by Ogawa and Chinnock (1979); it was also used in the integrating receivers of Ogawa et al. (1983), and its operation was discussed in Section IV.D. The feedback integrator is a common base stage, Q_1, with an extra capacitance, C_1, from the collector to the ground, that integrates the signal. The integrated feedback signal, v_f, is connected directly to the feedback resistor, R_F, and is ac-coupled by C_c to the bias side of the photodiode; thus, the photodiode capacitance, C_{pin}, is used as part of the feedback element.

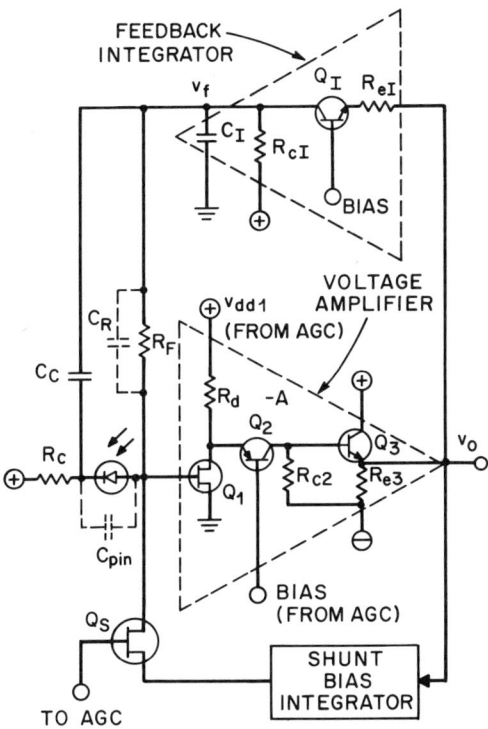

FIG. 19. Capacitive feedback HIC receiver circuit. [After Williams (1982).] (© 1982 IEEE.)

The AGC/dynamic-range-extender circuitry is identical to that of Fig. 18. At low photocurrents, Q_S is off and AGC is provided by a variable-gain post amplifier. The AGC circuit turns Q_S on and uses it as a variable-resistance input shunt for AGC at larger photocurrents where the extra Johnson noise due to Q_S does not matter. The bias voltage for the variable shunt resistance is provided by a slow feedback integrator. At yet higher photocurrents, the resistance of Q_S is at its minimum value, and the forward gain A is reduced to provide a third stage of AGC. The AGC circuit reduces A by reducing the input FET transconductance. It does this by reducing the drain voltage and drain current by reducing V_{dd1} and the bias voltage on the base of Q_2 together. (Alternatively, a dual-gate input FET could have been used.)

Now consider the frequency response of the photodiode capacitance feedback amplifier without AGC. The feedback element is the resistance R_F in parallel with the photodiode capacitance, C_{pin}, and the feedback resistor's parasitic feedback capacitance, C_R. The capacitances dominate over most of the bandwidth. Writing the total feedback capacitance as $C_F = C_{pin} + C_R$, the feedback current, i_f, is approximately

$$i_f = C_F \frac{dv_f}{dt}. \tag{53}$$

where v_f is the feedback voltage from the integrator. Writing

$$V_f = a \int v_o \, dt \tag{54}$$

where v_o is the output voltage and a is the feedback integrator gain constant, gives

$$i_f = aC_F v_o. \tag{55}$$

aC_F is the equivalent feedback conductance of the feedback element driven by the feedback integrator. Since $v_o = -Av_{in}$,

$$i_f = -AaC_F v_{in}, \tag{56}$$

and the equivalent input resistance r_e due to the feedback is

$$r_e = \frac{dv_{in}}{di_f} = (AaC_F)^{-1} \tag{57}$$

The receiver bandwidth is simply the input pole frequency due to r_e in parallel with the total input capacitance C_T:

$$f_p = \frac{1}{2\pi r_e C_T} = \frac{AaC_F}{2\pi C_T}. \tag{58}$$

If the receiver bandwidth is set equal to the bit rate, B, then

$$AaC_F = 2\pi C_T B. \tag{59}$$

This sets the minimum product AaC_F of the forward gain times the feedback integrator gain-constant times the feedback capacitance. Using C_{pin} as part of C_F typically increase C_F by about a factor of 5–10 and reduces the forward voltage gain and feedback integrator gain-constant needed accordingly.

By Eq. (51), the forward voltage gain of Fig. 19 is

$$A = g_{m1} R_{c2}, \tag{60}$$

where g_{m1} is the transconductance of the input FET Q_1 and R_{c2} is the collector resistor of Q_2. The gain constant of the common base feedback integrator is approximately

$$a = \frac{1}{R_{el} C_1}, \tag{61}$$

where R_{el} is the emitter resistor of the integrator transistor Q_I, and C_I is the integrating capacitor on the collector of Q_I. (This expression assumes that the Q_I emitter input impedance is much less than R_{el}.) a has the dimensions of inverse seconds, as required.

These circuits can readily be used up to $\sim 100 - 200$ Mb/s because the voltage gain needed from v_{in} to v_f at the high-frequency bandwidth limit (i.e., at the transimpedance response pole, f_p) is only $\sim 2 - 3$. The extra gain needed at lower frequencies is supplied by the feedback integrator. The feedback integrator has a very wide bandwidth because the dominant collector pole is used as the integrator pole by adding the extra capacitance C_I.

The gain needed at $f_p = 1/(2\pi r_e C_T)$ is derived by noting that, at that frequency, the current into the total front-end capacitance, C_T, is equal to the current through r_e, i.e., to the feedback current though the feedback capacitance, C_F. Thus, the voltage gain required from v_{in} to v_f at f_p is simply C_T/C_F. Now, $C_F = C_{pin} + C_R$ is essentially equal to C_{pin}. As before, $C_T = C_{pin} + C_{FET} + C_s$, where the stray capacitance C_s (which includes C_R) is small compared to C_{pin} and C_{FET}. For low noise, the size of the receiver input FET is scaled so that $C_{FET} = C_{pin} + C_s$ (Eq. 25a); thus, $C_T \cong 2C_{pin}$. Therefore C_T/C_F, the voltage gain needed from v_{in} to v_f at f_p, is only $\sim 2 - 3$.

At low frequencies, the parallel combination of R_F, C_R, and C_{pin} acts like a feedback resistor rather than a feedback capacitance; in addition, the feedback integrator acts like a feedback amplifier at low frequencies. Therefore, at low frequencies, the capacitive feedback receiver essentially reduces to a conventional resistive feedback transimpedance receiver. For a flat overall frequency response, the feedback zero frequency $f_z = 1/[2\pi R_F(C_{pin} + C_R)]$, below which the feedback is resistive rather than capacitive, must be equal to the feedback integrator pole frequency, $f_p = 1/(2\pi R_{CI} C_I)$, below which the feedback voltage is not integrated. Typically, R_{CI} is trimmed accordingly. This compensation is relatively insensitive to temperature.

The loop stability considerations are essentially identical to those discussed in Section V.B for the micro-FET feedback IC receivers; the phase shift due to poles in the forward voltage gain A (plus any extra poles in the feedback integrator) should be less than 45° over the bandwidth of the receiver. (Since the main pole of the feedback integrator and the capacitive part of the feedback element compensate each other, they contribute no net phase shift.)

F. IC Active-Feedback Receivers

This section first presents two IC active-feedback receiver examples, with a discussion of implementations in NMOS, CMOS, and GaAs IC technologies. Both examples use the micro-FET feedback basic receiver of Section V.B; they differ in the design of the voltage amplifier and in their dynamic-range-extender/AGC circuitry. The first example (Fig. 20) uses the input shunt AGC technique of Section V.D; this technique is presently preferred for most applications. The second example (Fig. 22) uses a transimpedance AGC circuit in which the AGC element is connected around the first stage only. This section concludes by discussing two design details concerning the amplifier response. The first extends the response discussion of Section V.B to include the parasitic gate-to-channel capacitance of the feedback FET, Q_F; the second describes how to approximately cancel both the effect of this capacitance and of the residual nonlinearities in Q_F, to first order. Both the capacitive effects and the nonlinear effects in Q_F are readily minimized in the design of optimized single-bit-rate receivers; canceling these effects offers useful but minor advantages. However, in "universal" receivers designed to operate at many different bit rates, canceling these effects is very helpful indeed.

Design scaling laws for these IC receivers are presented in Section V.G; sensitivity versus bit-rate calculations for present-technology implementations and for implementations in probable future technologies are presented in Section V.H

Figure 20 shows a simplified diagram of a typical active-feedback, high-sensitivity receiver IC with input shunt AGC. This analog circuit is compatible with both silicon MOSFET and GaAs MESFET fine-line digital IC technologies. From an IC production viewpoint, it looks like a very small memory.

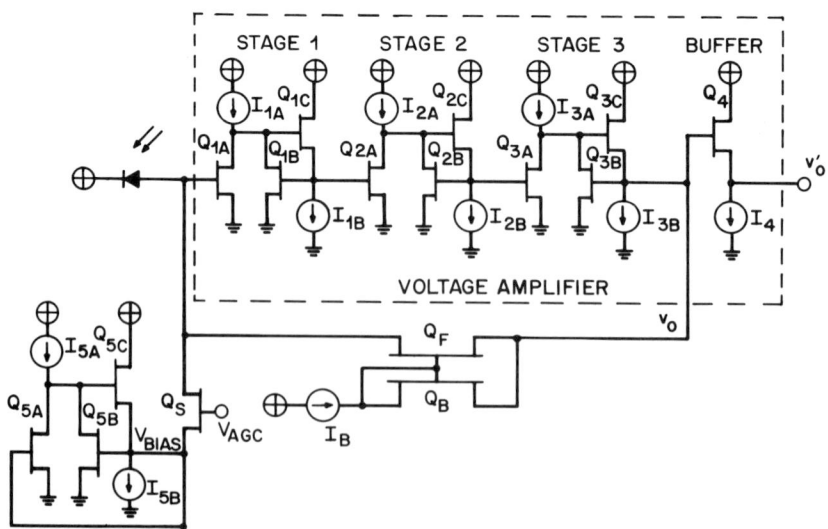

FIG. 20. Micro-FET feedback receiver IC with input shunt AGC. [After Williams (1982).] (© 1986 IEEE.)

This receiver circuit is a realization of the AGC receiver circuit of Fig. 17b discussed in Section V.D. The basic micro-FET receiver comprises a voltage amplifier of gain $-A$, formed by Q_{1A} through Q_4, plus the micro-FET Q_F, which acts like a feedback resistor (Section V.B); Q_B and I_B form the gate bias voltage supply for Q_F. The dynamic-range-extender/AGC circuit comprises the input shunt FET, Q_S, the bias voltage supply ($Q_{5A} - Q_{5C}$) for the source of Q_S, and an AGC control circuit (not shown). At low photocurrents, AGC is provided by a post amplifier (not shown); at higher photocurrents, AGC is provided by input shunt FET Q_S. The third AGC stage, in which the forward voltage gain A is reduced, is omitted from this example; stage three AGC techniques will be discussed later.

As mentioned in Section V.B, R_F is chosen to be large enough so that its Johnson noise (Eq. 8, Section III.B.1) is small compared to the input FET noise (Eq. 20), which is the majority of the noise in a well-designed receiver. The voltage gain, A, needed to keep the photocurrent signal from being integrated by the receiver input capacitance, C_T, is determined by R_F, C_T, and the bandwidth needed. If the receiver bandwidth is set equal to the bit rate, B, the minimum forward gain is then $A = 2\pi R_F C_T B - 1$ by Eq. (52) of Section V.B. Since R_F scales inversely as the bit rate squared (Section V.G), the required voltage gain A is *inversely* proportional to the bit rate; high-bit-rate designs need *less* voltage gain. Thus, this receiver design readily scales to high bit rates without requiring any increase in the number of stages and without stability problems. A bit rate of 1.7 Gb/s has been reported in a production version of this receiver (Dorman *et al.*, 1987).

The voltage amplifier in the lightwave receiver circuit of Fig. 20 has three similar stages of the shunt feedback type demonstrated by Hornbuckle *et al.* (1981) for GaAs wide-band voltage amplifiers. Stage one is transistors $Q_{1A} - Q_{1C}$; stage two is $Q_{2A} - Q_{2C}$; stage three is $Q_{3A} - Q_{3C}$. In silicon NMOS or GaAs MESFET ICs, the current source loads $I_{1A} - I_4$ are n-channel depletion FETs with gate shorted to source (Figs. 21a, 21b); in CMOS, the loads are p-FETs for the common-source FETs and n-FETs for the source-follower FETs (Fig. 21c). Consider stage one. Q_{1A} is a common-source stage driving the source follower, Q_{1C}. The source follower increases the gain-bandwidth product of the common-source stage because it isolates the collector of the common-source stage from the Miller-enhanced input capacitance of the following stage. Q_{1B} is a shunt feedback transistor. Without Q_{1B}, the low frequency gain of the stage would theoretically be infinite (assuming for now a perfect current-source load I_{1A} and infinite Q_{1A} drain resistance). With Q_{1B}, the stage gain is the ratio of the Q_{1A} size to the Q_{1B} size. For example, for a typical amplifier implemented in 1-micron gate-length technology (source to drain), if Q_{1A} is 500-μm wide and Q_{1B} is 100-μm wide, the gain of the first stage is approximately 5.

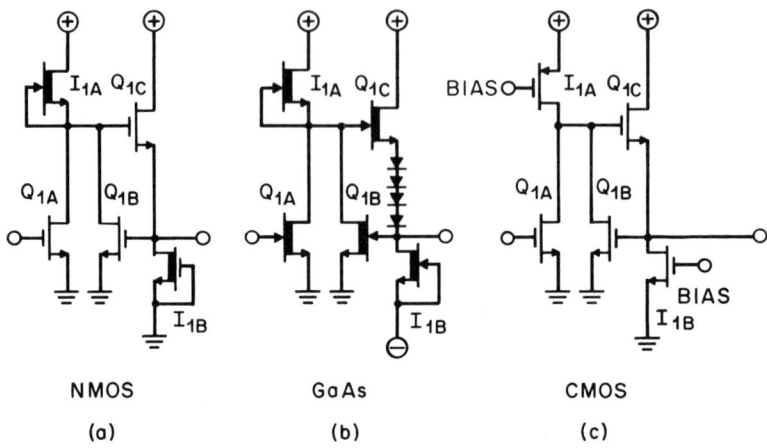

FIG. 21. Typical IC voltage gain stages.

In the lightwave receiver design of Fig. 20, the stages are all identical, except for size; they are all scale models of a common design. Thus, the closed-loop quiescent input and output voltages of all the stages are equal to the same voltage. This gives several advantages. First, since the quiescent input voltage of each stage is equal to the quiescent output voltage of the stage, the common-source input transistor of each stage (e.g., Q_{1A}) and the shunt feedback transistor of each stage (e.g., Q_{1B}) have the same quiescent gate voltage, hence the same transconductance per unit gate width, so that the stage gain is set by the size ratio, as desired. This is most important for depletion-mode devices, such as GaAs MESFETs, where the transconductance depends strongly on the gate voltage. Second, the dc bias calculations need be done for only a single stage rather than for a three-stage system; in addition, this type of receiver design reduces dc biasing problems caused by processing variations that change the dc parameters of the FETs on a given wafer. Third, a single such stage with input shorted to output gives a voltage equal to the quiescent amplifier input voltage. This is used as the bias source ($Q_{5A} - Q_{5C}$) for the AGC input shunt FET, Q_S, in this design. (At low bit rates ($B < 100$ Mb/s), a slow integrator is often preferred, as was discussed in Section V.D. This also can readily be integrated on the IC.)

An output source-follower buffer, Q_4, is used to avoid capacitive loading of the last stage, which would degrade the loop stability (Fraser et al., 1983).

The gate bias source for the resistive feedback FET, Q_F, in Fig. 20 is formed by a matched bias transistor, Q_B, and a current source, I_B. The ratio of I_B to the gate width (size) of Q_B determines how far Q_B, hence Q_F, is biased above threshold. If Q_F were operated in the constant drain current region, Q_B and Q_F

would form a FET current mirror; though Q_F operates instead in the linear drain-current versus drain-voltage region at low drain voltage, the biasing action of Q_B is the same.

Of course, I_B is typically a FET current source similar to these used as the drain loads $I_{1A} - I_{3A}$ for the common-source gain FETs. Note also that GaAs MESFET designs typically use a source-follower and level-shifting diodes between the drain of Q_B and the gates of Q_B and Q_F.

The scaling of the three stages is set by the noise considerations of Section III.B.2. The optimum-size input FET Q_{1A} is scaled so that its gate input capacitance is equal to the rest of the front-end capacitances combined (Eq. 25a). This sets the scale of the input stage; typically the input stage transistors are a few hundred micrometers wide, and the stage draws a few tens of milliamperes. The following stages can be made smaller to reduce both the chip area and the power consumption without much of a noise penalty; as usual, most of the noise comes from the first stage. Consider an amplifier with Q_{1A} 500-μm wide, Q_{2A} 100-μm wide, and Q_{3A} 50-μm wide, with a gain of 5 per stage. The mean-square equivalent input noise due to each stage is inversely proportional to both the stage size and the voltage gain squared preceding the stage. Thus, in the example, the second stage contributes $5/25 = 1/5$ the equivalent mean-square photocurrent noise of the first stage; the third stage contributes $10/625 = 0.016$ times as much as the first stage. The total root-mean-square equivalent photocurrent noise is only $\sqrt{1.22} = 1.10$ times that of the first stage alone. The second and third stage noise thus reduce the optical sensitivity by only 0.4 dB.

Simpler designs for voltage amplifier stages can be used, especially at lower bit rates. For example, the source-follower buffers may often be omitted in lower-bit-rate silicon MOS designs, as in the receiver of Fraser et al. (1983). (In GaAs MESFET versions, the source-follower buffer is still needed to drive the level-shifting diodes.) The shunt feedback transistor Q_B may often be omitted in GaAs or silicon NMOS stage designs, leaving an ordinary common-source stage. Most GaAs MESFETs have a relatively low drain resistance r_d, which may sometimes be used as the signal load in place of the shunt feedback FET Q_B (Hornbuckle et al., 1981). Similarly, in NMOS amplifier stages, the n-FET active loads have an output conductance equal to their back-gate transconductance; this conductance can be used as the signal load in place of Q_B, as in the receiver of Fraser et al. (1983). The IC yield then depends on how well the GaAs FET drain resistance or NMOS back gate effect are controlled.

The circuit of Fig. 20 is adequate for most LED-based systems. Typical LED transmitter optical outputs into a fiber are presently on the order of 0.05–0.1 milliwatts; at a 1.3-μm wavelength this corresponds to a maximum receiver photocurrent of 0.05–0.1 milliamperes, which the input shunt AGC circuit of Fig. 20 can easily handle.

Semiconductor laser transmitters presently put ~ 1 milliwatt optical into a fiber; at ~ 1.3 μm, this corresponds to a maximum receiver photocurrent of about a milliampere. For low-bit-rate receivers (e.g., 45 Mb/s), the minimum input shunt resistance of Q_S must then be only a few ohms; this means a large Q_S, extra input capacitance, and lower sensitivity, as was discussed in Section V.D. However, the minimum shunt resistance needed goes up proportional to the bit rate, as will be discussed in Section V.G. Thus, the circuit of Fig. 20 is presently adequate for most laser-based systems above a few hundred megabits per second.

Both laser and LED transmitter powers will increase; in addition, lasers will become cheaper and probably will be used in more low-bit-rate systems in the future. These future systems may well need more dynamic range than the two-stage AGC circuit of Fig. 20 can provide. A solution is to add a third stage of AGC to further extend the dynamic range.

One third-stage AGC technique is to reduce the forward voltage gain A, once Q_S has been servoed to its minimum resistance, as was discussed in Section V.D. One way to do this is to reduce the transconductance of the input FET, either by using a dual-gate input FET or (equivalently) by using a cascode input stage. The input FET transconductance is then reduced by reducing the input FET drain voltage by reducing the cascode FET gate bias and reducing the input FET drain current, e.g., by partially turning off the input stage load devices. This gain reduction technique was discussed for the hybrid IC designs of Section V.E. Other voltage-gain reduction techniques may also be used.

A typical transimpedance AGC receiver circuit is shown in Fig. 22. This circuit also increases the AGC range and in addition reduces the AGC threshold. In this circuit, a variable-resistance AGC FET, Q_{TZ}, is connected around the first amplifier stage. At low photocurrents, Q_{TZ} is off to avoid adding excess Johnson noise to the input photocurrent; AGC is provided by the postamplifier (not shown). Q_{TZ} is turned on only when the photocurrent is large enough so that the extra Johnson noise does not cause bit errors. Above this AGC threshold, Q_{TZ} is servoed to provide AGC, just as in the shunt AGC circuit. The voltage available to drive the photocurrent through Q_{TZ} is the input voltage times the first-stage gain plus one. Thus, if the first stage has a gain of 5, the maximum photocurrent is increased by a factor of 6 over the corresponding input shunt AGC circuit. In addition, the AGC threshold is reduced by a factor of $\sqrt{6} = 3.9$ dB because the resistance of Q_{TZ} at the AGC threshold is six times larger than that of a corresponding input shunt AGC FET; therefore, the RMS Johnson noise current is reduced by a factor of $\sqrt{6}$.

In the circuit of Fig. 22, the stages are all identical except for size, and operate with equal quiescent input and output voltages as before. This is the simplest way to avoid changing the amplifier operating point when Q_{TZ} is turned on. The same idea was used in the shunt AGC design of Fig. 20; note,

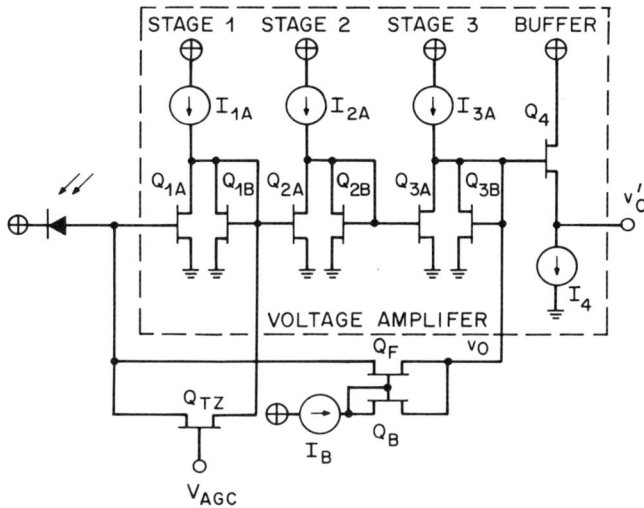

FIG. 22. IC receiver with transimpedance AGC.

however, that the output buffers on each stage are omitted in Fig. 22, which removes one high-frequency pole from inside the transimpedance AGC loop but reduces the high-frequency performance. (Note again that in GaAs designs, the buffers are needed for dc level shifting.)

The transimpedance AGC receiver of Fig. 22 might appear to be only marginally stable; for some value of the AGC feedback resistance, R_{TZ} of Q_{TZ}, the $R_{TZ}C_T$ feedback pole frequency will equal the first-stage collector pole frequency. However, if the low-frequency gain of the first stage is 5, the loop gain at the double-pole frequency is only 2.5 and the phase margin at unity loop gain is 53°. The feedback zero due to the input transistor gate-to-drain capacitance provides additional phase margin.

Of course, the receiver would be unstable for low values of R_{TZ} if Q_{TZ} were connected around all three stages. The same would apply if Q_F were used to provide AGC. (Q_F might be used to provide a limited amount of initial AGC in very-high-sensitivity, very-low-bit-rate receivers.)

In order to best design these IC receivers, the parasitic feedback capacitance effects must be considered in more detail than in Section V.B. The receiver may be represented by equivalent circuit of Fig. 23 in which the resistive feedback FET, Q_F, is represented by a feedback resistance, R_F, which is the source-to-drain resistance, shunted by a parasitic source-to-drain capacitance, C_{sd}, and with a distributed capacitance, C_{gc}, from the gate to the channel. The total input capacitance (photodiode plus input FET plus strays) is again represented by C_T.

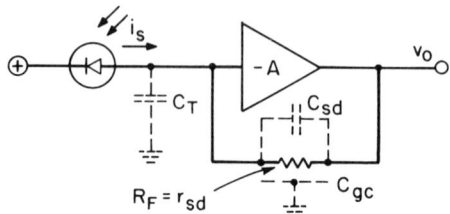

FIG. 23. IC receiver equivalent circuit. [After Williams, 1985.]

The parasitic source-to-drain capacitance, C_{sd}, appears across R_F as an alternate feedback path and can cause signal integration. If the $R_F C_{sd}$ feedback bandwidth is set at twice the bit rate B, then

$$C_{sd} < \frac{1}{4\pi R_F B}. \tag{62}$$

For the original 45-Mb/s receiver example, with $R_F = 1$ MΩ, C_{sd} must then be less than 0.002 pF. Experimentally, C_{sd} is negligible because the gate of Q_F acts as an electrostatic shield between the source and the drain; the flux lines go from source to gate and from drain to gate rather than from source to drain.

The gate-to-channel capacitance, C_{gc}, forms a distributed R-C delay line with the channel resistance, $r_{sd} = R_F$. If the gate is at ac ground, as in Fig. 23, the effect is to reduce the high-frequency feedback and to introduce an additional lagging phase shift into the feedback loop. Both effects increase the high-frequency response; the extra phase shift also reduces the phase margin against oscillation. Solving the R-C delay line equations and setting the extra phase shift less than 30°,

$$C_{gc} < \frac{3}{2\pi R_F B}. \tag{63}$$

For the 45-Mb/s receiver example, C_{gc} must then be less than 0.01 pF. This means the total gate area must be tiny; as mentioned, a typical feedback FET is 1-μm wide by 2–10 μm from source to drain.

The gate-to-channel capacitance effects can be canceled to first order in frequency by applying one third of the ac output voltage to the feedback transistor gate. (In MOS designs, the parasitic overlap capacitances of Q_B, Q_F, and I_B can be tailored to form a three-to-one capacitive voltage divider.) Then, if Eq. (63) is satisfied, the magnitude of the current feedback at the high-frequency end of the receiver bandwidth is 96% of the magnitude of the low-frequency current feedback. In addition, the extra phase shift due to C_{gc} is now leading rather than lagging, which increases the loop stability. The improved phase margin is the more important benefit in optimized single-bit-rate receiver designs and usually justifies using this technique. In "universal" receivers designed to operate at many different bit rates, this technique is important for both response and stability reasons.

Finally, although Q_F is used as a resistor, its current-voltage characteristic is not perfectly linear, even though the AGC circuit is used to restrict the range over which Q_F is used (Section V.B). If necessary, the residual nonlinearity can be canceled to first order by applying half the output signal voltage to the gate of Q_F so that

$$v_{gs} = V_{gso} + \frac{1}{2} i_{ph} R_F, \tag{64}$$

where V_{gso} is the quiescent gate bias and $i_{ph} R_F$ is the output signal voltage due to the photocurrent i_{ph} flowing through the feedback resistance R_F. Of course, this technique also approximately cancels the effect of the gate-to-channel capacitance; however, this technique is slightly more difficult to implement because it requires a modified gate bias circuit for Q_F. This technique is principally of interest for "universal" receivers that cannot be optimized for a single bit rate and for very-low-bit-rate receivers.

G. Design Scaling Laws for IC Receivers

This section discusses how the IC receiver designs of Section V.F scale to high bit-rates; the following section presents calculations of IC receiver sensitivities for bit-rates between 10 Mb/s and 4 Gb/s. These calculations include sensitivities of receiver designs using present pin-photodiode and FET IC technologies, plus sensitivities to be expected from probable future receiver designs, given expected advances in pin-photodiode and FET IC technologies.

These nonintegrating IC receiver designs can readily be extended to bit rates in excess of 1 Gb/s by appropriately scaling the design parameters. The feedback resistance R_F was chosen so that its Johnson noise was some small fraction of the input FET noise. Assume that the total input capacitance C_T due to the photodiode, the input FET, plus strays, is constant. The noise of the input FET scales as B^3 (Eq. 20); the Johnson noise of R_F scales as B/R_F (Eq. 8). If the ratio of these two noise sources is kept constant as the bit rate is increased, R_F scales as:

$$R_F \sim \frac{1}{B^2}. \tag{65}$$

By Eq. (52), the forward voltage gain A is proportional to R_F times the bit rate. Using Eq. (65) for the scaling of R_F, the voltage gain A scales as

$$A \sim \frac{1}{B}. \tag{66}$$

Thus, the forward voltage gain needed to avoid signal integration decreases as the bit rate increases because the noise considerations of Section III allow R_F to be decreased as the bit rate squared (Eq. 65). This permits these active-feedback IC receiver designs to be adapted to bit rates well in excess of 1 Gb/s.

The allowable Q_F parasitic feedback capacitances, C_{sd} and C_{gc}, are inversely proportional to $R_F B$ [Eq. (62) and (63)]. Since R_F scales as $1/B^2$, the allowable Q_F parasitic capacitances scale as

$$C_{sd}, \quad C_{gc} \sim B. \tag{67}$$

Thus, the parasitic feedback capacitance effects diminish at high bit rates, again because noise considerations allow R_F to scale as $1/B^2$.

If input shunt AGC is used, the size of the input shunt FET is inversely proportional to the bit rate. Thus, a third, gain reduction, AGC stage is not needed in high-bit-rate designs. Assume that the transmitter optical power is the same at all bit rates and assume that the maximum output voltage swing of the receiver is the same at all bit rates. Since A is inversely proportional to the bit rate, the input voltage swing available to drive the photocurrent through Q_S is proportional to the bit rate; therefore, the required conductance, hence size, of Q_S is inversely proportional to the bit rate.

H. Sensitivity Calculations for Present- and Future-Technology IC Receivers

Now consider the sensitivities achievable in optical receiver designs using present pin-photodiode and FET IC technologies and in receiver designs using probable future pin-photodiode and FET IC technologies. The receiver circuits used for these calculations are micro-FET feedback IC designs of the type illustrated in Fig. 20 in Section V.F; the sensitivities are calculated using the noise expression of Eq. (35), which includes all the noise sources in the circuit. For a given FET IC technology and total input capacitance, C_T, the different receiver designs for the different bit rates, B, are generated by scaling R_F as $1/B^2$ (Eq. 65), scaling A as $1/B$ (Eq. 66), and redesigning Q_F and the voltage amplifier accordingly.

The sensitivity versus bit-rate calculations for typical present-technology IC receivers were presented previously in Section III.B.3, along with sensitivity results from the literature and a brief comparison of GaAs and silicon FET IC technologies. The calculations assumed a typical commercial 1-μm channel-length GaAs MESFET or silicon MOSFET IC technology, and a total input capacitance, C_T, of 1 pF due to the receiver-input FET capacitance C_{FET}, the pin-photodiode capacitance, C_{pin}, plus any stray capacitance, C_s (C_s is typically due to the bonding pads, the shunt AGC FET, and the packaging.) For the silicon MOSFET IC designs, the ratio of the transconductance to the gate-input capacitance, g_m/C_{FET}, was taken as 70 mS/pF; for the GaAs MES FETs, g_m/C_{FET} was taken as 90 mS/pF. Since C_{FET} was 0.5 pF, the silicon IC input FETs had $g_{m1} = 35$ mS; the GaAs IC input FETs had $g_{m1} = 45$ mS. The channel-noise-factor, Γ, was taken as 1.2 for the silicon FETs and as 1.5 for the GaAs FETs. The photodiode leakage current was taken as 1 nA at 20°C and 15 nA at 85°C, which is typical of present commercial InGaAs photodiodes. As was discussed, the GaAs MESFET and silicon MOSFET receiver sensitivities were essentially equal in theory; the GaAs FETs have a higher transconductance, but the silicon FETs have a lower channel noise factor. In practice, the silicon FET receivers are presently a few decibels less sensitive; this performance gap should narrow in the future.

Future-technology receivers will offer superior performance due to improvements in both the FET IC technologies and the pin-photodiode technology. The principal FET IC technology improvement will be finer linewidths, which will permit shorter channel lengths. Now, to first order, the transconductance of an optimized short-channel FET is independent of the channel length from source to drain because the electrons in the channel move at the saturation velocity (Sze, 1983); however, the gate input capacitance, C_{FET}, is approximately proportional to the gate area and, therefore, is inversely proportional to the gate length. Thus, g_m/C_{FET} and f_T are inversely proportional to the gate length; therefore, using a finer-linewidth FET technology will reduce the principal noise terms is Eq. (35), thereby improving the receiver sensitivity. In addition, a higher f_T will allow higher-bit-rate receivers, and, in lower-bit-rate designs, will increase the achievable voltage gain per stage.

Since optimized 0.2-μm channel-length FETs were reported in 1983 (by Fichtner et al.), and 0.3-μm FETs are now commercially available, it is only a matter of time until 0.25-μm channel-length IC receivers are commercially practical, especially since the development of fine-line FET ICs is driven by the VLSI and VHSIC programs. The individual FETs in these future 0.25-μm IC technologies should have about four times the g_m/C_{FET} ratio and four times the f_T of present 1-μm IC FETs. Therefore, g_m/C_{FET} should be about 280 mS/pF for the 0.25-μm silicon technologies and about 360 mS/pF for the 0.25-μm GaAs technologies.

The principal pin-photodiode improvement will be low-doping or reduced-area devices with very small capacitances. Reducing the photodiode capacitance C_{pin} allows the total capacitance C_T to be reduced proportionately; as mentioned in Sections III.B.2 and V.F, the sensitivity is maximized by scaling the size of the input FET so that $C_{FET} = C_{pin} + C_s = 1/2 C_T$; since the stray capacitance C_s can be made negligible, $C_T \cong 2C_{pin}$. By inspection, reducing C_T reduces all of the principal noise terms in Eq. (35), thereby improving the receiver sensitivity.

As discussed in Section III.B.2, the photodiode capacitance could be reduced immediately from the present ~0.4 pF to ~0.1 pF, simply by reducing the photodiode diameter from the present ~75 μm to ~30 μm. This means that C_T would be reduced from the present ~1 pF to ~0.25 pF, simply by reducing the size of the input FET to match C_{FET} to $C_{pin} + C_s$, and taking care to minimize C_s. Note that a reduced-area photodiode can easily be coupled to present single-mode fiber by use of a microlens or a GRIN lens; the technology to efficiently couple a small-area optical device to an optical fiber has already been developed for semiconductor laser transmitters. The coupling optics should become inexpensive as laser transmitters ride down their learning curve. Eventually, the doping density almost certainly will be reduced to the point that full-size 0.1-pF InGaAs photodiodes become available; if the present rate of progress continues, this may well happen in the next five years.

Future pin photodiodes will also have lower leakage currents, which will allow better receiver sensitivities at low bit rates. Kim *et al.* (1985) and Campbell *et al.* (1985) have already demonstrated long-wavelength photodiodes with a 50-pA leakage current at 20°C; it is only a matter of time until commercial InGaAs photodiodes match this performance.

Figure 24 shows sensitivity-versus-bit-rate calculations for present-technology IC receiver designs, next-technology IC receiver designs, and future-technology IC receiver designs. The present-technology receiver calculations are taken directly from Section III.B.2 and, as mentioned, assume a 1-μm channel-length GaAs or silicon FET IC technology, a C_T of 1 pF, and a leakage current of 1 nA at 20°C and 15 nA at 85°C. The next-technology

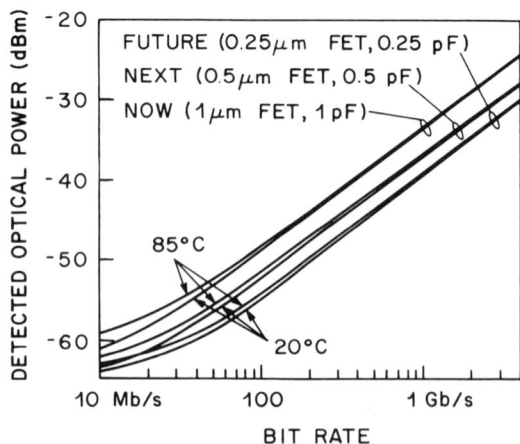

FIG. 24. Projected evolution of pin-FET receiver sensitivities. $\lambda = 1.3$ μm, 10^{-9} BER.

receiver calculations assume a 0.5-μm channel-length FET IC technology, a C_T of 0.5 pF, and a leakage current of 0.2 nA at 20°C and 3 nA at 85°C. The future-technology receiver calculations assume a 0.25-μm channel-length FET IC technology, a C_T of 0.25 pF, and a leakage current of 50 pA at 20°C and 0.75 nA at 85°C.

The sensitivity improvements calculated for the "next" and "future" receivers of Fig. 24 can be understood intuitively on the basis of the figure-of-merit discussion and analysis of Section III.B.2. For these two examples, the projected sensitivity improvements are equally due to the reduced gate length of the FETs and to the lower photodiode capacitances. The FET-technology figure of merit is $M_{FET} = g_m/(C\Gamma)$ by Eq. (27); the optical receiver sensitivity is approximately proportional to the square root of M_{FET}. Since g_m/C is inversely proportional to the FET channel length, the receiver sensitivity is inversely proportional to the square root of the FET channel length. Thus, using 0.5-μm FETs rather than present 1-μm FETs should increase the sensitivity by a factor of about the $\sqrt{2}$ or 1.5 dB; using 0.25-μm FETs should give a 3-dB improvement.

Similarly, the pin photodiode figure of merit is $M_{pin} = 1/[4(C_{pin} + C_s)] = 1/[2C_T]$ by Eq. (28) and thus is approximately inversely proportional to C_{pin} and exactly inversely proportional to C_T. The receiver sensitivity is approximately proportional to the square root of M_{pin} and thus inversely proportional to the square root of C_T. Reducing C_T from 1 pF to 0.5 pF should increase the sensitivity by about 1.5 dB; reducing C_T to 0.25 pF should give a 3-dB improvement.

Thus, adding the sensitivity improvements due to the projected photodiode advances to those due to the projected FET advances indicates that the next-technology receivers (0.5-μm FETs, $C_T = 0.5$ pF) should give about 3 dB better sensitivity than present-technology receivers; the future-technology receivers (0.25-μm FETs, $C_T = 0.25$ pF) should give about 6 dB better sensitivity. This agrees reasonably well with the complete calculations of Fig. 24.

Note that the calculated sensitivities in Fig. 24 scale quite accurately as $B^{-3/2}$, except at low bit rates. For a well-designed receiver, the principal terms in Eq. (35) for the mean-square receiver noise are the input FET noise, the noise of the load device for the input FET, and the feedback resistor Johnson noise. The mean-square noise of the input FET and its load device are both proportional to B^3; the mean-square noise of the feedback resistor usually is also proportional to B^3 because R_F is usually scaled as $1/B^2$, following Eq. (65) of Section V.G. If these were the only noise terms, the mean-square noise would be exactly proportional to B^3; the root-mean-square noise would be exactly proportional to $B^{3/2}$, and the receiver sensitivity would be exactly proportional to $B^{-3/2}$. In fact, except at low bit rates, summing over these terms usually give the sensitivity to within 0.3–0.4 dB of the exact calculations; the sum over the following stages gives a minor mean-square noise contribution (which scales approximately as B^3); the leakage current noise is proportional to B and becomes important only at low bit rates.

At low bit rates, the calculated sensitivities are less than predicted by the $B^{-3/2}$ law because of the leakage current noise and because R_F cannot be increased indefinitely as $1/B^2$ as the bit rate is decreased.

References

Abidi, A. A., Kasper, B. L., and Kushner, R. A. (1984). In "1984 IEEE International Solid State Circuits Conference Digest of Technical Papers," p. 76. IEEE, New York.
Baechtold, W. (1972). *IEEE Trans. Electron Dev.* **ED-19**, 674.
Brooks, R. (1980). *Electron Lett.* **16**, 458.
Campbell, J. C., Dentai, A. G., Holden, W. S., and Kasper, B. L. (1983). *Electron. Lett.* **19**, 818.
Campbell, J. C., Dentai, A. G., Qua, G. J., Long, J., and Riggs, V. G. (1985). *Electron Lett.*, to be published.
Davenport, W. B., and Root, W. L. (1958). "Introduction to the Theory of Random Signals and Noise." McGraw-Hill, New York.
Dorman, P. W., Yoder, J. D., Tatsuguchi, I., Gibson, W. C., Wemple, S. H., and Owen, B. (1987). To be published. IEEE, New York.
Fichtner, W., Fuls, E. N., Johnston, R. L., Watts, R. K., Weick, W. W. (1983). In "1982 International Electron Devices Meeting Technical Digest," p. 384. IEEE, New York.
Forrest, S. R., Williams, G. F., Kim, O. K., and Smith, R. G. (1981). *Electron. Lett.* **17**, 917.
Fraser, D. L., Williams, G. F., Jindal, R. P., Kushner, R. A., and Owen, B. (1983). In "1983 IEEE International Solid-State Circuits Conference Digest of Technical Papers," p. 80. IEEE, New York.
Gloge, D. C., and Ogawa, K. (1985). In "1985 Conference on Optical Fiber Communication, Digest of Technical Papers," p. 84. Optical Society of America, Washington DC.
Gloge, D., Albanese, A., Burrus, C. A., Chinnock, E. L., Copeland, J. A., Dentai, A. G., Lee, T. P., Li, T., and Ogawa, K. (1980). *Bell Syst. Tech. J.* **59**, 1365. Short Hills, N.J.
Goell, J. E. (1974). *Bell Syst. Tech. J.* **53**, 629. Short Hills, N.J.
Hooper, R. C., Rejman, M. A. Z., Ritchie, S. T. D., Smith, D. R., and White, B. R. (1980). In "Sixth European Conference on Optical Communications, Proceedings," p. 222.
Hornbuckle, D. P., and van Tuyl, R. L. (1981). *IEEE Trans. Electron Devices.* **ED-28**, 175.
Kanbe, H., Susa, N., Nakagome, H., and Ando, H. (1980). *Electron. Lett.* **16**, 163.
Kasper, B. L., Campbell, J. C., Gnauck, A. H., Dentai, A. G., and Talman, J. R. (1985). *Electron Lett.* **21**, 982.
Kim, O. K., Dutt, B. V., McCoy, R. J., and Zuber, J. R. (1985). *IEEE J. Quantum Electron.* **QE-21**, 138.
Kroemer, H. (1982). *Proc. IEEE* **70**, 13.
Kroemer, H. (1983). *J. Vac. Sci. Technol.* **B1**, 126.
Linke, R. A., Kasper, B. L., Ko, J.-S., Kaminow, I. P., and Vodhanel, R. S. (1983). *Electron Lett.* **19**, 775.
Linke, R. A., Kasper, B. L., Campbell, J. C., Dentai, A. G., and Kaminow, I. P. (1984). *Electron. Lett.* **20**, 489.
Maione, T. L., Sell, D. D., and Wolaver, D. H. (1978). *Bell Syst. Tech. J.* **57**, 1837.

Matsushima, Y., Akiba. S., Sakai, K., Kushiro, Y., Noda, Y., and Utaka, K. (1982). *Electron. Lett.* **18**, 945.
McIntyre, R. J. (1966). *IEEE Trans. Electron Dev.* **ED-13**, 164.
Melchior, H., Hartman, A. R., Schinke, D. P., and Seidel, T. E., (1978). *Bell Syst. Tech. J.* **57**, 1791.
Mikawa, T., Kagawa, S., Kaneda, T., Sakurai, T., Ando, H., and Mikami, O. (1981). *IEEE J. Quantum Electron* **QE-17**, 210.
Miller, S. E. (1979). *In* "Optical Fiber Telecommunication," (S. E. Miller and A. G. Chynoweth, eds.), Chap. 21. Academic Press, New York.
Morrison, D. P. (1984). FOCLAN-84 Post-Deadline Paper.
Nishida, K., Taguchi, K., and Matsumoto, Y. (1979). *Appl. Phys. Lett.* **35**, 251.
Ogawa, K. (1981). *Bell Syst. Tech. J.* **60**, 923.
Ogawa, K., and Chinnock, E. L. (1979). *Electron. Lett.* **15**, 650.
Ogawa, K., Owen, B. and Boll, H. J. (1983). *Bell Syst. Tech. J.* **62**, 1181.
Paski, R. M. (1980). Private communication.
Personick, S. D. (1973a). *Bell Syst. Tech. J.* **52**, 843.
Personick, S. D. (1973b). *Bell Syst. Tech. J.* **52**, 875.
Rousseau, M. (1976). *Electron. Lett.* **12**, 478.
Ruch, J. G., and Fawcett, W. (1970). *J. Appl. Phys.* **41**, 3843.
Smith, D. R., Hooper, R. C., Ahmad, K., Jenkins, D., Mabbitt, A. W., Nicklin, R. (1980). *Electron. Lett.* **16**, 69.
Smith, D. R., Hooper, R. C., Smythe, P. P. and Wake, D. (1982). *Electron. Lett.* **18**, 453.
Smith, R. G., and Personick, S. D. (1980), *In* "Semiconductor Devices for Optical Communication," (H. Kressel, ed.), Chap. 4. Springer-Verlag, New York.
Smith, R. G., Brackett, C. A., and Reinbold, H. W. (1978). *Bell Syst. Tech. J.* **57**, 1809.
Snodgrass, M. L., and Klinman, R. (1984). *IEEE J. Lightwave Tech.* **LT-2**, 968.
Steininger, J. M., and Swanson, E. J. (1986). *In* "1986 International Solid-State Circuits Conference Digest of Technical Papers," p. 60. IEEE, New York.
Susa, N., Nakagome, H., Mikami, D., Ando, H. and Kanbe, H. (1980). *IEEE J. Quantum Electron.* **QE-16**, 864.
Sze, S. M. (1981). *Physics of Semiconductor Devices*, John Wiley, New York.
Takasaki, Y., Tanaka, M., Maeda, N., Yamashita, K., and Nagano, K. (1976). *IEEE Trans. Commun.* **COM-24**, 404.
Williams, G. F. (1982). *In* "1982 IEEE International Solid-State Circuits Conference Digest of Technical Papers," p. 160. IEEE, New York.
Williams, G. F. (1985). U.S. Patent #4,540,952 (Filed 1981).
Williams, G. F., and LeBlanc, H. P. (1986), *IEEE J. Lightwave Technol.* **LT-4**, 1502.
Williams, G. F., Capasso, F., and Tsang, W. T. (1982). *IEEE Electron Dev. Lett.* **EDL-3**, 71.

…

Frequency and Phase Modulation of Semiconductor Lasers

SOICHI KOBAYASHI

Photonic Integration Research, Inc. (PIRI)
Columbus, Ohio

YOSHIHISA YAMAMOTO AND TATSUYA KIMURA

NTT Basic Research Laboratories
Musashino-shi, Tokyo, Japan

I. Introduction 151
II. Direct Frequency Modulation 153
   A. Principle of Frequency Modulation 153
   B. Frequency Modulation Measurement 159
   C. Frequency Characteristics of Direct Frequency Modulation . . . 161
   D. Carrier Density Modulation Effect 163
   E. Temperature Modulation Effect 167
   F. Phase Relationship between FM and AM Signals 171
   G. Theoretical Analysis 174
   H. Advanced Technology of Direct Frequency Modulation . . . 180
III. Phase Modulation by Injection Locking 186
   A. Phase Modulation Mechanism by Injection Locking 186
   B. Phase Modulation Measurement 189
   C. Frequency Characteristics of Induced PM Signal 191
   D. Phase Modulator for Transmission Applications 195
IV. Summary 197
   Acknowledgments 199
   References 199

I. Introduction

Optical fiber transmission systems developed so far have modulated the intensity of the transmitted optical energy. However, optical frequency modulation and optical phase modulation using homodyne and heterodyne detection are more sensitive than conventional intensity modulation with direct detection. Angle modulation of a coherent laser beam, single-polarization-mode fiber, and optical heterodyne (or homodyne) detection are expected to extend repeater spacing and improve transmission capacity in coherent optical fiber transmission systems (Yamamoto, 1980; Yamamoto and Kimura, 1981). Direct frequency and phase modulation in a conventional semiconductor laser using injection current modulation also show promise for coherent transmission systems, since it is easily possible to achieve an FM transmitter and a tunable local oscillator (Saito et al., 1980, 1981).

Optical frequency modulation is inherently accomplished by internal or external modulators, determined by whether the modulators are placed within the laser cavity or outside of it. (Kaminow, 1974; Kubota et al., 1978). The mechanisms of these modulators are based on electro-optic and acousto-optic effects in the dielectric medium. An electro-optic modulator is used for internal modulation and can respond with a bandwidth of up to several gigahertz. An acousto-optic modulator is used for external modulation, but the modulation bandwidth is limited to several hundred megahertz. The modulation frequency is limited in this case by the transit time of the acoustic wave in the modulator cell.

Direct frequency modulation of a semiconductor laser has been achieved by three different methods: (1) modulation of the injection current in a conventional semiconductor laser (Nakamura et al., 1978; Osterwalder and Rickett, 1980; Kobayashi et al., 1981a, 1982), (2) monolithic integration of an electro-optic modulator and laser amplifier (Reinhart and Logan, 1975), and (3) photoelastic modulation by an acoustic wave (Ripper et al., 1966; Whitney and Pratt, 1970). All these methods use the refractive-index change in the laser waveguide induced by an electric input signal.

Electro-optic frequency modulation in semiconductor lasers by method 2 uses an intracavity phase modulator, monolithically fabricated in the laser cavity and directly modulated by current. Injection current modulation and electro-optic modulation by methods 1 and 2 include a spurious AM component. This is because of free-carrier absorption in method 1, and the Franz–Keldish effect in method 2. Photoelastic modulation of method 3 cannot be applied to the baseband modulation scheme because the acoustic wave is generated through a transducer attached to the semiconductor laser. Direct current modulation of method 1 is useful not only for intensity modulation, but also for frequency modulation.

This chapter specifically focuses on injection current frequency modulation in AlGaAs and InGaAsP semiconductor lasers. The most important feature of direct current modulation is the unified structure of the oscillator and modulator. A semiconductor laser used as a frequency modulator is convenient for controlling the oscillation frequency by bias current and by laser mount temperature. A semiconductor laser modulator can also be integrated with other optical components.

Frequency modulation by injection currect is due to the refractive-index change induced by injected carrier change and temperature change. Refractive-index change in semiconductor lasers biased below the threshold has been measured by the frequency shift of Fabry–Perot modes by

Nathan et al. (1963). This change has two causes. Nash (1973) showed that a refractive-index change is caused by a free-carrier plasma dispersion effect, and Thompson (1972) showed that band-to-band interaction also causes such a change. The thermal effect, induced by current injection in semiconductor lasers, consists of a linear thermal expansion and thermal refractive-index change.

Optical phase modulation is achieved by a linear electro-optic or Pockels effect in external modulators and also in reverse-biased GaP p–n junctions (Reinhart, 1968) and GaAs double heterostructures (DH) (Reinhart and Miller, 1972). In a DH modulator, the high cutoff frequency because of series resistance and depletion layer capacitance is expected to be as high as several GHz. Another method of optical phase modulation, using an injection locking technique with a CO_2 ring laser, has been proposed by Buczek and Freiberg (1972). This technique achieves injection-locked phase modulation by injecting coherent cw laser light into a frequency-modulated laser. In a semiconductor laser, injection locking can also be applied to the direct frequency modulated laser to generate a phase modulation signal (Kobayashi and Kimura, 1982a,b). Phase modulation is induced and frequency modulation is simultaneously suppressed by injection locking, thus decreasing the carrier density in the locked laser. Injection-locked phase modulation in an AlGaAs semiconductor laser will also be treated in this chapter.

Section II of this chapter describes frequency modulation by injection current modulation in semiconductor lasers. The frequency characteristics of frequency deviation caused by carrier and thermal effects are detailed. The structural dependence for both effects is analyzed by rate and thermal equations. Advanced technologies for FM semiconductor lasers are also discussed. Section III describes direct phase modulation induced by injection locking in AlGaAs semiconductor lasers. Frequency characteristics for phase deviation restricted by the locking bandwidth are discussed. An analytical discussion of injection-locked phase modulation is also represented by the van der Pol equation. Finally, Section IV summarizes several problems for future applications using direct frequency modulation and phase modulation by injection locking in semiconductor lasers.

II. Direct Frequency Modulation

A. Principle of Frequency Modulation

A frequency-modulated optical wave is represented by

$$E = E_0 \exp[j\{2\pi f_0 t + \beta \sin(2\pi f_m t)\}],$$
$$\beta = \Delta F / f_m. \tag{1}$$

Here, f_0 is the carrier center frequency, ΔF is the maximum frequency deviation (MFD), f_m is the modulation frequency, and β is the frequency modulation index. Typical optical frequency spectra of directly modulated FM signals from an AlGaAs semiconductor laser observed with a Fabry–Perot interferometer are shown in Fig. 1. Particular spectra from (a) to (h) are obtained at a modulation frequency of 150 MHz and a bias current 1.4 times the threshold. The optical field spectrum of a frequency-modulated wave can be expanded in terms of modulation sidebands with amplitudes given by

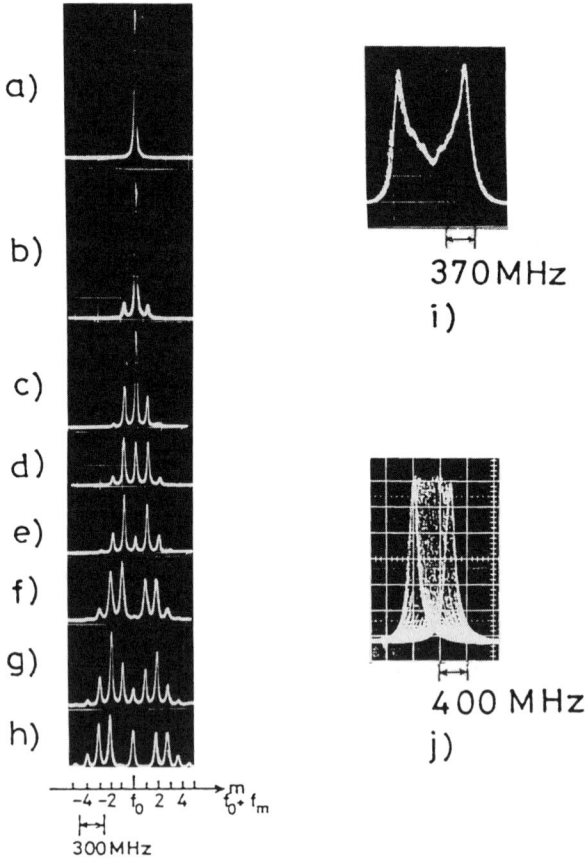

FIG. 1. Power spectra for a directly modulated CSP laser measured with a Fabry–Perot interferometer. (a)–(h): $f_m = 150$ MHz, $I = 1.4 \times I_{th,C}$, $I_{th,C} = 69$ mA. (a): $\Delta I_{0-p} = 0$ mA, (b): 0.76 mA, (c): 1.4 mA, (d): 1.84 mA, (e): 2.24 mA, (f): 2.88 mA, (g): 3.44 mA, (h): 4 mA. Free spectral range (FSR) is 1.5 GHz. (i): $f_m = 100$ KHz, $I = 1.5 \times I_{th,C}$, $\Delta I_{0-p} = 0.43$ mA, and FSR = 1.9 GHz. (j): $f_m = 400$ KHz, $I = 1.5 \times I_{th,C}$, $\Delta I_{0-p} = 0.11$ mA, and FSR = 1.9 GHz. [After Kobayashi et al. (1982).] (©1982 IEEE.)

Bessel functions $J_l(\beta)$ of the first kind:

$$E = J_0(\beta)E_0 \sin(2\pi f_0 t) + J_1(\beta)E_0 \sin\{2\pi(f_0 + f_m)t\}$$
$$- J_1(\beta)E_0 \sin\{2\pi(f_0 - f_m)t\} + \cdots\cdots\cdots\cdots$$
$$+ J_l(\beta)E_0 \sin\{2\pi(f_0 + lf_m)t\}$$
$$+ (-1)^l J_l(\beta)E_0 \sin\{2\pi(f_0 - lf_m)t\}. \quad (2)$$

Each sideband intensity corresponds to the square of the coefficient of the corresponding term in Eq. (2). The frequency modulation index β can be obtained from the ratio of the carrier to the first sideband intensity. The normalized amplitudes for the carrier and various sidebands from Fig. 1 are plotted in Fig. 2 as a function of modulation current (bottom) and modulation index β (top). Marks corresponding to sidebands were experimentally

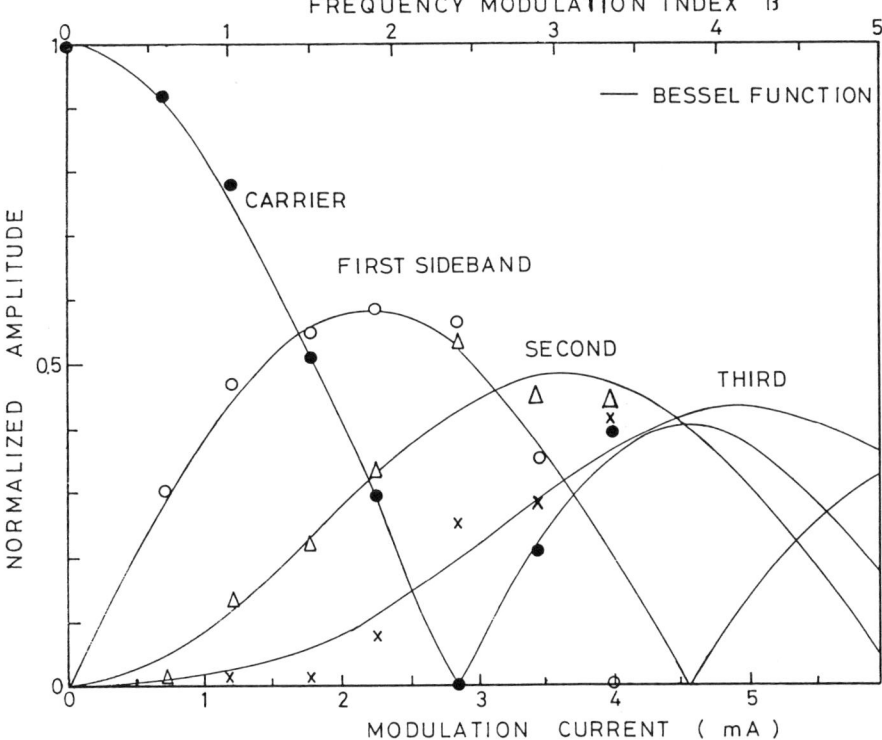

FIG. 2. Normalized amplitudes for various order sidebands observed in the power spectra of Fig. 1(a)–(h) vs. modulation current and frequency modulation index. ●, carrier, ○, first sideband, △, second sideband, ×, third sideband. Solid lines show theoretical values of Bessel functions of the first kind. [After Kobayashi et al. (1982).] (© 1982 IEEE.)

obtained by averaging the upper and lower sideband amplitudes to eliminate the intensity modulation effect. Solid curves indicate theoretical values of Bessel functions of the first kind. The most remarkable feature of frequency-modulated signals is that the amplitude of the carrier frequency vanishes at modulation index $\beta = 2.4$. Each result agrees well with the Bessel function value.

Frequency characteristics of a directly modulated FM signal are important for optical transmission applications. Frequency characteristics of frequency deviation normalized by modulation current are shown in Fig. 3. The frequency deviation is negative. Experimental results measured by various methods are shown. These results have two remarkable features. There is a flat response and a resonant peak in the region from 10 MHz to 5.2 GHz, as indicated by the solid line. The curve shows a frequency shift due to the refractive-index change itself, induced by carrier density change with modulation current injection. There is also a monotonous decrease in proportion to the modulation frequency from 200 Hz to 10 MHz, as indicated by the broken line. This curve corresponds to a frequency shift resulting from the change in the laser cavity length caused by thermal expansion, and the refractive-index change induced by the temperature increase accompanying modulated current injection.

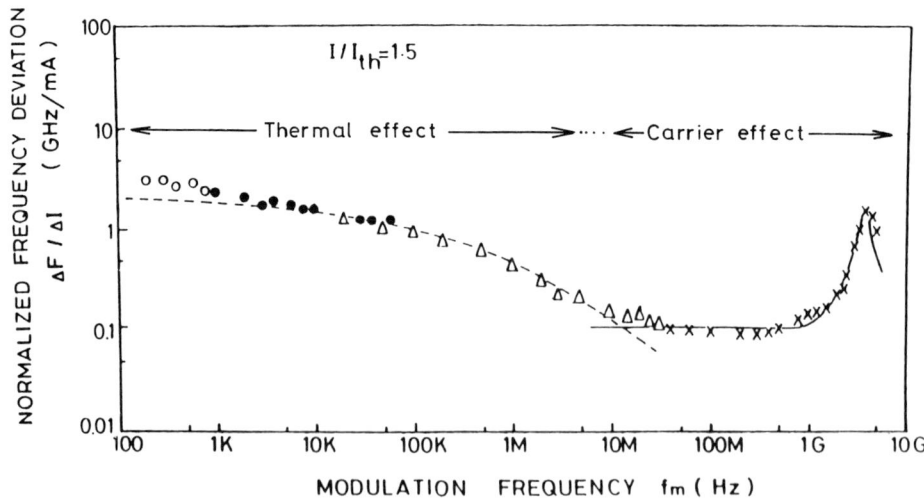

FIG. 3. Frequency deviation per unit sinusoidal modulation current amplitude in CSP laser. Measurement methods indicated by (a)–(c) below correspond to (a)–(h) ∼ (j) in Fig. 1. See text for detailed procedure. $I = 1.5 \times I_{th,C}$. Experimental: × × ×, Fabry–Perot (a); △ △ △, Fabry–Perot (b); ○ ○ ○, Fabry–Perot (c); ● ● ●, birefringent filter. Theoretical: ----, thermal effect; ——— carrier effect. [After Kobayashi et al. (1982).] (©1982 IEEE.)

Carrier density and temperature, in a semiconductor laser active layer, are changed by the injection current modulation. The normalized Fabry–Perot mode frequency deviation in the semiconductor laser can be expressed as (Ito and Kimura, 1980)

$$\delta f/f_0 = -(\alpha_c/\bar{n})\Delta N_{\text{eff}} - (\kappa + \gamma)\Delta T_{\text{eff}}. \tag{3}$$

The first term on the right-hand side represents a frequency deviation induced by carrier density change, and the second represents the deviation due to temperature change. Oscillation frequency increases as the carrier density increases, but decreases as the temperature increases. The refractive index varies with changes in carrier density and temperature. The cavity length varies through thermal expansion with temperature change. Here, $\bar{n}$ is the refractive index, ΔN_{eff} is the effective carrier density change, f_0 is the oscillation frequency, ΔT_{eff} is the effective temperature change, α_c is the coefficient relating the carrier change to the refractive-index change, κ is the linear thermal expansion coefficient, and γ is the refractive-index change per unit temperature rise normalized by the refractive index.

The normalized oscillation frequency deviation due to carrier density modulation is

$$\delta f_c/f_0 = -[(C_{\text{BB}} + C_{\text{FC}})/\bar{n}]\Delta N_{\text{eff}}$$
$$\Delta N_{\text{eff}} = \Gamma_y \int_{-\infty}^{\infty} \Delta N_e(x)|\varepsilon(x)|^2\, dx, \tag{4}$$

where Γ_y is the mode confinement factor along the vertical direction, $\Delta N_e(x)$ is the distribution of carrier density modulation along the lateral direction, and $\varepsilon(x)$ is the normalized optical lasing field that satisfies $\int_{-\infty}^{\infty}|\varepsilon(x)|^2\, dx = 1$. The function $|\varepsilon(x)|^2$ is calculated for a specific diode structure with the values of effective core width W_{eff} and effective refractive-index difference Δ_{eff}. Coefficients C_{BB} and C_{FC}, relating the carrier density change to the refractive-index change, represent the anomalous dispersion effect due to band-to-band transition and the free carrier plasma dispersion effect (Nash, 1973), respectively.

The value of C_{BB} lies in a certain range for the shapes of band tail in conduction and valence bands (Selway et al., 1974). C_{BB} can be obtained by calculating the gain spectrum when carrier N_e is injected into the active layer, as follows (Casey and Panish, 1978):

$$g(E) = (\pi e^2 h/\varepsilon_0 m^2 cnE) \int_{-\infty}^{\infty} \rho_c(E')\rho_v(E' - E)|M(E', E' - E)|^2$$
$$\times [f_v(E' - E) - f_c(E')]\, dE'. \tag{5}$$

Here, $E(=h\nu)$ is the photon energy, and ρ_c and ρ_v are the density of states in the conduction and valence bands, respectively. These are calculated by extrapolating the Kane function (Kane, 1963) to the Halperin–Lax band tail (Halperin and Lax, 1966). $|M(E', E' - E)|^2$ is the transition matrix element between the conduction and valence bands (Casey and Stern, 1976), f_v and f_c are the quasi-Fermi level distributions, e is the electron charge, ε_0 is the dielectric constant in vacuum, m is the effective mass in the conduction band, and c is the light velocity in the semiconductor laser. The calculated gain spectra and the related refractive-index-anomalous dispersion as a function of minority carrier density are shown in Fig. 4 (Yamamoto et al., 1983). The anomalous dispersion is calculated by the following Kramers–Kronig integral of theoretical gain spectra:

$$\Delta n(E) = -(P/2\pi^2) \int_0^\infty [1/(E' + E)]\{[\Delta g(E') - \Delta g(E)]/(E' - E)\} \, dE'. \quad (6)$$

Here, Δg shows the gain difference with and without carrier injection. P is the principal value integral (Landau and Lifshitz, 1960; Henry et al., 1981). The refractive index decreases in proportion to the minority carrier increase at the maximum gain wavelength. The C_{BB} value is -5 and $\sim 15 \times 10^{-21}$ cm^3 and decreases in proportion to the minority carrier density increase.

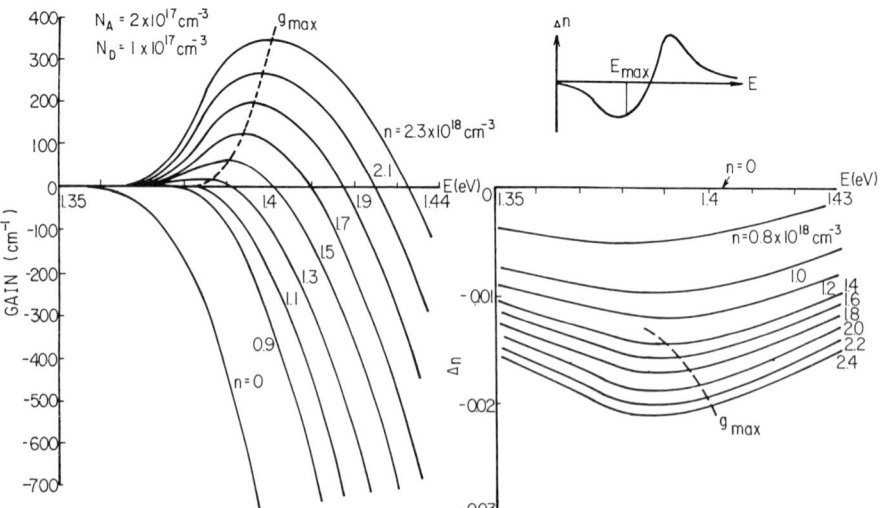

FIG. 4. Calculated gain spectra and refractive-index dispersion due to anomalous dispersion. [After Yamamoto et al. (1983).] (©1983 IEEE.)

The value $C_{FC} = -1.8 \times 10^{-21}$ cm^3 is obtained from (Nash, 1973)

$$C_{FC} = -e^2/(2m\omega^2\varepsilon_0 \bar{n}), \tag{7}$$

where m is the effective electron mass and ω is the angular frequency. The discrepancy between some experimental values of frequency shift per unit current in AlGaAs heterostructure lasers biased below the threshold is mainly due to the confinement factor Γ_y and the value of C_{BB} (Ito and Kimura, 1980; Selway et al., 1974; Matthews et al., 1981; Manning and Olshansky, 1980). The effective carrier density change, ΔN_{eff}, is obtained from the rate equation for lateral carrier diffusion (Kobayashi and Kimura, 1982b).

The normalized oscillation frequency deviation due to temperature modulation in an active layer is

$$\delta f_T/f_0 = -(\kappa + \gamma)\Delta T_{eff} \tag{8}$$

$$\Delta T_{eff} = \int \Delta T_a(x)|\varepsilon(x)|^2 \, dx,$$

where $\kappa = (1/L)\,dL/dT = 6 \times 10^{-6}$(deg^{-1}) is the linear thermal expansion coefficient (Novikova, 1961) of GaAs, $\gamma = (1/\bar{n})\,d\bar{n}/dT = 0.45 \times 10^{-4}$(deg^{-1}) is the thermal refractive-index coefficient (Cardona, 1960) of GaAs, and ΔT_a is the distribution of temperature modulation in an active layer. Thickness, thermal conductivity, and thermal diffusivity are different in each layer of the semiconductor laser. The dimensions of the submount and heat sink also differ from those of the semiconductor laser chip. To understand the three-dimensional thermal diffusion behavior in the laser chip, ΔT_a must be obtained precisely, requiring complex calculations. In Section II.G, the temperature modulation, $\Delta T_a(x)$, is calculated by the Fourier and Laplace transform method for a time-dependent thermal equation.

B. Frequency Modulation Measurement

The frequency deviation of direct frequency modulation in a semiconductor laser was measured with a spectrometer, scanning Fabry–Perot interferometer, birefringent crystal, and Michelson interferometer, and by heterodyne detection, and FM–AM conversion using a Fabry–Perot interferometer. Many researchers, using spectrometers, have the measured refractive index change caused by a frequency deviation of the Fabry–Perot mode-induced injection current change in semiconductor lasers biased below the oscillation threshold (Nathan et al., 1963; Ito and Kimura, 1980; Selway et al., 1974; Matthews et al., 1981; Manning and Olshansky, 1980; Fenner, 1964; Hurley et al., 1979). Under the lasing condition, Olsson and Tang (1981) measured the refractive index change in AlGaAs semiconductor lasers for light injection from external mirrors. Stubkjaer et al. (1980) measured the static frequency deviation in buried heterostructure GaInAsP/InP lasers at 1.6 μm below threshold, and Kishino et al. (1982) measured the frequency characteristics of the dynamic frequency deviation up to 5 GHz with the bias current above threshold.

Power spectra of directly modulated FM signals were observed with the scanning Fabry–Perot interferometer and by heterodyne detection. Nakamura et al. (1978) observed the power spectra of directly modulated FM signals on a channeled-substrate-planar (CSP) AlGaAs Laser with the Fabry–Perot etalon, where the central carrier peak disappeared at a current of 17 mA_{0-p}. Power spectra were measured at resonance frequency, and the refractive index change induced by carrier injection was discussed by Osterwalder and Rickett (1980) for an AlGaAs laser with stripe contact geometry. Frequency characteristics of frequency deviation per unit current in AlGaAs directly frequency-modulated lasers were measured and clarified so that the thermal and the carrier effects occupy low and high frequency regions from 0 to 5.2 GHz, respectively (Kobayashi et al., 1982).

As indicated in Fig. 1, the output modes from the scanning Fabry–Perot interferometer were measured to be different for various modulation frequency ranges. Above 40 MHz, the mode shown in Fig. 1a–h is used. Applied sinusoidal modulation currents vary from 0 to 4 mA_{0-p}. The frequency modulation index β can be obtained from the ratio of the carrier to the sideband intensity. Between 10 kHz and 30 MHz, a large frequency deviation is expected with a low modulation drive current; when the modulation frequency falls in this range, an FM spectrum with a large frequency deviation is used to measure β, as shown in Fig. 1i. For a frequency modulation index larger than 20, twice the maximum frequency deviation $2\Delta F$ is directly observed from the separation between two sideband peaks. The spectrum shown in Fig. 1i is obtained at a modulation frequency of 100 kHz, a modulation current of 0.43 mA_{0-p}, and a maximum frequency deviation $\Delta F = 442$ MHz_{0-p}.

At a modulation frequency lower than the scanning frequency of the Fabry–Perot interferometer, the output shows an instantaneous frequency shift on an oscilloscope display, as shown in Fig. 1j. This spectrum corresponds to a maximum frequency deviation of 340 MHz_{0-p} at a modulation current of 0.11 mA_{0-p} and a modulation frequency of 400 Hz. The Fabry–Perot scanning frequency is 885 Hz. Heterodyne detection with a frequency-stabilized local oscillator was also used to measure the power spectra of frequency-shift-keying (FSK) signals in an AlGaAs laser (Saito et al., 1981).

Frequency deviations of frequency modulated signals were also detected by FM–AM conversion. Ito and Kimura (1981) used birefringent crystals to measure frequency characteristics of frequency deviation due to thermal effects using current modulation in the low-frequency region below 10 MHz. In the intermediate modulation frequency range between 1 kHz and 50 kHz, FM-AM conversion by birefringent crystals is useful, as shown in Fig. 3. The high-frequency limit of this method is imposed by the spurious AM component in the direct modulated FM output. Using the unbalanced Michelson interferometer, Dandridge and Goldberg (1982) measured the transient response of frequency deviation due to the thermal effect by applying a square current pulse in AlGaAs lasers. Tsuchida et al. (1983), also using a Michelson interferometer to measure visibility, proposed a method of measuring the frequency deviation of a directly modulated signal without any intensity modulation effect, and obtained frequency characteristics up to 10 MHz. Since this method focuses on the carrier component of an FM signal with visibility measurement of the interference fringe, the frequency deviation at high modulation frequencies of up to several gigahertz can be measured without high-speed photodetectors.

C. Frequency Characteristics of Direct Frequency Modulation

Double-heterostructure AlGaAs semiconductor lasers with three different structures were tested. The three structures were channeled-substrate-planar (CSP), buried-heterostructure (BH), and transverse-junction-stripe (TJS) lasers. A dc bias current above threshold and a sinusoidal modulation current in the frequency range of 200–5.2 GHz was applied to them.

Frequency deviation normalized by modulation current is shown in Fig. 5 as a function of modulation frequency for CSP, BH, and TJS lasers (Kobayashi et al., 1982). Bias levels are 1.17, 1.17, and 1.25 times the threshold for the TJS, CSP, and BH lasers, respectively. The operating conditions were as follows: (1) The CSP laser had a threshold current $I_{th,C} = 69$ mA and a single longitudinal mode wavelength at $\lambda_C = 840$ nm (Aiki et al., 1977). The laser chip was mounted so that the p-side electrode was in contact with a submount (p-side down). (2) The BH laser was mounted p-side up with a threshold current $I_{th,B} = 36$ mA, and a single-frequency operation at wavelength $\lambda_B = 820$ nm (Chinone et al., 1979). (3) The TJS laser was mounted p-side up with a threshold current $I_{th,T} = 39$ mA and an emitting wavelength $\lambda_T = 840$ nm at a single frequency (Namizaki, 1976). The three kinds of lines in the figure show theoretical values for the three kinds of lasers. These values were obtained by theoretical analyses considering the carrier density modulation effect and temperature modulation effect.

Above 1 GHz, all the lasers show similar resonant phenomena. The frequency deviation values for TJS and BH lasers at resonant frequencies are

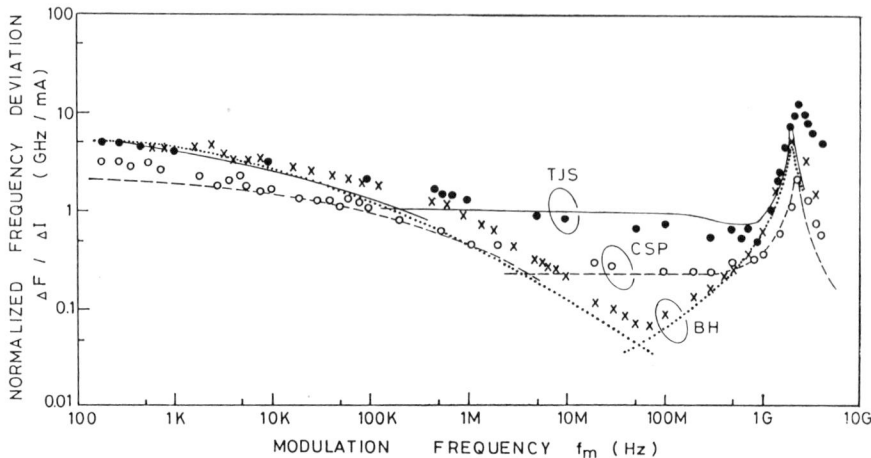

FIG. 5. Normalized frequency deviation for TJS, CSP, and BH lasers vs. modulation frequency. Experimental: ●, TJS laser, $I = 1.17 \times I_{th,T}$, bias current is $I_{th,T} = 39$ mA; ○, CSP laser, $I = 1.17 \times I_{th,C}$, $I_{th,C} = 69$ mA; ×, BH laser, $I = 1.25 \times I_{th,B}$, $I_{th,B} = 36$ mA. Theoretical: ——, TJS laser; - - - -, CSP laser; ......, BH laser. [After Kobayashi et al. (1982).] (©1982 IEEE)

the same, while the value for the CSP laser is smaller. There is a shallow dip in the frequency deviation value in the TJS laser at 1 GHz. The TJS and CSP lasers show a flat response in the frequency range between 10 MHz and 1 GHz, while V-shaped characteristics are observed in the BH laser. The frequency deviation values for the lasers decrease in the order TJS, CSP, and BH, except for those values at the resonant peak. In the frequency region below 10 MHz, frequency characteristics show monotonically decreasing characteristics with modulation frequency. The frequency deviation values for TJS and BH lasers are nearly the same, while those for the CSP laser are slightly lower. Low-frequency characteristics are governed by the thermal effect. The difference in the frequency deviations is caused by different thermal impedances. The CSP laser is mounted p-side down and has a low thermal impedance, while the TJS and BH lasers have high impedance values due to their p-side-up mounts.

The frequency deviation responses of an InGaAsP laser diode emitting at 1.3μm have been measured with a Fabry–Perot interferometer by Jacobsen et al. (1982). Figure 6 shows the experimental results biased at 1.33 and $1.5 \times I_{th}$. The BH laser (Hitachi HLP 5400) was tested, and its threshold current was 30 mA. At low modulation frequencies, the two curves in Fig. 6 show no significant difference, and thermal effects are expected to dominate the behavior. The frequency deviation of 500 MHz/mA for $f_m = 1$ kHz is about half of the dc sensitivity (1.03 GHz/mA). For $f_m \approx 1$ MHz, a minimum of 60–80 MHz/mA is obtained in the two cases of bias currents. The V-

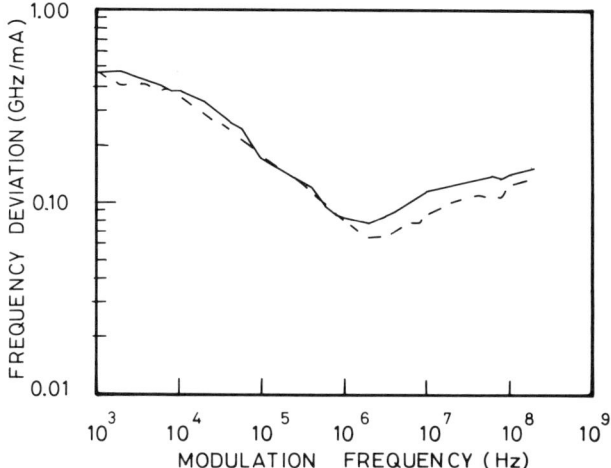

FIG. 6. Modulus of current/frequency-modulation transfer function (CF-MTF) for a Hitachi HLP 5400 laser. ----------, I/I_{th} = 1.33; ———, I/I_{th} = 1.50. [After Jacobsen et al. (1982).]

shaped characteristics between 100 kHz and 100 MHz agree qualitatively with observations on a BH laser emitting at 830 nm.

D. Carrier Density Modulation Effect

Frequency deviation in the AlGaAs CSP laser, normalized by modulation current, is shown in Fig. 7 as a function of modulation frequency. The spurious AM modulation index is kept below 5 percent. Frequency characteristics of the laser drive and receiver circuits are calibrated by the intensity modulation of the laser at a bias current 1.35 times the threshold. Solid line curves are theoretical values obtained by the rate equation analysis for photon and carrier distribution in the active layer and lateral carrier diffusion effect (Kobayashi et al., 1982). The frequency deviation at a unit modulation current is constant in the modulation frequency range of 50 MHz to 1 GHz. The flat FM response in the range of modulation frequency between 10 MHz and 500 MHz decreases as the dc bias current increases, because an increase in lasing optical field intensity tightly clamps the quasi-Fermi level and reduces $\Delta N_e(x)$. The steeper decrease in FM response at a modulation frequency higher than 500 MHz arises from the resonant enhancement of the FM response due to relaxation oscillation. The relaxation oscillation resonant frequency increases and peak FM response decreases as the dc bias current increases, as shown in Fig. 7. These are the same characteristics as in the intensity modulation

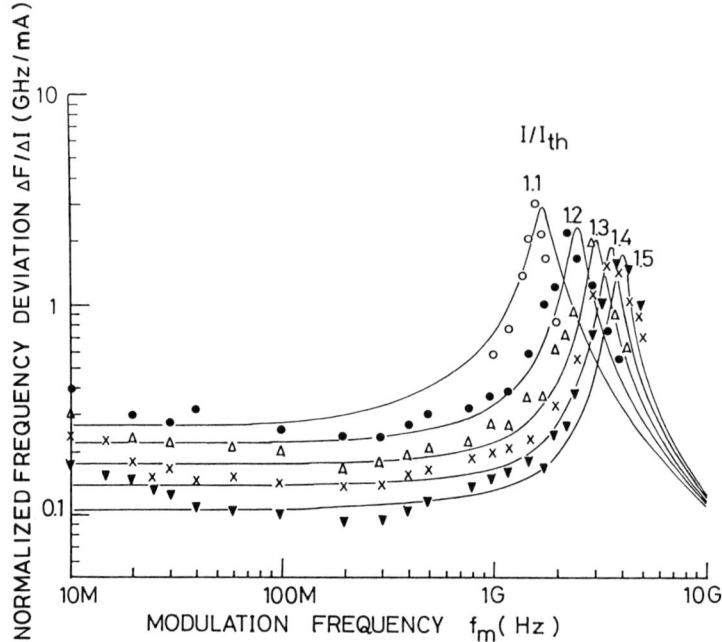

FIG. 7. Normalized frequency deviation for CSP laser vs. modulation frequency. Solid lines show calculated values that take into account lateral carrier diffusion. [After Kobayashi et al. (1982).] (©1982 IEEE.)

performance. Figure 8 shows maximum frequency deviation ΔF in the CSP laser as a function of modulation current ΔI at a bias current 1.4 times the threshold $I_{th,C}$. The frequency deviation increases linearly with the modulation current. These are very important characteristics for the FM modulators in transmission systems. These gradients increase as the modulation frequency increases because of the resonant effect in carrier density modulation. For InGaAsP laser structures emitting at 1.3 μm, the measured chirp widths of a longitudinal mode as a function of the modulating current at 100 MHz are shown in Fig. 9 (Dutta et al., 1984). The gain-guided lasers exhibit considerably greater frequency chirping than the index-guided lasers (CSPH and DCPBH). The small difference between the strongly index-guided lasers is due to a variation in the dimension of the active region. Comparison of frequency modulation and intensity modulation characteristics is interesting. Figure 10 shows the normalized modulation efficiency of intensity modulation as a function of modulation frequency in the same CSP laser. Experimental results are given for various bias currents from 1.04 to 1.14 times the threshold in the modulation frequency region of 10 MHz to 2 GHz. The frequency character-

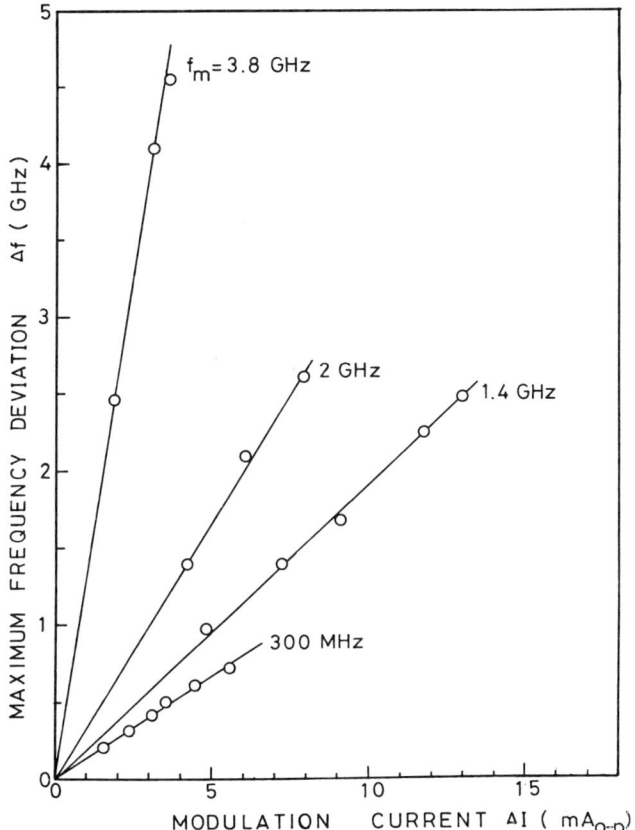

FIG. 8. Maximum frequency deviation for CSP laser vs. modulation current. [After Kobayashi et al. (1982).] (©1982 IEEE.)

istics of the driver and receiver circuits are calibrated by the method described in Kobayashi et al. (1982). The IM frequency characteristics at a frequency less than 500 MHz are independent of the bias current. The resonant frequency and peak values agree well with the theoretical values calculated by the first-order rate equation (Ito et al., 1979). The measured resonant frequencies for FM and IM obtained from Figs. 7 and 10 are plotted in Fig. 11. Open and closed circles show FM and IM resonant frequencies, respectively. The solid line represents the theoretical value calculated from the rate equation. The IM resonant frequency has been measured up to $1.14 \times I_{th,C}$, and the FM resonant frequency has been measured up to the modulation frequency limit imposed by the free-spectral-range (FSR) of the Fabry–Perot interferometer.

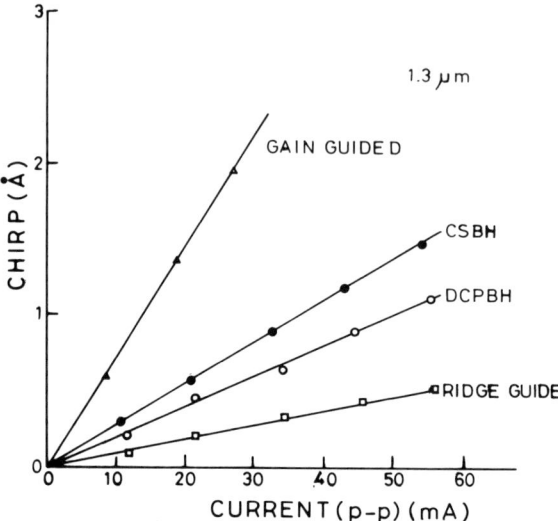

Fig. 9. Measured chirp width of the longitudinal mode as a function of the *p–p* modulating current at 100 MHz of InGaAsP lasers emitting at 1.3 μm. [After Dutta *et al.* (1984).]

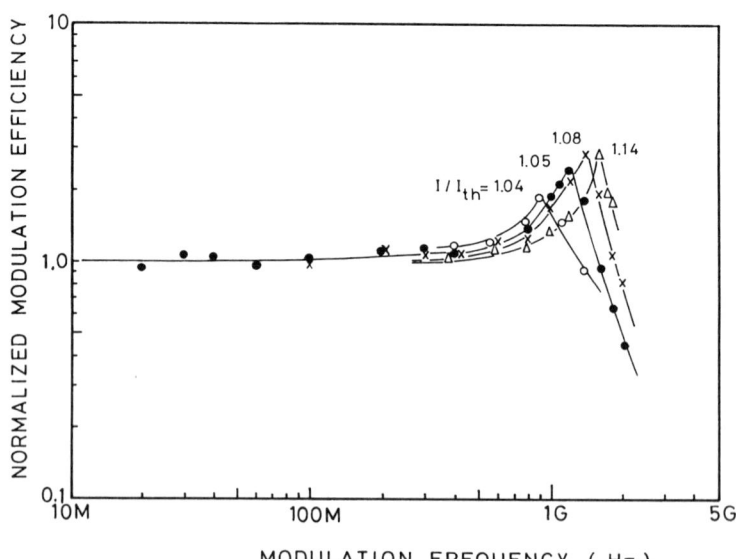

Fig. 10. Normalized modulation efficiency of intensity modulation for a CSP laser vs. modulation frequency. ○, $I = 1.04 \times I_{th,C}$; ●, $1.05 \times I_{th,C}$; ×, $1.08 \times I_{th,C}$; △, $1.14 \times I_{th,C}$. [After Kobayashi *et al.* (1982).] (©1982 IEEE.)

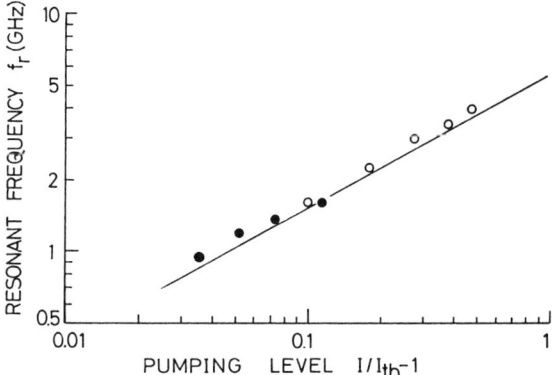

FIG. 11. Resonant frequency of FM and IM modulation vs. dc drive current. ○, FM modulation; ●, IM modulation. Solid line indicates the calculated results as shown in Fig. 7. [After Kobayashi et al. (1982).] (©1982 IEEE.)

E. Temperature Modulation Effect

Frequency deviation decreases monotonically in the modulation frequency region below 10 MHz in Figs. 3 and 5. This shows a frequency shift caused by the change in cavity length due to thermal expansion, and a refractive-index change caused by the change in temperature due to modulation of current injection. Experimental results are almost completely independent of the dc bias current, as shown in Fig. 12. This is because the ohmic heat generated, which depends on the dc bias current, is less than a quarter of the heat generated in an active layer, which is almost independent of the dc bias current because the junction voltage V_{th} is clamped above the threshold. Solid lines show theoretical values calculated by the thermal equation for the laser and submount structures shown in Fig. 13 (Kobayashi et al., 1982; Ito and Kimura, 1981). The laser chip is assumed to have a stripe-geometry double heterostructure. The GaAs active layer is bounded with p- and n-type GaAlAs cladding layers. The p-GaAs cap layer and the n-GaAs substrate are typically metallized with Au–Cr and Au–Ge–Ni as electrodes. The indium layer attaches the chip to the submount. The active layer is defined by stripe-width A and cavity length L. The thermal properties are assumed to be uniform along the laser cavity axis, and heat flow is treated as two-dimensional. The laser chip is mounted on a submount. The output facet is set coplanar with the submount side surface for good optical coupling with the fiber. Heat flow in the submount must be treated three-dimensionally. The submount is attached to a heat sink, the bottom of which is kept at a constant temperature.

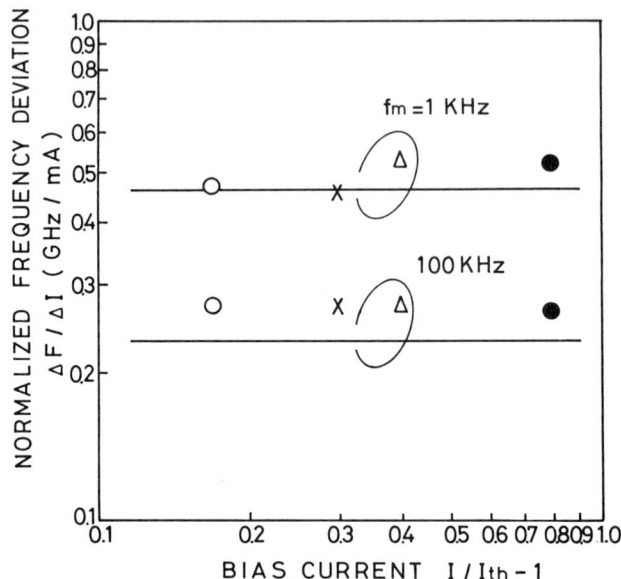

FIG. 12. Bias current dependence of frequency deviation in the TJS laser low-frequency region vs. dc drive current. $I_{th,T} = 39$ mA. ○, $1.17 \times I_{th,T}$; ×, $1.3 \times I_{th,T}$; △, $1.4 \times I_{th,T}$; ●, $1.5 \times I_{th,T}$. [After Kobayashi et al. (1982).] (©1982 IEEE.)

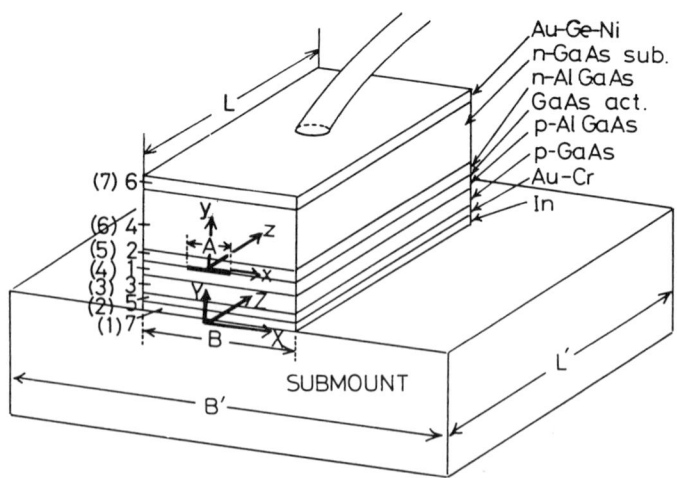

FIG. 13. Laser diode structural model used for thermal analysis. [After Kobayashi et al. (1982).] (©1982 IEEE.)

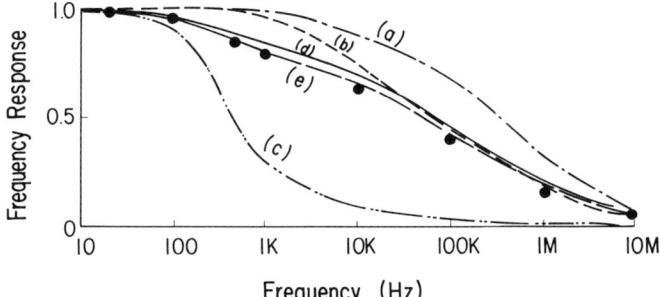

FIG. 14. Theoretical and experimental frequency responses of active layer temperature rise. (a) chip, (b) submount, (c) heat sink, (d) total, and (e) theoretical curve based on parameters obtained by step pulse response measurement. [After Ito and Kimura (1981).] (©1981 IEEE.)

When a sinusoidal current is superposed on the dc bias to the AlGaAs CSP laser diode, the temperature of the active layer varies periodically. A comparison between theoretical and experimental frequency characteristics is shown in Fig. 14. The curves show the response due to heat conduction: (a) in the chip, (b) in the submount, (c) in the apparent heat sink. Curve (d) is the overall frequency response. The thermal resistances corresponding to (a), (b), and (c) are 20.4, 6.9, and 7.7 deg/W, respectively. Curve (e) is the overall frequency response obtained using experimental thermal resistance values to the step pulse response measurement. Here, 19.1, 6.5, and 9.4 deg/W are assumed for the three thermal resistances, respectively. The theoretical curves agree well with the experimental values denoted by the closed circles.

The frequency deviation values for the TJS and BH lasers are nearly the same, while those for the CSP laser are lower, as shown in Fig. 5. Low-frequency characteristics are governed by the thermal effect. The difference in the frequency deviations is due to different thermal impedances. The CSP laser was mounted *p*-side down and had a low thermal impedance, while the TJS and BH lasers had high impedance values due to their *p*-side-up mounts. Step pulse response are measured to verify the difference of thermal impedances depending on the distance between the active layer and the heat sink.

The calculated step drive current responses in frequency deviation for CSP, BH, and TJS lasers are shown in Fig. 15. The *p*-side-up-mounted TJS and BH lasers have time constants in step pulse response about one order of magnitude larger than the *p*-side-down-mounted CSP laser. Therefore, TJS and BH lasers have lower cutoff modulation frequencies than the CSP laser, as shown in Fig. 5. The time constant for a step pulse response is independent of stripe width and decreases as the thickness $T(1)$ in Fig. 13 decreases. Experimental results for step response were measured by Goldberg *et al.* (1981)

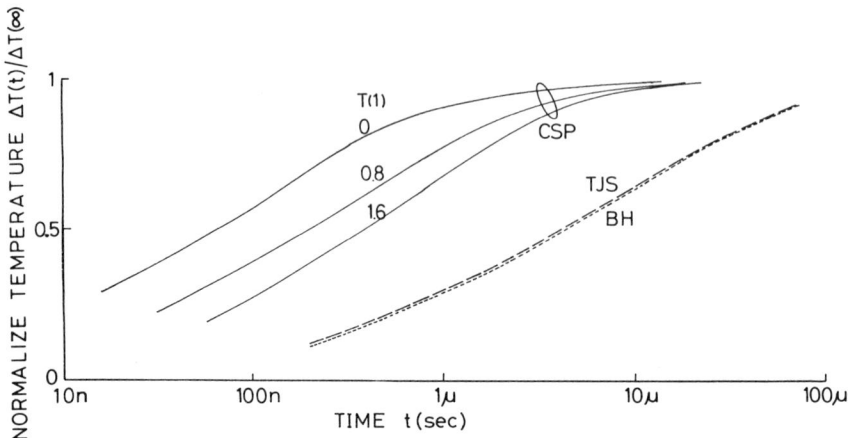

FIG. 15. Step pulse response of effective temperature change for CSP, TJS, and BH lasers. [After Kobayashi et al. (1982).] (©1982 IEEE.)

and Dandridge and Goldberg (1982) with a scanning Fabry–Perot interferometer and Michelson interferometer. Experimental results for four kinds of AlGaAs semiconductor lasers are shown in Fig. 16. Step responses of CSP and TJS lasers show the same trends as the calculated value show in Fig. 15. These experimental results also show that the low-frequency responses of

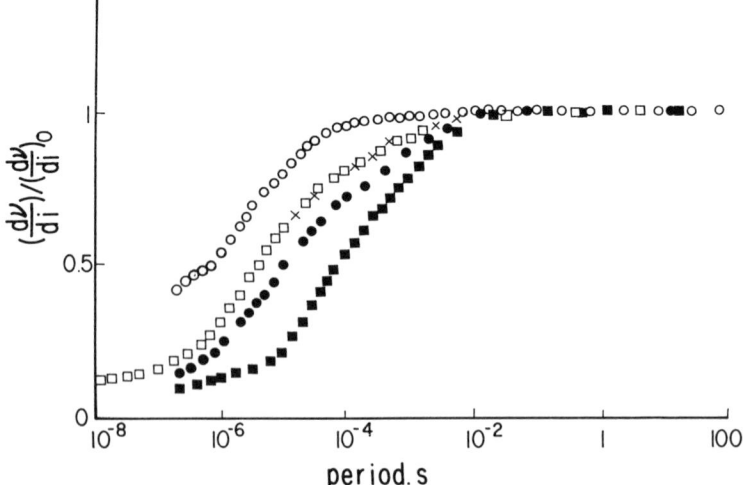

FIG. 16. Variation of dv/di as a function of modulation frequency for four lasers. □, Hitachi CSP; ■, Mitsubishi TJS; ●, Laser Diode Labs Gaussian CSP; ○, Laser Diode Labs Gaussian CSP = laser pigtailed to single-mode fiber. [After Dandridge and Goldberg (1982).]

direct modulation in semiconductor lasers are due to the thermal response, which depends on the distance between the active layer and the heat sink.

F. Phase Relationship between FM and AM Signals

In the power spectra of the FM signal shown in Figs. 1a–1h, there are asymmetric power spectra for upper and lower sidebands according to the phase difference between FM and AM components. The upper sideband amplitude is smaller than that of the lower, because an AM modulation is superimposed on the FM modulation. A simultaneously FM and AM modulated optical wave is represented as follows:

$$E = E_0\{1 + M\cos(2\pi f_m t)\}\exp[j\{2\pi f_0 t + \beta\sin(2\pi f_m t)\}], \tag{9}$$

$$f = f_0 + \Delta F\cos(2\pi f_m t). \tag{10}$$

Here, f is the instantaneous frequency and M the amplitude modulation index. Expanding Eq. (9) gives the carrier and first sideband amplitudes as follows:

carrier:
$$J_0 E_0 \exp\{j(2\pi f_0 t)\}, \tag{11}$$

upper first sideband:
$$[J_1(\beta) - (M/2)\{J_0(\beta) + J_2(\beta)\}]E_0 \exp\{j2\pi(f_0 + f_m)t\}, \tag{12}$$

lower first sideband:
$$-[J_1(\beta) + (M/2)\{J_0(\beta) + J_2(\beta)\}]E_0 \exp\{j2\pi(f_0 - f_m)t\}. \tag{13}$$

These equations show that FM and AM components are out of phase and the phase difference is π. These equations also show that the FM modulation index can be obtained from the average value of the upper and lower sideband amplitudes. Spurious intensity modulation is encountered when the injection current directly modulates the frequency. It should be minimized for FM laser applications. Intensity modulation depth values, accompanied with a 100-MHz frequency deviation for CSP, BH, and TJS lasers, are summarized in Table I. Frequency responses of the phase delay measured by Jacobsen *et al.* (1982) and Olesen and Jacobsen (1982) with a Michelson interferometer are shown in Fig. 17 and Fig. 18. Figure 17 shows the characteristics of the CSP type AlGaAs laser whose FM frequency characteristics are shown in Fig. 5. Figure 17 supports the FM spectrum that is shown to be almost out of phase at 150 MHz in Fig. 1. Figure 18 shows frequency responses of the BH-type InGaAsP semiconductor laser emitting at 1.3 μm, whose characteristics are different from the CSP characteristics as shown in Fig. 6. A phase change of

TABLE I

Direct Frequency Modulation Characteristics and Numerical Parameters for CSP, TJS, and BH Lasers (after Kobayashi et al., 1982; ©1982 IEEE.)

	CSP	TJS	BH
$\Delta F/\Delta I$ (MHz/mA)	250	750	85
IM(%) at $\Delta F = 100$ MHz ($I/I_{th} = 1.2$, $f_m = 50$ MHz)	3.3	2	13
Threshold current	69 mA	39 mA	36 mA
Effective core width W_{eff}	6 μm	2 μm	2.5 μm
Refractive-index difference Δ_{eff}	1×10^{-3}	2×10^{-3} $(p-p)$ $5-10 \times 10^{-3}$ $(p-n)$	1.1×10^{-1}
Stripe width A	$A > W_{eff}$	$A > W_{eff}$	$A = W_{eff}$
Mode confinement factor Γ_y	0.13	0.52	0.17

Layers	h_i (μm)	K_i (W/μm) $\times 10^{-4}$	κ_i (μm^2/s) $\times 10^8$	h_i (μm)	K_i (W/μm) $\times 10^{-4}$	κ_i (μm^2/s) $\times 10^8$	h_i (μm)	K_i (W/μm) $\times 10^{-4}$	κ_i (μm^2/s) $\times 10^8$
First	1.6	0.1	0.05	3.0	0.1	0.05	1.0	0.1	0.05
Second	0.4	0.1	0.05	1.5	0.1	0.05	1.5	0.1	0.05
Third	0.8	0.4	0.27	60	0.45	0.27	50	0.45	0.27
Fourth	50	0.45	0.27	1.5	0.4	0.27	1.5	0.1	0.05
Fifth	0.8	3.2	1.0	1.0	3.18	1.26	1.0	3.18	1.26
Sixth	0.3	3.2	1.0	0.5	3.18	1.26	0.5	3.18	1.26
Seventh	0.2	0.87	0.6	2.0	0.87	0.5	2.0	0.87	0.5

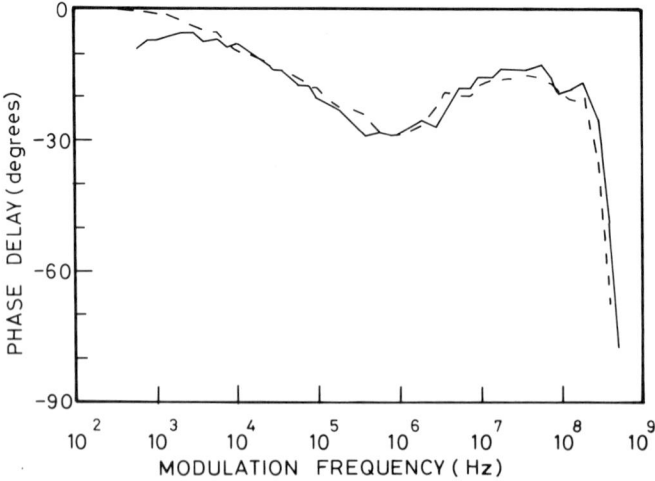

FIG. 17. Comparison between measured phase delay of CF-MTF and values obtained from a Hilbert transform of measured modulus for a Hitachi HLP 1400 laser. ——, $I/I_{th} = 1.11$ (experimental); ----------, theoretical. [After Jacobsen et al. (1982).]

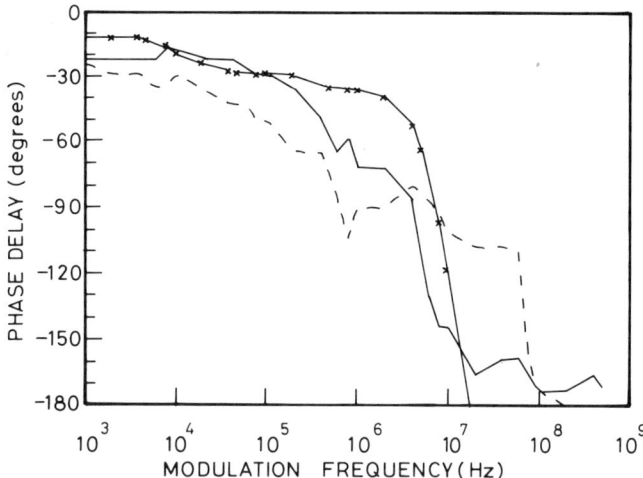

FIG. 18. Phase delay of CF-MTF for a Hitachi HLP 5400 laser. Theoretical curve for $I/I_{th} = 1.33$ is obtained from Hilbert transform of modulus. ———, $I/I_{th} = 1.33$ (experimental); ——— × ———, theoretical; ----------, $I/I_{th} = 1.50$ (experimental). [After Jacobsen et al. (1982).]

π is observed in the range 1 kHz–1 GHz. This discrepancy in phase delay characteristics between CSP and BH lasers can be explained from the difference between gain-guided and index-guided laser structures. These explicit differences in phase delay responses between gain-guided and index-guided lasers more easily explain the flat response in frequency deviation characteristics and phase delay due to red shift or blue shift between FM and AM components. Red shift, which means a downward frequency shift of the oscillation frequency, occurs even in the high-frequency region where the carrier modulation effect dominates. This is because the effective refractive index for the lasing mode increases due to the inhomogeneous gain saturation in the CSP-type gain-guided semiconductor lasers. On the other hand, blue shift, which means the frequency upper shift of the oscillating frequency caused by increase in injection current occurs in the BH laser. This is because the effective refractive index for the lasing mode decreases as a result of the homogeneous gain saturation, i.e., carrier increase by the current injection in the BH type index-guided semiconductor lasers. Theoretical analyses of these discussions are given in Section II.G. Phase delay in the sinusoidal frequency response in frequencies less than 10 MHz was also measured by Ito and Kimura (1981), as shown in Fig. 19. The closed circles show experimental results measured with birefringent crystals. Various solid curves show theoretical values in the chip (a), in the submount (b), and in the heat sink (c). Curve (d) is the overall phase delay, and curve (e) is the value obtained using

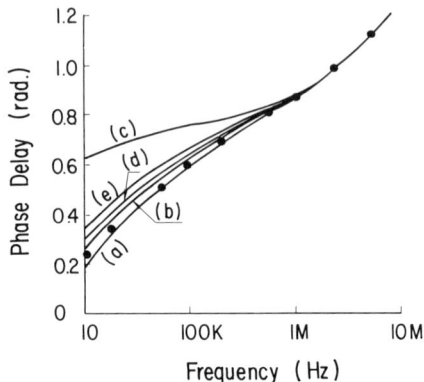

FIG. 19. Theoretical and experimental phase delays in sinusoidal response. Meanings of (a)–(e) are the same as in Fig. 14. [After Ito and Kimura (1981).] (©1981 IEEE.)

the step pulse response measurement. This phase delay has the opposite sign to that in Fig. 17. Both phase delay characteristics show the same trend in the frequency region less than 10 MHz.

G. *Theoretical Analysis*

The carrier effect in frequency modulation is represented by rate equations. The rate equations, ignoring spontaneous emission coupled to the lasing mode and lateral diffusion of carriers, are represented as follows:

$$dN/dt = J - N/\tau_s - nA(N), \tag{14}$$

$$dn/dt = n[A(N) - 1/\tau_p], \tag{15}$$

where N is the number of carriers, n is the number of photons, J is the pumping rate, τ_s is the carrier lifetime, τ_p is the photon lifetime, and $nA(N)$ is the number of stimulated emissions per second. Introducing small, time-dependent perturbations ΔN, Δn, and ΔJ around the stationary solution of Eqs. (14) and (15), these can be written as follows:

$$j\Omega \Delta N = \Delta J - \Delta N/\tau_s - \Delta n A(N) - n\Delta N(\partial A/\partial N), \tag{16}$$

$$j\omega \Delta n = n\Delta N(\partial A/\partial N), \tag{17}$$

where we have used the stationary quantities

$$A = 1/\tau_p. \tag{18}$$

The laser oscillation frequency shift is obtained using $\Delta\omega/\omega = \Delta n_r/n_r$ as follows:

$$\Delta\omega = \tfrac{1}{2}\alpha\,\Delta N(\partial A/\partial N), \tag{19}$$

where α is a parameter (Henry, 1982) defined as

$$\alpha = (\partial n'_r/\partial N)/(\partial n''_r/\partial N), \tag{20}$$

where n'_r and n''_r are the real part and the imaginary part of the complex refractive index n_r. The α parameter can be experimentally obtained from the intensity modulation index m and the frequency (phase) modulation index β as follows (Harder et al., 1983).

$$\alpha_m = -2(\beta/m). \tag{21}$$

For a buried optical guide laser (BOG, Hitachi 3400, $\lambda = 816$ nm), $|\alpha_m| = 4.5$ has been measured at a modulation frequency of 2 GHz, a bias level of 1.3 times threshold, and an intensity modulation depth of 10% by Harder et al. (1983). The measured α values did not have any modulation frequency dependency in the range 1 to 3.5 GHz. ΔN is calculated from Eqs. (16) and (17), and $\Delta\omega$ is obtained as follows:

$$\Delta\omega = (1/2n)\alpha B\,\Delta J/\{1/\tau_s + B + j(\Omega - AB/\Omega\}, \tag{22}$$

where B is the small-signal carrier decay rate due to stimulated emission,

$$B = n(\partial A/\partial N). \tag{23}$$

Figure 20 shows the schematic FM response and phase delay characteristics due to the carrier effect. Dotted lines show the characteristics represented by Eq. (22). The FM response curve shows a monotonic increase proportional to Ω in the lower-frequency region, and a decrease proportional to Ω^{-1} in the region higher than the resonant frequency. The phase delay between FM and AM(IM) modulation signals is $\pi/2$ above the resonant in the lower-frequency region, and $-\pi/2$ in the higher-frequency region above the resonant frequency. These trends in the FM response and the phase delay calculated from Eq. (22) can explain the experimental results of the BH laser frequency characteristics shown in Figs. 5, 6, and 18.

The frequency characteristics of CSP and TJS lasers, however, have different shapes from those of the BH laser. The flat carrier-induced FM response, the red-shift frequency chirping in the higher-frequency region, and the absence of phase reversal at the thermal cutoff frequency are mysterious characteristics compared with the above theoretical considerations. Nilsson and Yamamoto (1985) proposed a new model for a semiconductor laser carrier distribution with two regions, as shown in the schematic figure of Fig. 21, with different α parameter values and different gain values to explain the characteristics of CSP and TJS lasers.

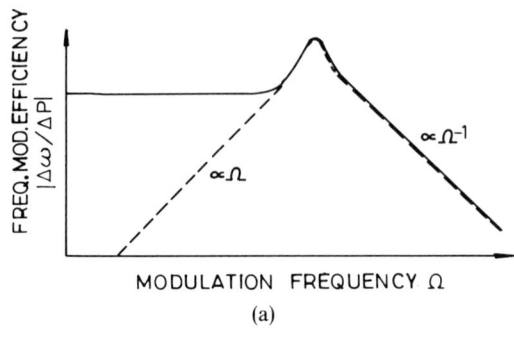

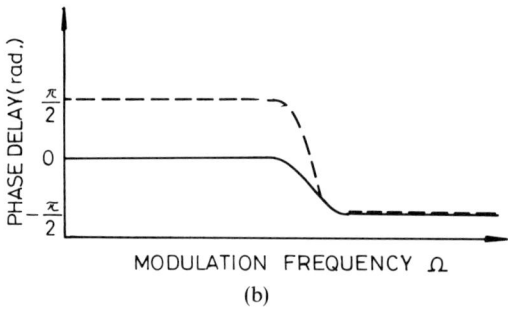

FIG. 20. Schematic figures of (a) FM response and (b) phase delay in a semiconductor laser. ----------, with homogeneous α parameter; ———, with inhomogeneous α parameters.

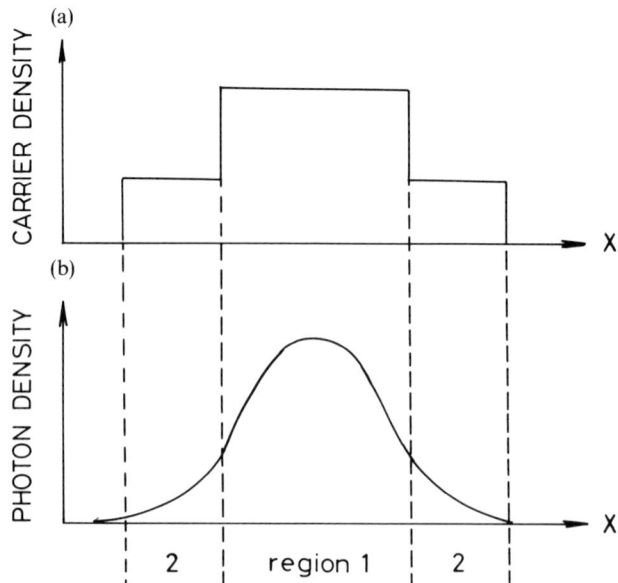

FIG. 21. (a) Carrier density and (b) optical field distributions of the two-region model with different values of α.

The rate equations can then be written as follows, omitting the spontaneous emission term:

$$dN_1/dt = J_1 - N_1/\tau_{s1} - nA_1(N_1), \quad (24)$$

$$dN_2/dt = J_2 - N_2/\tau_{s2} - nA_2(N_2), \quad (25)$$

$$dn/dt = n[A_1(N_1) + A_2(N_2) - 1/\tau_p], \quad (26)$$

when n, N_i, J_i, τ_i, and A_i have the same meaning as in Eqs. (14) and (15), and the subscript i means regions 1 and 2 in Fig. 21. The perturbation theory as used in Eqs. (16) and (17) gives the angular frequency deviation as follows:

$$\Delta\omega = (1/2n)[\Delta J_1\{A_2C_2(\alpha_1 - \alpha_2)/P_2 + j\Omega\alpha_1\}$$
$$\cdot \{A_1 + A_2C_2P_1/C_1P_2 + j\Omega P_1/C_1\}^{-1}$$
$$+ \Delta J_2\{A_1C_1(\alpha_2 - \alpha_1)/P_1 + j\Omega\alpha_2\}.$$
$$\cdot \{A_2 + A_1C_1P_2/C_2P_1 + j\Omega P_2/C_2\}^{-1}], \quad (27)$$

$$P_{1,2} = 1 + C_{1,2} + j\Omega\tau_{s1,2} \quad (28)$$

$$C_{1,2} = n(\partial A_{1,2}/\partial N_{1,2})\tau_{1,2} = B_{1,2}\tau_{s1,2}.$$

The optical density in region 1 is considerably higher than in region 2, so A_1 is much larger than A_2, where the vertical confinement factors are considered for each region. Frequency deviation mainly depends on the ΔJ_2 in the low-modulation-frequency region because ΔJ_1, ΔP_1, and C_1 nearly equal ΔJ_2, ΔP_2, and C_2, respectively. When the ΔJ_1 is applied, the optical density becomes high, and the current injection of ΔJ_1 suppresses the frequency deviation due to the gain saturation effect in region 1, where the optical density is fairly high. On the other hand, as the gain saturation is not so complete in region 2 when a current of ΔJ_2 is injected, the carrier density increases in region 2 and decreases in region 1. As the α parameter is larger in the center of the cavity of region 1 than in region 2, frequency modulation is mainly governed by the carrier density change in region 1. The increase in injected carriers ($\Delta J_2 > 0$) induces a decrease in carrier density and an increase in refractive index in the center part of the laser cavity. Consequently, there is a frequency red shift in the FM laser opposite to the blue shift induced by the carrier injection in the semiconductor lasers biased under threshold. The solid lines in Fig. 20 show the characteristics of FM response and phase delay represented by the two-region model. Experimental results for CSP and TJS shown in Figs. 5 and 17 can be explained by the two-region model induced from the α parameter depending on the carrier density. Frequency characteristics with a flat response and no phase delay arise from the second term proportional to $\Delta J_2(\alpha_2 - \alpha_1)$ in Eq. (27). This term is limited by the carrier lifetime, and instead the $j\Omega\alpha_1\Delta J_1$ term dominates in the high-frequency region, where the resonant peak and the characteristics are proportional to Ω^{-1}.

Temperature modulation $\Delta T_a(x)$ is calculated by the Fourier and Laplace transform method for a time-dependent thermal equation. The structure of the laser diode used for thermal analysis is shown in Fig. 13. Thickness h_i, thermal conductivity k_i, and thermal diffusivity κ_i in each layer are assumed to be constant for a small modulation in temperature. Heat is also assumed to be generated uniformly within the stripe of width A and cavity length L. Two-dimensional heat flow in the chip and three-dimensional heat flow in the submount will be discussed here, including the heat radiation effect from the top layer to the ambient atmosphere and the dominant heat transfer to a heat sink. The thermal equation in each layer i is given by

$$\partial T_i/\partial t = \kappa_i \nabla^2 T_i + W_i/C, \tag{29}$$

where C is heat capacity per unit volume, and W_i is heat generation power. Temperature modulation $\Delta T_a(x)$ due to heat conduction in the diode chip itself and that due to the heat transfer in the submount are obtained separately. First, $\Delta T_a(x)$ is derived. The time-dependent heat generation function W_{act} (W/cm^2) in an active layer is given by

$$W_{act} = V_D[J - \eta_L(J - J_{th}) - \eta_{sp}J_{th}] - W_{sp}, \tag{30}$$

where V_D is the junction voltage, J is a time-varying drive current density, η_L is the external quantum efficiency above the threshold current, η_{sp} is the external quantum efficiency below the threshold current, and W_{sp} is the spontaneous emission power absorbed in the GaAs substrate and cap layer. Here, the junction voltage V_D is assumed to be independent of dc bias current because the quasi-Fermi level is fully clamped in the central portion of the stripe, where the lasing optical field intensity is sufficiently high.

The Laplace transform of the temperature modulation in the ith layer, caused by the heat generation in an active layer, is given by

$$\hat{T}_i(x, y) = \sum_n \hat{B}_{i,n}[\cosh(\hat{\beta}_{i,n} y) - \hat{e}_{i,n} \sinh(\hat{\beta}_{i,n} y)] \cos(k_n x), \tag{31}$$

where $\hat{\beta}_{i,n}^2 = p/\kappa_i + k_n^2$, $K_n = 2\pi n/B$, B is the diode chip width, and n is an integer. P is the Laplace operator, and the sign $\hat{}$ denotes a Laplacian. Coefficients $B_{i,n}$ and $e_{i,n}$ are determined by solving equations for temperature and heat flow continuity at the layer boundaries. They are expressed as closed form functions of h_i, $\beta_{i,n}$, κ_i, and W_{act} and have been summarized by Ito and Kimura (1981).

The time-dependent heat generation function W_r (W/cm^2) in an ohmic resistance is given by

$$W_r = J^2 r, \tag{32}$$

where r is the total ohmic resistance in a diode chip.

Ohmic heat is mainly generated in the p-AlGaAs cladding layer for GaAs lasers with good ohmic contacts (Newman et al., 1978). Therefore, the temperature modulation due to ohmic heat generation can be calculated by Eq. (31). First the p-AlGaAs layer is divided into two. Then the layers are renumbered so that the heat-generation surface is bounded by layers 1 and 2. The real-time temperature modulation is evaluated by a numerical inverse Laplace transform of Eq. (31).

Next, temperature modulation due to the heat conduction in the submount will be derived. The new coordinate system (X, Y, Z) is shown in Fig. 13. Each layer is renumbered as shown in the parentheses in Fig. 13. The Laplace transform of temperature variation in the submount is expressed by

$$T_s(X, Y, Z) = \sum_{m,l} \hat{B}_{ml}\{\cosh(\hat{\beta}_{ml}Y) - e_{ml}\sinh(\hat{\beta}_{ml}Y)\}\cos(k_n X)\cos(k_l Z), \quad (33)$$

where $\hat{\beta}_{ml}^2 = p/\kappa_9 + k_m^2 + k_l^2$, $k_n = 2n\pi/B'$, and $k_l = l\pi/L'$. κ_9 is the thermal diffusivity of the submount. B' and L' are the width and length of the submount, respectively. The coefficients $\hat{B}_{ml}^2$ and e_{ml} have been summarized by Ito and Kimura (1981). The Laplace transform of temperature modulation in the ith layer due to heat conduction in the submount is given by

$$\hat{T}_i(x) = \hat{B}_{i,0} + \sum \hat{B}_{i,n}\cos(k_n X), \quad (34)$$

where coefficients $\hat{B}_{i,0}$ and $\hat{B}_{i,n}$ are summarized in Ito and Kimura (1981). The total temperature modulation is given by the simple sum of the temperature modulations caused by each heat source and due to each thermal conduction because of the linearity in thermal equation (29).

The experimental results of frequency modulation characteristics at low modulation frequencies for CSP, BH, and TJS lasers are successfully explained by the present thermal analysis, as shown in Fig. 5. Numerical parameters used for thermal analysis are determined by microscopic observation of etched facets and from manufacturer's data. These parameters are summarized in Table I. TJS and BH lasers have a higher FM response than the CSP laser because their p-side-up mounting produces a large distance between their active layers and submount surfaces. Several factors influencing the temperature modulation effect are shown in Fig. 22. The most important parameter is the static normalized temperature modulation $\Delta T_{\text{eff}}/\Delta W$ as a function of factor-of-two variations in several dimensions with respect to their nominal values for the CSP laser. It is important because the normalized frequency deviation is inversely proportional to the diode length. Stripe width is the next most-important parameter and should be increased to suppress the temperature modulation effect. The increase in spot size, as well as the increase in the stripe width A, effectively suppresses the temperature modulation effect, although this suppression is partially canceled out by the threshold current

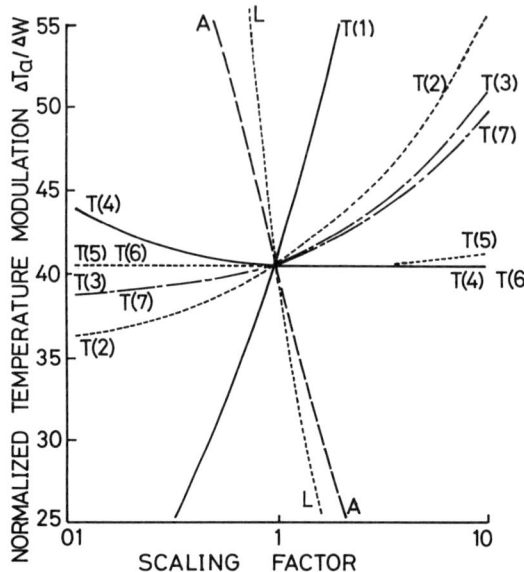

FIG. 22. Normalized effective temperature modulation $\Delta T_{\text{eff}}/\Delta W = \int \Delta T_a(x)|\varepsilon(x)|^2\,dx/\Delta W$ (deg/W) at zero modulation frequency vs. factor-of-two changes in several dimensions with respect to their nominal values for the CSP laser. [After Kobayashi et al. (1982).] (©1982 IEEE.)

increase. FM response due to carrier density modulation also is reduced as L and A increase. Among the film thickness values for each layer, $T(1)$ and $T(3)$ are important and should be minimized to suppress the temperature modulation effect. To eliminate this effect entirely, the cutoff modulation frequency should be decreased, as well as the static frequency deviation value.

H. Advanced Technology of Direct Frequency Modulation

Let us now discuss a new type of FM semiconductor laser. Two-electrode diode lasers giving a flat FM response, low spurious intensity modulation components, and suppressed temperature effect are proposed from the theory based on the two-region model (Nilsson et al., 1987). A schematic figure of a two-electrode laser is shown in Fig. 23, where the α parameters in region 1 and region 2 are different. Theoretical analyses are based on the rate equations with perturbation treatment represented by Eqs. (24), (25), and (26). The frequency deviations with two different α values are represented by Eq. (27). The terms of ΔJ_1 and ΔJ_2 have opposite signs, so a push–pull operation for phase reversal between ΔJ_1 and ΔJ_2 can enhance the FM response. The optical

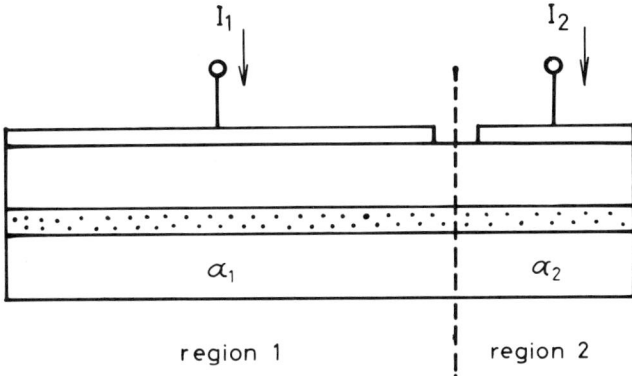

FIG. 23. Two-electrode model with different α parameters in a Fabry–Perot-type semiconductor laser. [After Nilsson and Yamamoto (1985).]

angular frequency deviation for the two-region model is

$$\Delta\omega = \tfrac{1}{2}[\alpha_1 \Delta N_1(\partial A_1/\partial N_1) + \alpha_2 \Delta N_2(\partial A_2/\partial N_2)] \\ - (\Delta J_1 + \Delta J_2)a_T/(1 + j\Omega\tau_T), \quad (35)$$

where a_T is a measure of the thermal resistance and the temperature dependence of the refractive index, and τ_T is the thermal time constant. The push–pull operation of ΔJ_1 and ΔJ_2 can also suppress the thermal effects in the two regions, making them cancel each other out in Eq. (35). Spurious intensity modulation components are represented as follows:

$$\Delta n = \Delta J_1[A_1 + (A_1 C_2 P_1/C_1 P_2) + j\Omega P_1/C_1]^{-1} \\ + \Delta J_2[A_2 + (A_1 C_1 P_2/C_2 P_1) + j\Omega P_2/C_2]^{-1}. \quad (36)$$

In Eq. (36), the push–pull operation that satisfies the condition of $\Delta J_2/\Delta J_1 = -C_1(1 + C_2)/C_2(1 + C_1)$ can completely suppress the spurious intensity modulation component Δn in the lower-frequency region in FM response. The frequency deviation and phase delay characteristics of two-electrode FM lasers are shown in Figs. 24a and 24b, respectively. Curve (a) is for a hypothetical laser without gain saturation or inhomogeneities in the α parameter. Curve (b) is for the same laser, but with substantial gain saturation included and a π phase shift in thermal rolloff. If the α parameters differ in the two regions, an FM response is obtained as shown by curve (c). Curve (d) is obtained by adding a phase-reversing and pre-equalization network to the laser driver.

A three-electrode InGaAsP DFB laser with a buried heterostructure has been developed that emits at 1.5 μm. It is shown in Fig. 25 (Yoshikuni and

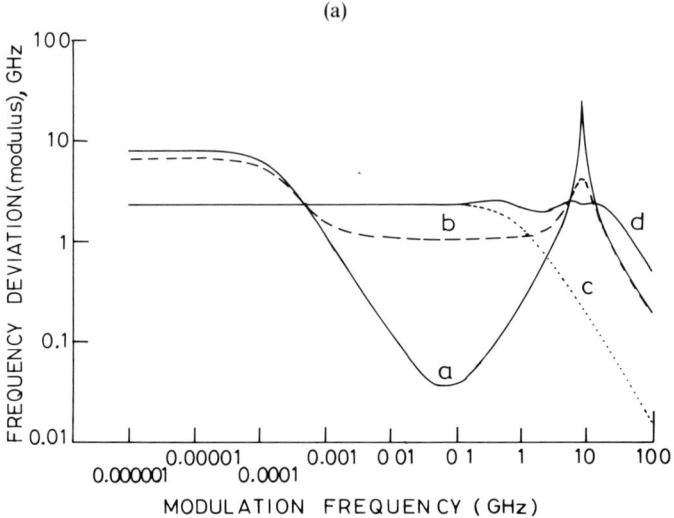

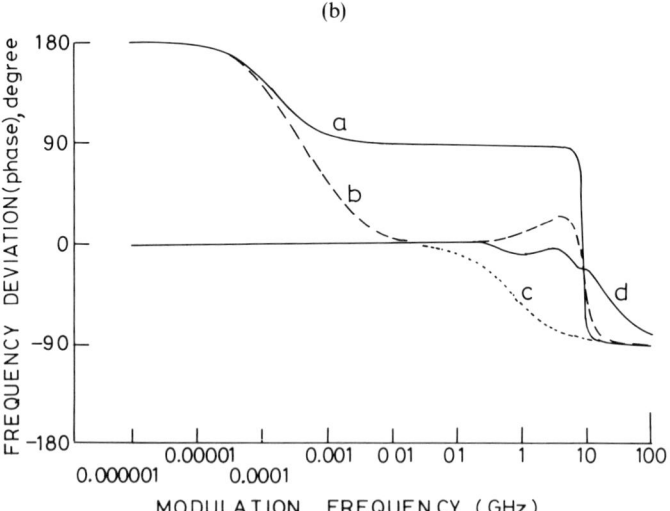

FIG. 24. Modulus of frequency deviation (a) and phase delay (b) as functions of modulation frequency (calculated with two-region model). a: $\alpha_1 = \alpha_2 = 5.5$, $n_{si} = \infty$, $\Delta J_1 = \Delta J_2$. b: $\alpha_1 = \alpha_2 = 5.5$, $n_{si} = 1.88 \times 10^7$, $\Delta J_1 = \Delta J_2$. c: $\alpha_1 = 6$, $\alpha_2 = 5$, $n_{si} = 1.88 \times 10^7$, $\Delta J_1 = -\Delta J_2$. d: As in c, with the phase-reversing and pre-equalization network. [After Nilsson et al. (1987).]

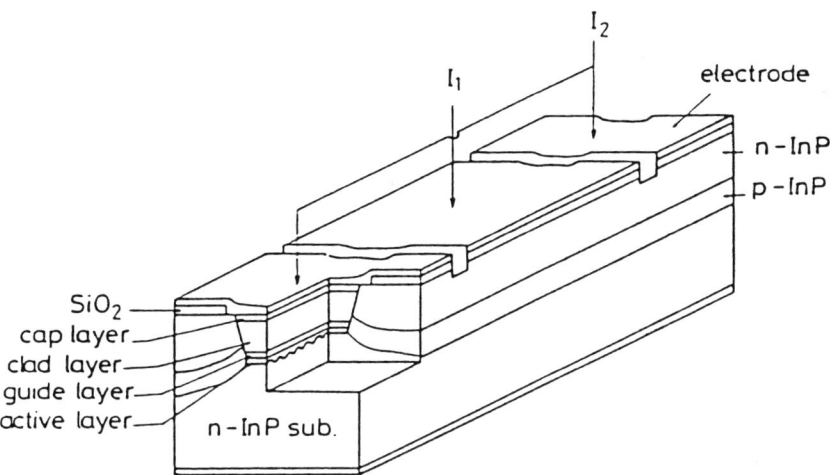

FIG. 25. Schematic diagram of a multielectrode DFB laser. Total current was distributed to center (I_1) and side electrodes (I_2). [After Nakano et al. (1987).]

Motosugi, 1986; Nakano et al., 1987). The side electrodes of the laser are connected together. The currents applied to the center (I_1) and side (I_2) electrodes are adjusted so that carrier density could be artificially created to be nonuniform between the side and center regions of the laser cavity. The phase delay and FM response of the three-electrode DFB laser are shown in Fig. 26. In the measurement, the total current was set at 100 mA with a constant current ratio of $I_1/I_2 = \frac{3}{2}$. In the frequency region from 5 Hz to 200 MHz, the multielectrode DFB laser did not exhibit any change in phase delay. The FM response is flat up to 200 MHz, and characteristics are observed in a current ratio range of $\frac{3}{2} \leq I_1/I_2 \leq 9$. The linewidth decreased linearly up to $3I_{th}$, and the minimum linewidth was 4.5 MHz. The conventional DFB laser exhibited a decrease in the phase delay from π to 0 as the modulation frequency increased. This result indicates that the frequency shift changed from a red shift (due to the thermal effect) to a blue shift (due to the carrier effect). Conversely, the multielectrode DFB laser exhibited no change in phase delay at all. These results suggest that frequency shift by carrier density nonuniformity is larger than that by the thermal effect even in the low-frequency region. When the modulation current is applied only to the center electrode, the lasing frequency moved out of phase with the modulation current (red shift). However, modulation to the front electrode caused an in-phase frequency shift (blue shift). This is because the increse in the modulation current into the front electrode causes a decrease in the carrier density in the center. On the other hand, amplitude modulations are in phase with the current in both cases, as

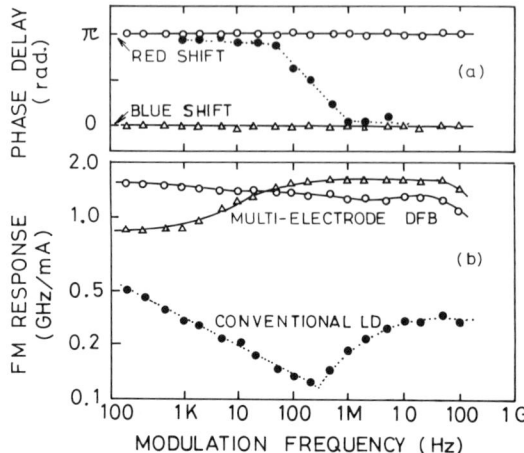

FIG. 26. (a) Phase delay of the frequency shift to the modulation current as a function of the modulation frequency. Results in the multielectrode DFB laser are shown for two different bias points with circles ($I_c = 0.6I_t$) and triangles ($I_c = 0.5I_t$). Dots represent a conventional DFB laser. (b) Modulation frequency dependence of frequency modulation response. [After Yoshikuni and Motosugi (1987).] (©1987 IEEE.)

expected. When the red-shift effect of the center electrode modulation is combined with the blue-shift effect of the front electrode modulation, the amplitude and frequency can be modulated independently. When the modulation currents applied to the front and the center electrode are nearly in phase, the amplitude modulation has suppressed chirping. On the other hand, when the modulation currents are out of phase and their amplitudes are controlled to cancel any modulation in the lasing amplitude, the amplitude change is completely suppressed. This push–pull operation is similar to the two-region model proposed by Nilsson and Yamamoto, and the experimental results are also explained by the two α parameters in the different regions. However, the DFB laser is more sensitive than the Fabry–Perot laser because the Bragg condition must also be satisfied when the refractive index changes.

An FM laser for a coherent optical communication system requires high FM efficiency with low spurious IM component, a flat and broadband FM response, and a narrow spectral linewidth or low FM noise. Recently designed DFB lasers have useful characteristics for FM light sources that satisfy these conditions. Figure 27 shows the improved DFB FM lasers for coherent optical communication systems. Figure 27a shows the phase-tunable (PT) GaInAsP/InP DFB laser emitting at 1.5 μm (Murata *et al.*, 1987). It has a 250-μm-long DFB region and a 130-μm-long phase-tuning region. The two regions are electrically isolated from each other by 200-μm-wide etched grooves, formed on both sides of the center mesa stripe area. The feature of

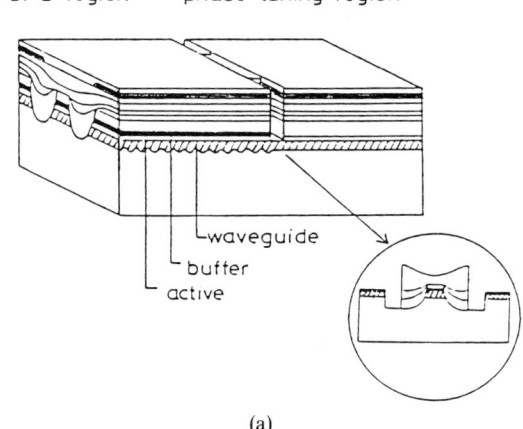

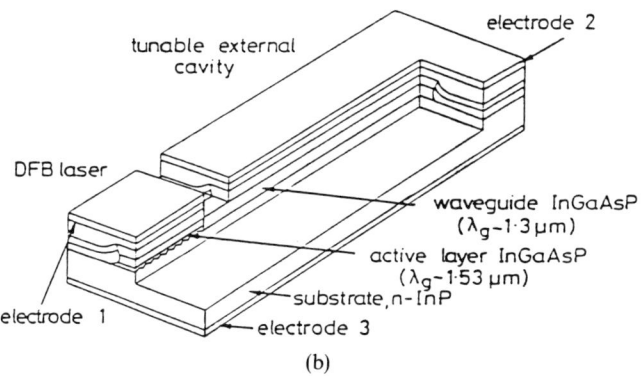

FIG. 27. (a) Phase-tunable DFB laser device structure. Inset: cross-sectional view of boundary region. [After Murata et al. (1987).] (b) Schematic diagram of DFB laser integrated with a tunable external cavity. [After Lee and Menocal (1987).]

this laser is the frequency deviation controlled by the facet phase change. The continuous-wavelength tuning range is over 1.2 nm (150 GHz). The FM response is flat up to 100 MHz, and to 200 MHz at 3 dB down bandwidth. The FM efficiency measured at 10 MHz also changes periodically from a few gigahertz per milliamp to 16 GHz/mA as the tuning current is increased. These values are more than one order of magnitude larger than for a normal DFB laser. In the PT-DFB laser, a parasitic intensity modulation accompanying the 1 GHz frequency deviation is suppressed to less than 1%. The minimum linewidth is 20 MHz. Figure 27b shows an integrated InGaAsP DFB laser with a tunable external cavity (Lee and Menocal, 1987). The DFB laser has a 0.4-mm-long cavity with a first-order grating with a Bragg wavelength at 1.56 μm. The length of the external cavity ranged from 1.0 to 3.51 mm. The minimum linewidth measured is 18 MHz. The lasing frequency deviation shows a flat response up to 500 MHz/mA without any broadening of the linewidth.

III. Phase Modulation by Injection Locking

Optical phase modulation by injection locking was suggested by Buczek and Freiberg (1972), who theoretically explained phase change by external mirror vibration in an injection-locked CO_2 laser. This has not yet been achieved in a semiconductor laser. Injection locking in these lasers has been studied for coherent optical system applications. The bandwidth-versus-gain relationship of injection-locked semiconductor laser amplifiers (Kobayashi and Kimura, 1980a, 1981), coherence of phase-locked lasers (Kobayashi and Kimura, 1980b), and single mode operations in directly modulated semiconductor lasers with cw light injection (Kobayashi et al., 1980; Yamada et al., 1980, 1981) were reported.

Here, we discuss phase modulation induced in an AlGaAs semiconductor laser by modulating the locked-laser cavity frequency by cw coherent light injection.

A. Phase Modulation Mechanism by Injection Locking

When the frequency of coherent light injected into a semiconductor laser is swept, the phase of the locked-laser-output wave changes by the frequency shift, as schematically represented in Fig. 28a. This is an inherent injection locking phenomenon (Kobayashi and Kimura, 1981). The locking bandwidth is determined by the cavity Q of a locked semiconductor laser, and by the input power and output power. The full phase change corresponding to the locking bandwidth is π. There is no phase modulation in this model, however, because of the accompanying frequency shift.

On the other hand, when cw light injection changes the cavity (or resonant) frequency of the locked laser, the phase of the locked-laser output wave changes without frequency deviation. This is shown in Fig. 28b, where the locked-laser-output frequency is locked to the injected light frequency. The phase change corresponding to the locking bandwidth is also π. This is the basic principle of phase modulation by optical injection locking.

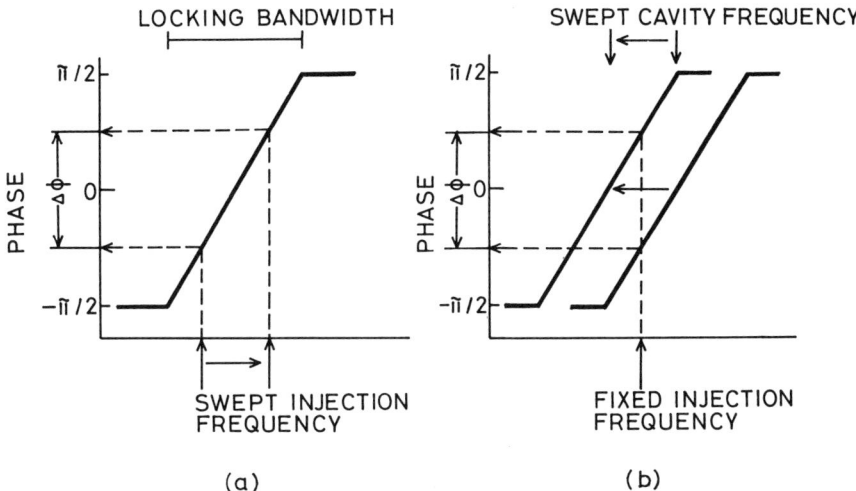

FIG. 28. A phase modulation model by cw light injection to a frequency-modulated laser: (a) conventional injection locking, (b) induced phase modulation.

The static phase shift obtained by changing the injected light frequency is shown in Fig. 29. The cavity frequency of the locked laser is shifted by changing the bias current indicated by the bottom abscissa. The total phase shift can be as large as π radians, corresponding to the frequency shift of locking bandwidth $2\Delta f$. The locking bandwidth is determined by cavity Q, input power P_{in}, and output power P_{out} as follows (Kobayashi and Kimura, 1980a, 1981):

$$2\Delta f = (f_0/Q)\sqrt{P_{in}/P_{out}}, \quad (37)$$
$$Q = 2\pi f_0 \tau_p.$$

Here, f_0 is the oscillation frequency and τ_p is the photon lifetime in the semiconductor laser. One of the features of phase shift by injection locking is that the phase changes in direct proportion to the cavity frequency change (top abscissa). By small-signal analysis, the phase shift is represented as follows (Kobayashi and Kimura, 1982b):

$$\Delta\Phi = (f_c - f_{in})/\Delta f. \quad (38)$$

Here, f_c is the cavity center frequency (or resonant frequency) of the locked laser, f_{in} is the injected light frequency, and Δf is the locking half bandwidth. Another feature is that the gradient of the phase shift depends on the locking bandwidth, as indicated by Eq. (38). Therefore, as the locking bandwidth becomes large, the phase shifts slowly.

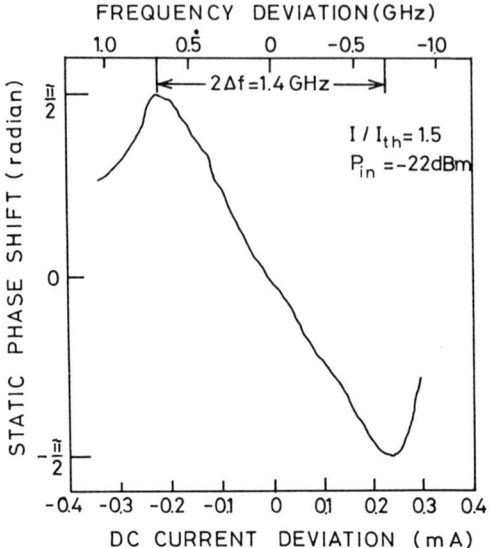

FIG. 29. Static phase shift obtained by changing the locked laser frequency with respect to the master laser through the bias current. [After Kobayashi and Kimura (1982a).] (©1982 IEEE.)

Dynamic phase modulation is performed by modulating the injection current in a locked laser by cw light injection. Power spectra of phase-modulated signals with injection and of frequency-modulated signals without injection are shown in the insets in Fig. 30. Suppression of the sideband component means that phase modulation is induced by injection locking rather than by frequency modulation suppression (Kobayashi et al., 1981b; Kobayashi and Kimura, 1982c). The phase modulation signal is represented as

$$E_{\text{PM}} = E_0 \exp\{j[2\pi f_0 t + \Delta\Phi \sin(2\pi f_m t)]\}. \tag{39}$$

Here, $\Delta\Phi$ is the maximum phase deviation (MPD). In ideal PM, $\Delta\Phi$ does not change when the modulation frequency f_m is changed. On the other hand, the frequency modulation index β varies inversely with the modulation frequency. Frequency characteristics of induced phase modulation by injection locking are represented (Kobayashi and Kimura, 1982b)

$$\Delta\Phi = \beta\sqrt{1/[(\Delta f/f_m)^2 + 1]},$$
$$\beta = \Delta F/f_m. \tag{40}$$

Here, ΔF is the frequency deviation and f_m is the modulation frequency. Maximum phase deviation $\Delta\Phi$ is a constant, $\Delta F/\Delta f$, in the region below the locking half bandwidth Δf. On the other hand, $\Delta\Phi$ changes inversely with the

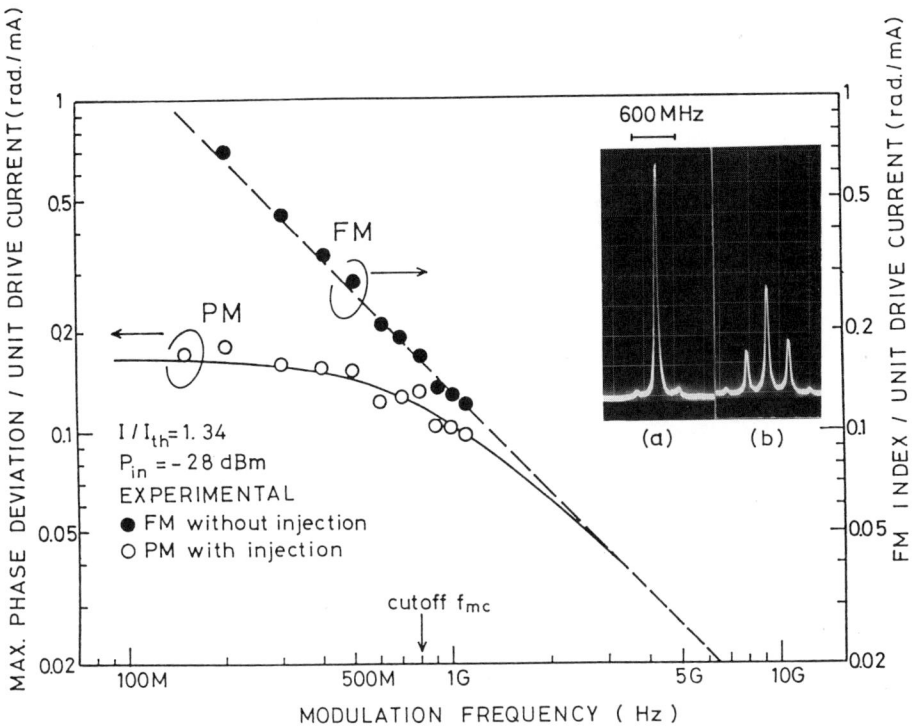

FIG. 30. Maximum phase deviation with injection and FM index without injection vs. modulation frequency. Solid and broken lines show calculated values using the van der Pol equation. Photograph (a): PM spectrum with −28 dBm input power injection. Maximum phase deviation $\Delta\Phi$ is 0.17 rad/mA. Photograph (b): FM spectrum without injection. FM index $\Delta F/f_m$ is 0.47 rad/mA. Locking half-bandwidth Δf is 800 MHz. $I/I_{th} = 1.34$. $P_{in} = -28$ dBm. Experimental results: ●, FM without injection; ○, PM with injection. [After Kobayashi and Kimura (1982a).]

modulation frequency f_m in regions above Δf. These characteristics were verified in an injection locking experiment, shown in Fig. 30, where the cutoff frequency was equal to Δf.

B. Phase Modulation Measurement

The experimental setup for studying phase modulation is shown in Fig. 31. Two double-heterostructure AlGaAs semiconductor lasers (master laser and locked laser) with identical cavity lengths operate continuously in single longitudinal mode. These lasers have channeled-substrate-planar (CSP)

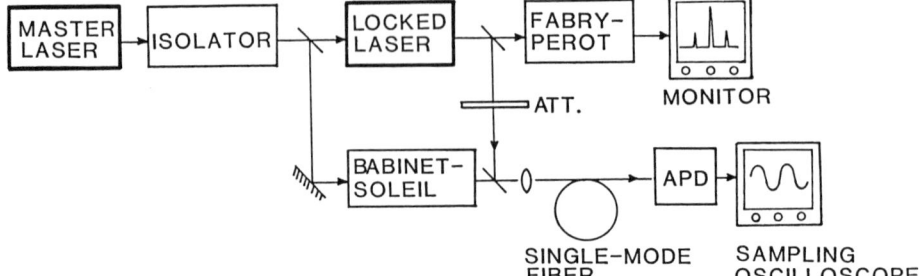

FIG. 31. Configuration for optical phase modulation experiment using optical injection locking. A Mach–Zehnder interferometer and scanning Fabry–Perot interferometer measure the phase modulation.

geometry (Aiki *et al.*, 1977). They are individually mounted on copper heat sinks whose temperatures are stabilized to an accuracy of $\pm 0.05°C$ by thermoelectric elements. The master laser output is collimated with antireflection-coated microlenses and injected into the locked laser. Reflected light from the locked laser to the master laser is eliminated by an optical isolator with 40 dB isolation (Kobayashi and Seki, 1980). The longitudinal mode frequency tuning and coupling efficiency of the two lasers were measured by the method given by Kobayashi and Kimura (1980c, 1981). The locking bandwidth can be measured by changing the master laser current as reported in Kobayashi and Kimura (1981).

The scanning Fabry–Perot interferometer, shown in Fig. 31, measured the modulation sideband spectra. The Mach–Zehnder interferometer configuration measured the phase shift. The attenuator balanced the light amplitudes in the two branches of the interferometer, and the Babinet–Soleil compensator acted as a phase shifter to calibrate the phase deviation. The single-mode fiber ensured wavefront matching of cw and phase modulation (PM) light at the surface of the Si-APD.

For static characteristics measurement, the two lasers were oscillated continuously without an rf signal. A change in the locked-laser-resonant frequency caused by changes in its drive current results in a static phase shift with respect to the injected light wave. This static phase shift is shown in Fig. 30 versus the locked laser resonant frequency shift, measured by homodyne detection using the Mach–Zehnder interferometer configuration. The master laser and the locked laser are dc-driven. Injected cw light power in the locked laser is $P_{in} = -22$ dBm. A static phase shift of π takes place when the locked laser bias current is changed by 0.48 mA (Kobayashi and Kimura, 1982a). The full locking bandwidth $2\Delta f$ is 1.4 GHz. The phase shift curve between $-\pi/2$ and $\pi/2$ is nearly linear, rather than sinusoidal. The visibility of the Mach–Zehnder interferometer output is more than 95 percent when the Babinet–Soleil compensator and attenuator are well adjusted to match wavefronts at the cavity center frequency in a locking state. At the locking boundary, the visibility was reduced to between 74 and 89 percent because of undesired spontaneous power emitted from the other longitudinal mode. The phase change per unit drive current becomes large when the locking bandwidth becomes narrow.

For dynamic phase modulation measurement, an rf sinusoidal signal with frequency f_m is superimposed on the locked laser bias current for direct current modulation. The induced PM signal is observed on the display of the sampling oscilloscope through the Mach–Zehnder interferometer. The optical phase bias is adjusted by the Babinet–Soleil compensator to give a sinusoidal signal on the display. The maximum phase deviation (MPD) is calibrated by the known phase shift of the Babinet–Soleil compensator. Frequency characteristics of the Si-APD have been calibrated with a beat note between the two lasers, both under cw drive conditions. The 3-dB down cutoff frequency was about 500 MHz. Frequency characteristics of the locked laser drive circuit were obtained from these Si-APD characteristics, assuming that the locked laser has flat frequency characteristics for small current modulation. The cutoff frequency was measured as 600 MHz. The frequency characteristics of the drive and detection circuits have been compensated for to give intrinsic PM performance over a wide modulation frequency range.

When the locked laser is driven by an rf current with cw light injection, the phase modulation outputs are as shown in Fig. 30. Frequency spectra show (a) PM modulation with, and (b) FM modulation without, injection measured by the scanning Fabry–Perot interferometer (Kobayashi and Kimura, 1982a, b). The measured frequency deviation is 130 MHz/mA. The FM index, which is defined by $\beta = \Delta F/f_m$, is suppressed from 0.47 to 0.17 per milliamp by injecting -28 dBm cw light power. The maximum phase deviation (MPD) per unit drive current is 0.17 rad/mA for the locking half bandwidth of $\Delta f = 800$ MHz. The MPD measured from carrier and sideband amplitudes coincides with that of the detected PM signal measured by the Mach–Zehnder interferometer.

C. Frequency Characteristics of Induced PM Signal

Injection locking can be represented by the van der Pol equation in the rotating wave approximation, as follows (Sargent *et al.*, 1974; Shimoda and

Yajima, 1972; Hirota and Suematsu, 1979):

$$d\Phi/dt = (E_{in}/E_L)\sin(\Psi - \Phi)/2\tau_p - \Delta\omega + d\Theta/dt \tag{41}$$

$$P_L = E_L^2,$$

$$P_{in} = E_{in}^2,$$

$$\Delta\omega = \omega_{in} - \Omega_0 (= \omega_L - \Omega_0).$$

Here, P_L is the locked photon density, τ_p is the photon lifetime, P_{in} is the injected photon density, E_L is the locked electric field amplitude, E_{in} is the injected electric field amplitude, ω_{in} is the injection light angular frequency, ω_L is the locked angular frequency, and Ω_0 is the cavity angular frequency of the resonator of the locked laser. In an injection-locked state, $\omega_{in} = \omega_L$. Φ is the phase of the locked signal, Θ is the phase of the locked laser under a free-running state, and Ψ is the phase of the injection signal from the locked laser.

The static phase change taken from Eq. (41) with $d\Phi/dt = 0$ and $\Psi_0 = 0$ is

$$\Phi_0 = \sin^{-1}[(E_0/E_{i0})4\pi\tau_p(f_0 - f_{in})]. \tag{42}$$

Here, the 0 means direct-current operation. As f_0 is shifted by the drive current, Φ_0 is changed because f_{in} is fixed to the cavity center. Φ_0 is changed from 0 by $\pm\pi/2$ when the frequency shift $f_{in} - f_0$ is equal to the locking-half bandwidth, $\Delta f = (E_{i0}/E_0)/4\pi\tau_p$. At around $f_{in} = f_0$, the phase shift can be approximated by a linear function represented by Eq. (38). The static phase shift is linear in the locked-laser characteristics near the cavity center frequency, as shown in Fig. 29. Dynamic frequency characteristics are represented by the first-order perturbation equation from Eq. (41) as follows:

$$j\omega\Delta\Phi = \{[E_{i1}\sin(\Psi_0 - \Phi_0) + E_{i0}(\Delta\Psi - \Delta\Phi)\cos(\Psi_0 - \Phi_0)]/2\tau_p \\ - \Delta\omega E_1 - j\omega\Delta\Theta E_0\}/E_0. \tag{43}$$

Here, subscripts 0 and 1 mean dc and first-order perturbation.

At the center frequency of the locked laser cavity, $\Delta\omega = 0$ and $\Psi_0 = \Phi_0$. For a small signal phase deviation $\Delta\Psi$ compared with the modulated laser phase deviation $\Delta\Theta$, Eq. (43) is approximated by

$$\Delta\Phi = \Delta\Theta\sqrt{(f_m/\Delta f)^2/[1 + (f_m/\Delta f)^2]}. \tag{44}$$

Here, $\Delta f = (E_{i0}/E_i)/4\pi\tau_p = (P_{in}/P_L)^{1/2}/4\pi\tau_p$ is the locking half-bandwidth (LHB). When cw light is injected into the directly modulated FM laser, the modulation index is suppressed by the factors (Kobayashi et al., 1981b; Kobayashi and Kimura, 1982c).

$$\beta_{out}/\beta_a = \Delta\Phi/\Delta\Theta = \sqrt{(f_m/\Delta f)^2/[1 + (f_m/\Delta f)^2]}. \tag{45}$$

Here, β_{out} is the modulation index of locked output with cw light injection, and β_a is the modulation index without injection. FM index suppression by injection locking is measured as represented in Fig. 32. The frequency modulation component is strongly suppressed as the locking half-bandwidth increases. The small bar in each curve shows this bandwidth.

The phase deviation can be represented with $\Delta\Theta = \beta_a = \Delta F/f_m$ as

$$\Delta\Phi = \Delta F \sqrt{1/(\Delta f^2 + f_m^2)}. \tag{46}$$

Frequency characteristics of phase modulation induced by injection locking can be represented by the characteristics of frequency deviation of FM without injection, locking bandwidth, and modulation frequency. Typical modulation frequency characteristics between 100 MHz and 1.4 GHz of induced PM signals are shown in Fig. 33. The ordinate on the left shows the maximum phase deviation (MPD) per unit drive current with injection; the ordinate on the right shows FM index per unit current without injection. Closed-circles curves and brokenline curves show the values of experimental and theoretical FM indices per unit current. Other symbols and lines show the respective values of MPD per unit current for each locking bandwidth. The small vertical bars on the curves show the locking half-bandwidth for each set of experimental conditions.

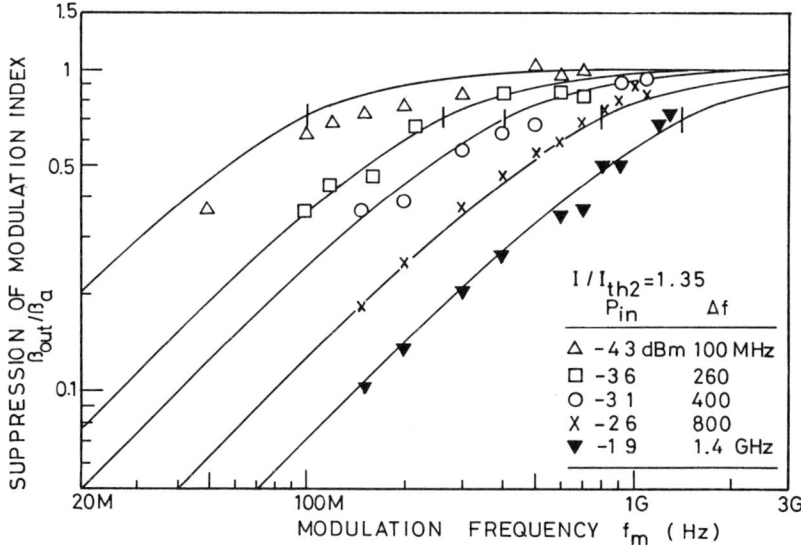

FIG. 32. Frequency characteristics of modulation index suppression. [After Kobayashi and Kimura (1982b).] (©1982 IEEE.)

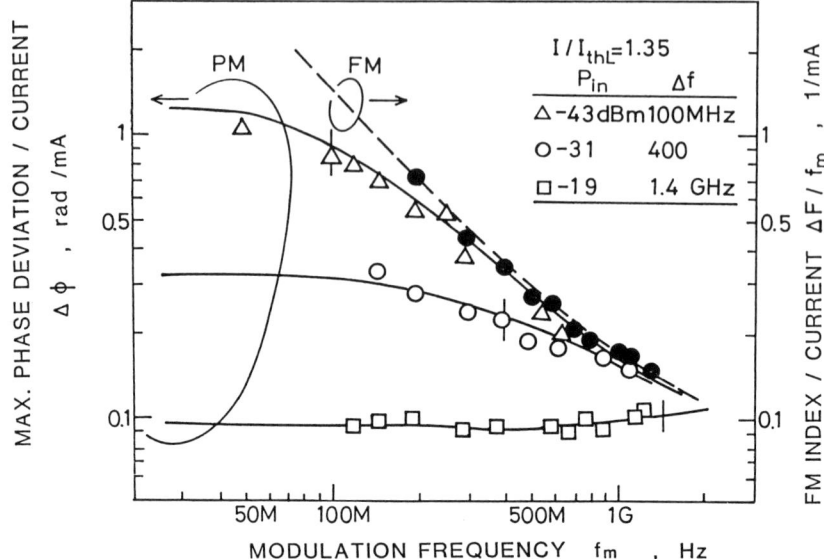

FIG. 33. Maximum phase deviation characteristics. Bias current is $1.35 \times I_{th,L}$. Closed circles and broken lines show experimental and calculated results of maximum frequency deviation without injection. Solid lines show theoretical values for PM. [After Kobayashi and Kimura (1982b).] (©1982 IEEE.)

The FM index without injection decreases inversely with the modulation frequency as represented by Eq. (40). The slope of the FM index curve becomes gradual at frequencies higher than 500 MHz because of carrier resonance (Kobayashi et al., 1981a, 1982). The MPD curves tend to approach the FM index curve at high modulation frequencies, and the MPD becomes constant at modulation frequencies less than the locking half-bandwidth. Theoretical solid lines calculated by Eq. (46) fit well with the experimental results. As the locking bandwidth becomes broad, the MPD per unit rf current becomes small, and the cutoff frequency f_{mc}, at which the MPD decreases to $1/\sqrt{2}$, becomes high. The MPD becomes small when the bias current increases at a large locked-laser bias current.

The frequency characteristics of MPD per unit current from 1 kHz to 1.4 GHz are shown in Fig. 34. The broken and dot-and-broken lines show theoretical thermal and carrier effect curves calculated from the FM deviation. The solid line in Fig. 34 is calculated from Eq. (46) by substituting the measured maximum frequency deviation values and the locking half-bandwidth. The broken line shows the relation $\Delta\Phi = \Delta F/\Delta f$, where ΔF is obtained from the thermal refractive-index change by calculating the thermal conduction in each layer of the AlGaAs laser (Kobayashi et al., 1981a, 1982).

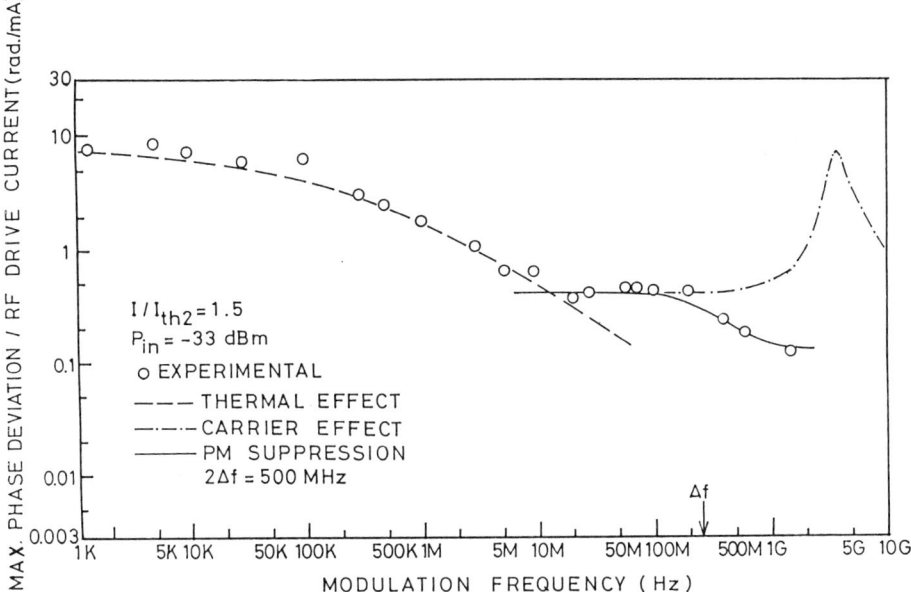

FIG. 34. Frequency characteristics of maximum phase deviation. [After Kobayashi and Kimura (1982b).] (© 1982 IEEE.)

The dot-and-broken line shows the same relation $\Delta\Phi = \Delta F/\Delta f$, where ΔF is obtained from the carrier refractive index change considering lateral carrier diffusion in a CSP laser (Kobayashi et al., 1981a, 1982).

D. Phase Modulator for Transmission Applications

Homodyne or heterodyne coherent optical transmission requires a phase modular with (1) a large PM response, (2) a large bandwidth, and (3) flat frequency characteristics. The first point requires a large FM response. As detailed in Section II, the vertical mode confinement factor Γ_y must be large for a large FM response (Kobayashi et al., 1982). In operational techniques, a low bias-current level is also effective. The bias-current dependence of the product of normalized maximum phase deviation and locking half-bandwidth is represented in Fig. 35 as a function of normalized modulation frequency. The restriction of the phase deviation by the locking bandwidth is the same on each curve. However, as bias current affects frequency deviation ΔF, the PM responses have different values on each curve. These are theoretically represented by Eq. (46). Since the total phase shift is determined by the locking

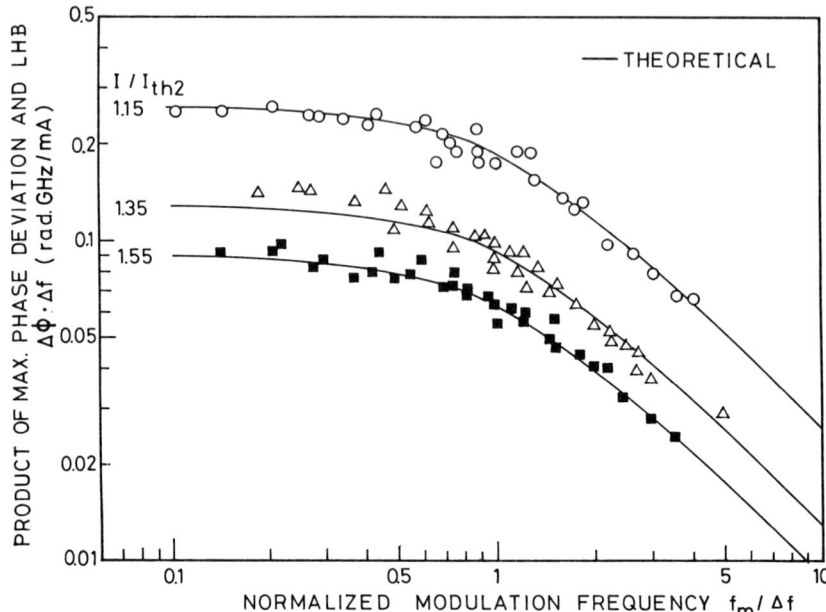

FIG. 35. Product of maximum phase deviation and locking half bandwidth as a function of normalized modulation frequency. ———, theoretical. [After Kobayashi and Kimura (1982b).] (© 1982 IEEE.)

bandwidth as π, another way to obtain a large PM response is to reduce the locking bandwidth.

A large bandwidth requires a large locking bandwidth and trades off with a large PM response, as shown in Fig. 36. A large input power, small output power, low bias current, and low Q cavity satisfy this condition. The flatness of the frequency characteristics of the FM response are discussed in Section II. To satisfy requirement 2 above, the carrier density should be distributed outside the photon distribution in a lateral direction.

A specific feature of phase modulation by injection locking is that when the locking half bandwidth is set at the carrier resonant frequency, the phase deviation becomes flat, as shown in Fig. 33. This is because the decrease in the PM response restricted by the locking bandwidth is compensated by the frequency deviation increase due to the resonant characteristics of carrier density. For coherent optical transmission, the linewidth should be as narrow as possible (Saito et al., 1983). Phase modulation by injection locking has an advantage here, since a large locking bandwidth requires a large input power, which can be obtained at a high bias level in the master laser (Saito and Yamamoto, 1981; Nilsson et al., 1981). An interesting feature of phase

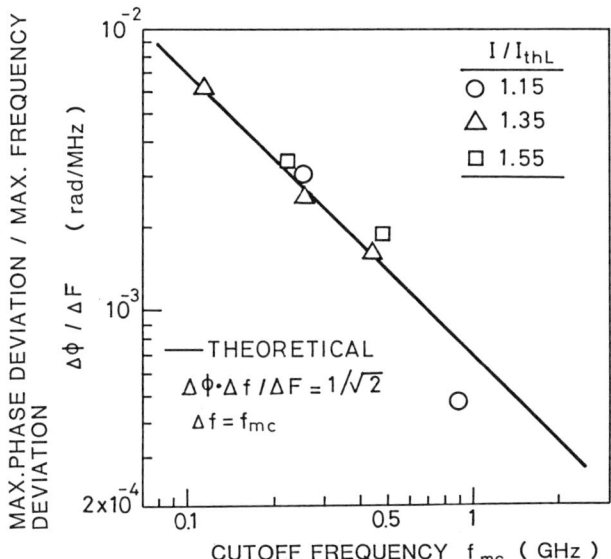

FIG. 36. Maximum phase modulation normalized by maximum frequency deviation as a function of the cutoff frequency. [After Kobayashi and Kimura (1982b).] (©1982 IEEE.)

modulation characteristics by injection locking is that phase deviation is determined only by the locking bandwidth. This is true even in the low-frequency region due to the thermal effect and in the high-frequency region due to the carrier effect.

The other problems to be solved are spurious intensity modulation due to free-carrier absorption in semiconductor lasers, and the phase relation between the thermal and carrier effect at a modulation frequency of 10 MHz. These are the same problems that will be encountered in applying direct frequency modulation to future coherent transmission systems.

IV. Summary

This chapter discussed frequency modulation and injection-locked phase modulation by modulated current injection in semiconductor lasers. For frequency modulation, the frequency characteristics of FM response were given from 0 to 5.2 GHz. On the border of 10 MHz, the carrier modulation effect is dominant in the high-frequency region, and the temperature modulation effect is dominant in the low-frequency region. The frequency modulation due to the carrier modulation effect exhibits a resonant peak because of relaxation oscillation and cutoff characteristics at a modulation frequency of several gigahertz. The FM response decreases as the dc bias level increases. A flat FM response is obtained in lasers such as CSP and TJS, where lateral carrier confinement is weak. However, there is a V-shaped decrease in the FM response at about 100 MHz in a laser such as BH, where the carrier and optical field are well confined in the active region. These characteristics arise from structural differences between gain-guided and index-guided lasers. The phase delay between FM and AM components shows a red shift and a blue shift in the higher-modulation-frequency regions of CSP and BH lasers, respectively. The two-region model with different α parameters in FM lasers can explain the flat response and phase delay characteristics of the red shift in the high-modulation frequency range. The frequency modulation due to the temperature modulation effect depends strongly on the distance between an active layer and a heat sink. A p-side-up-mounted laser has a larger thermal FM response and lower cutoff frequency than a p-side-down-mounted laser. Multielectrode semiconductor lasers with push−pull operations offer a flat, efficient FM response with small spurious intensity modulation. DFB FM lasers with phase control by facet phase change also usefully satisfy the flat response and high FM efficiency with small spurious intensity modulation.

Phase modulation is obtained by injection locking in directly modulated FM AlGaAs semiconductor lasers. Phase modulation by injection locking clarifies the fundamental characteristics of phase shift in relation to frequency shift and locking bandwidth. The phase shift is directly proportional to the frequency shift of the locked laser cavity. The maximum phase shift is π radians for a frequency shift corresponding to the locking bandwidth, an essential characteristic in the injection-locked phase modulator. Phase modulation frequency characteristics by injection locking can be represented by three parameters: (1) locking bandwidth, (2) maximum frequency deviation (MFD) in direct frequency modulation, and (3) maximum phase deviation (MPD). The frequency characteristics of phase deviation are similar to those in direct frequency modulation in the frequency region below the locking half-bandwidth. Maximum phase deviation increases in direct proportion to the maximum frequency deviation and in inverse proportion to the locking bandwidth. Frequency deviation caused by the carrier and thermal effects in direct frequency modulation is suppressed uniformly and transformed into phase modulation by injection locking. The small-signal van der Pol equation is useful in explaining FM suppression and PM induction characteristics.

Transmission in an optical phase modulator requires sufficient maximum phase deviation, wide bandwidth, and flat response. Unfortunately, the absolute maximum phase deviation is limited to π radians with an injection-locked modulator. However, synchronized cascade injection-locked phase modulators would give sufficient phase deviation larger than π. The key parameter in wide-band modulation is the locking bandwidth, which is determined from the cavity Q, output power, and input power. Since these are trade-offs between the locking bandwidth and the maximum phase deviation, the maximum frequency deviation is expected to become large. The structural design of an FM laser with a large frequency deviation was discussed in this chapter. A flat PM modulation response would be accomplished through considerations similar to those for direct frequency modulation.

Acknowledgments

The authors wish to thank Dr. T. Mukai and Dr. S. Saito for stimulating discussions and suggestions about coherent optical fiber transmission systems, and Dr. M. Ito for calculating the thermal effect in semiconductor lasers.

References

Aiki, K., Nakamura, M., Kuroda, T., and Umeda, J. (1977). *Appl. Phys. Lett.* **30**, 649.
Buczek, C. J., and Freiberg, R. J. (1972). *IEEE J. Quantum Electron.* **QE-18**, 64.
Cardona, M. (1960). *Int. Conf. Semiconductor Phys., Prague*, 388.
Casey, H. C., Jr., and Panish, M. B. (1978). *Heterostructure Lasers Part A: Fundamental Principles*, Academic Press, New York.
Casey, H. C., Jr., and Stern, F. (1976). *J. Appl. Phys.* **47**, 631.
Chinone, N., Saito, K., Ito, R., Aiki, K., and Shige, N. (1979). *Appl. Phys. Lett.* **35**, 513.
Dandridge, A., and Goldberg, L. (1982). *Electron. Lett.* **18**, 302.
Dutta, N. K., Olsson, N. A., Koszi, L. A., Besoni, P., Wilson, R. B., and Nelson, R. J. (1984). *J. Appl. Phys.* **56**, 2167.
Fenner, G. E. (1964). *Appl. Phys. Lett.* **15**, 198.
Goldberg, L., Tailor, H. F., and Weller, J. F. (1981). *Electron Lett.* **17**, 498.
Halperin, B. I., and Lax, M. (1966). *Phys. Rev.* **148**, 722.
Harder, C., Vahara, K., and Yariv, A. (1983). *Appl. Phys. Lett.* **42**, 328.
Henry, C. H. (1982). *IEEE J. Quantum. Electron.* **QE-18**, 259.
Henry, C. H., Logan, R. A., and Bertness, K. A. (1981). *J. Appl. Phys.* **52**, 4457.
Hirota, O., and Suematsu, Y. (1979). *IEEE J. Quantum Electron.* **QE-15**, 142.
Ito, M., and Kimura, T. (1980). *IEEE J. Quantum Electron.* **QE-16**, 910.
Ito, M., and Kimura, T. (1981). *IEEE J. Quantum Electron.* **QE-17**, 787.
Ito, M., Ito, T., and Kimura, T. (1979). *J. Appl. Phys.* **50**, 6168.
Jacobsen, G., Olesen, H., and Birkedahl, F. (1982). *Electron. Lett.* **18**, 874.
Kaminow, I. P. (1974). *An Introduction to Electrooptic Devices*, Academic Press, New York and London.
Kane, E. Q. (1963). *Phys. Rev.* **131**, 79.
Kishino, K., Arai, S., and Suematsu, Y. (1982). *IEEE J. Quantum Electron.* **QE-18**, 343.
Kobayashi, K., and Seki, M. (1980). *IEEE J. Quantum Electron.* **QE-16**, 11.
Kobayashi, S., and Kimura, T. (1980a). *IEEE J. Quantum Electron.* **QE-16**, 915.

Kobayashi, S., and Kimura, T. (1980b). *Electron. Lett.* **16**, 668.
Kobayashi, S., and Kimura, T. (1980c). *Electron. Lett.* **16**, 230.
Kobayashi, S., and Kimura, T. (1981). *IEEE J. Quantum Electron.* **QE-17**, 681.
Kobayashi, S., and Kimura, T. (1982a). *Electron Lett.* **18**, 210.
Kobayashi, S., and Kimura, T. (1982b). *IEEE J. Quantum Electron.* **QE-18**, 1662.
Kobayashi, S., and Kimura, T. (1982c). *IEEE J. Quantum Electron.* **QE-18**, 575.
Kobayashi, S., Yamada, J., Machida, S., and Kimura, T. (1980). *Electron. Lett.* **16**, 746.
Kobayashi, S., Yamamoto, Y., and Kimura, T. (1981a). *Electron Lett.* **17**, 350.
Kobayashi, S., Yamamoto, Y., and Kimura, T. (1981b). *Electron. Lett.* **17**, 849.
Kobayashi, S., Yamamoto, Y., Ito, M., and Kimura, T. (1982). *IEEE J. Quantum Electron.* **QE-18**, 582.
Kubota, K., Minakata, M., Saito, S., and Uehara, S. (1978). *Opt. Quantum Electron.* **10**, 205.
Landau, L. D., and Lifshitz, E. M. (1960). *Electrodynamics of Continuous Media*, Addison-Wesley, Reading, Massachusetts.
Lee, T. P., and Menocal, S. G. (1987). *Electron. Lett.* **23**, 1090.
Manning, J. S., and Olshansky, R. (1980). *Electron. Lett.* **17**, 506.
Matthews, M. R., Dyott, R. B., and Carling, W. P. (1981). *Electron. Lett.* **8**, 570.
Murata, S., Mito, M., and Kobayashi, K. (1987). *Electron. Lett.* **23**, 12.
Nakamura, M., Aiki, K., Chinone, N., Ito, R., and Umeda, J. (1978). *J. Appl. Phys.* **49**, 4644.
Nakano, Y., Itaya, Y., Fukuda, M., Noguchi, Y., Yasaka, H., and Oe, K. (1987). *Electron. Lett.* **23**, 1372.
Namizaki, H. (1976). *Trans. IECE Japan* **E59**, 8.
Nash, F. R. (1973). *J. Appl. Phys.* **44**, 4696.
Nathan, M. I., Fowler A. B., and Burns, G. (1963). *Phys. Rev. Lett.* **11**, 152.
Newman, D. H., Bond, D. J., and Stefani, J. (1978). *Solid State Electron. Rev.* **2**, 41.
Nilsson, O., and Yamamoto, Y. (1985). *Appl. Phys. Lett.* **46**, 223.
Nilsson, O., Saito, S., and Yamamoto, Y. (1981). *Electron. Lett.* **17**, 589.
Nilsson, O., Gillrer, L., and Goobar, E. (1987). *Electron. Lett.* **23**, 1372.
Novikova, S. I. (1961). *Soviet Physics—Solid State* **3**, 129.
Olesen, H., and Jacobsen, G. (1982). *8th ECOC, Cannes, Tech. Digest*, 291.
Olsson, A. and Tang, C. L. (1981). *Appl. Phys. Lett.* **39**, 24.
Osterwalder, J. M., and Rickett, B. J. (1980). *IEEE J. Quantum Electron.* **QE-16**, 250.
Reinhart, F. K. (1968). *Appl. Phys.* **39**, 3426.
Reinhart, F. K., and Logan, R. A. (1975). *Appl. Phys. Lett.* **27**, 532.
Reinhart, F. K., and Miller, B. I. (1972). *Appl. Phys. Lett.* **20**, 36.
Ripper, J. E., Pratt, G. W., Jr., and Whitney, C. G. (1966). *IEEE J. Quantum Electron.* **QE-2**, 603.
Saito, S., and Yamamoto, Y. (1981). *Electron. Lett.* **17**, 325.
Saito, S., Yamamoto, Y., and Kimura, T. (1980). *Electron. Lett.* **16**, 826.
Saito, S., Yamamoto, Y., and Kimura, T. (1981). *IEEE J. Quantum Electron.* **QE-17**, 935.
Saito, S., Yamamoto, Y., and Kimura, T. (1983). *IEEE J. Quantum Electron* **QE-19**, 180.
Sargent, M. S., III, Scully, M. O., and Lamb, W. E., Jr. (1974). *Laser Physics*, Addison-Wesley, Reading, Massachusetts.
Selway, P. R., Thompson, G. H. B., Henshall, G. D., and Whiteaway, J. E. A. (1974). *Electron. Lett.* **10**, 455.
Shimoda, K., and Yajima, T. (1972). *Quantum Electronics*, Shokabo, Tokyo. In Japanese.
Stubkjaer, K., Suematsu, Y., Asada, M., Arai, S., and Adams, A. K. (1980). *Electron. Lett.* **16**, 896.
Thompson, G. H. (1972). *Opto-Electron.* **4**, 257.
Tshuchida, H., Tako, T., and Ohtsu, M. (1983). *Japan J. Appl. Phys.* **22**, L19.
Turley, S. E. H., Thompson, G. H. B., and Lovelace, D. F. (1979). *Electron. Lett.* **15**, 257.

Whitney, C. G., and Pratt, G. W., Jr. (1970). *IEEE J. Quantum Electron*, **QE-6**, 352.
Yamada, J., Kobayashi, S., Machida, S., and Kimura, T. (1980). *Japan J. Appl. Phys.* **19**, L689.
Yamada, J., Kobayashi, S., Nagai, H., and Kimura, T. (1981). *IEEE J. Quantum Electron.* **QE-17**, 1006.
Yamamoto, Y. (1980). *IEEE J. Quantum Electron.* **QE-16**, 1251.
Yamamoto, Y., and Kimura, T. (1981). *IEEE J. Quantum Electron.* **QE-17**, 919.
Yamamoto, Y., Saito, S., and Mukai, T. (1983). *IEEE J. Quantum Electron.* **QE-19**, 47.
Yoshikuni, Y., and Motosugi, G. (1986). *OFC '87, Atlanta*, TUF1.
Yoshikuni, Y., and Motosugi, G. (1987). *J. Lightwave Tech.* **LT-5**, 516.

Coherent Optical Fiber Transmission Systems

SHIGERU SAITO, YOSHIHISA YAMAMOTO, AND
TATSUYA KIMURA

NTT Basic Research Laboratories
Musashino-shi, Tokyo, Japan

I. Introduction 203
II. System Operations and Configurations 206
III. Advantages of Optical Heterodyne or Homodyne Detection 211
    A. Receiver Sensitivity 212
    B. Frequency Selectivity 220
IV. System Applications 225
    A. Long-Distance High-Speed Transmission Systems 225
    B. Optical Amplifier Repeater Systems 230
    C. FDM Systems 232
V. Essential Technology for Developing Coherent Systems 233
    A. Coherent Laser Sources 234
    B. Modulation–Demodulation Technology 238
    C. Optical Fibers 244
    D. Optical Amplifiers 249
VI. System Experiments 251
VII. Conclusion 257
    References 258

I. Introduction

Since the successful fabrication of the first low-loss fiber in 1970, optical fiber communications have evolved rapidly through three system generations (Henry, 1985). The first generation uses multimode fibers and an operating wavelength of about 0.85 μm. The second uses a wavelength of 1.3 μm, which allows optical fibers to show lower loss characteristics. This wavelength shift enables larger separation between regenerative repeaters in long-distance transmission systems. Most second-generation systems use single-mode fibers, rather than multimode fibers, and enjoy reduced dispersive effect since typical single-mode fibers show zero chromatic dispersion at around 1.3 μm wavelength. The first two generations have been manufactured and installed on a commercial basis. Some 1.3-μm single-mode fiber systems have already been commissioned as long-distance trunk lines. Development of third-generation single-mode fiber systems, in which it is intended to utilize the minimum-loss characteristics of silica fibers at about 1.55 μm wavelength, is now underway.

Present optical systems, even of the first generation, have surpassed conventional coaxial cable and radio relay systems in terms of repeater spacing and channel capacity.

The successful evolution of current fiber communication systems is primarily due to progressive improvement in fiber loss characteristics. Simultaneous advances in semiconductor device technology, such as AlGaAs and Si technology in the shorter wavelength region and InGaAsP and Ge technology in the longer region, have also promoted this evolution. Improvement in laser lifetime, after the continuous operation of a semiconductor laser at room temperature in 1970, has played an especially important role in the practical applications of current systems.

Modulation and demodulation technology has contributed less to the evolution of today's fiber communications than fiber and device technology. Simple methods of intensity modulation (IM) and direct detection (DD) are used commonly throughout the three generations. Lightwave intensity, not field amplitude, is modulated linearly with respect to the input signal voltage in the transmitter, and directly converted into a demodulated electrical output by a photodetector in the receiver. Essentially, no attention is paid to optical carrier frequency and phase, so the transmitted signal spectrum is much wider than the modulation input spectrum and, in some cases, spreads over 1 THz. Spectral coherence, which is the most important property of laser light, is not fully utilized. The optical energy of a noisy carrier wave is used instead to convey information. This indicates that present optical fiber communication systems are as primitive as radio engineering was in Marconi's time.

There is still room for further development in modulation–demodulation technology, and this is expected through heterodyne or homodyne detection of optical signals, resulting in a significant improvement in receiver sensitivity (Oliver, 1961; Haus and Townes, 1962; Yamamoto, 1980). The intensity modulation method is not compatible with these detection schemes. An advanced method such as amplitude modulation (AM), frequency modulation (FM), or phase modulation (PM) of a lightwave field, or their digital formats, amplitude-shift keying (ASK), frequency-shift keying (FSK), or phase-shift keying (PSK), is used. Around 1980, research started on optical fiber systems using such sophisticated modulation–demodulation methods applied to transmission lines with long repeater spacing and large information capacity (Yamamoto and Kimura, 1981; Favre et al., 1981; Okoshi and Kikuchi, 1981). They are commonly referred to in literature as coherent optical fiber communication systems because of the strict requirement for a highly coherent lightwave. Coherent optical fiber communications may be regarded as the fourth generation of fiber systems.

In nonfiber systems, coherent optical modulation–demodulation techniques are not new topics. Just after the invention of lasers in 1960, research efforts were made to utilize coherent laser light for optical communications.

Signal transmission using an optical frequency or phase modulation scheme instead of intensity modulation—that is, the lightwave analogue of radio-frequency communications—was one of the initial focuses of research interest. The signal-to-noise ratio (SNR) of optical heterodyne and homodyne detection was theoretically shown in 1962 to be higher than that of direct detection (Oliver, 1961; Haus and Townes, 1962). The information capacity of various optical communication systems, including heterodyne and homodyne receiver systems, was discussed (Gordon 1962). A heterodyne detection transmission experiment using 3.39-μm He–Ne lasers demonstrated a significant SNR improvement in 1967 (Goodwin, 1967). Based on the strict frequency selectivity of heterodyning or homodyning, the concept of frequency-division multiplexing (FDM) using coherent detection schemes was proposed in 1970 (DeLange, 1970). Research into important coherent communication systems, however, was not evolved further because of such obstacles as short lifetime and insufficient stability of lasers, as well as severe turbulence and instability of transmission media.

As mentioned earlier, advances in fiber and device technology in the 1970s led to successful development of currently used optical systems, rather than coherent systems. Low-loss and broadband fiber characteristics enabled long repeater spacing and large information capacity. Although primitive in modulation–demodulation technology, both direct intensity modulation of semiconductor lasers or light-emitting diodes (LEDs), and direct power detection using avalanche photodiodes (APDs) or p–i–n photodiodes, are very simple and reliable, giving conventional systems the advantage of a simple configuration.

The latter half of the 1970s witnessed further development in semiconductor optical device technology for providing semiconductor lasers that could operate in a single longitudinal mode. This led to growing interest in reviving coherent techniques in fiber communications. An interference experiment of AlGaAs laser light propagating through a 4.15-km single-mode fiber cable reported in 1979 revealed two important features (Machida *et al.*, 1979). First, the semiconductor laser is capable of producing, under appropriate conditions, a sufficiently narrow linewidth output to exhibit clear interference fringes. Second, the single-mode fiber is capable of maintaining a stable polarization direction over several kilometers. These features strongly suggested the possibility of coherent optical communication through fibers using compact and convenient semiconductor lasers. Encouraged by this experiment, the receiving signal levels in various optical demodulation schemes were evaluated to confirm the usefulness of frequency or phase modulation and heterodyne or homodyne demodulation (Yamamoto, 1980).

At present coherent optical fiber communication systems are being studied on a worldwide scale (Yamamoto and Kimura, 1981; Favre *et al.*, 1981; Okoshi and Kikuchi, 1981; Kimura, 1987; Nosu and Iwashita, 1988;

Okoshi, 1987; Hooper *et al.*, 1983; Emura *et al.*, 1987; Gimlett *et al.*, 1987; Chikama *et al.*, 1990). Because of improved receiver sensitivity, the main research and development interest is in the field of long-distance and high-speed transmission lines, which will be used as terrestrial trunk lines or undersea systems. Recently, increased attention has been directed to FDM techniques, which will be applied to large-capacity multiterminal local networks, rather than to long-haul point-to-point transmission links (Nosu and Iwashita, 1988; Stanley *et al.*, 1987; Wagher *et al.*, 1987). Coherent optical modulation–demodulation methods are being discussed extensively in the field of intersatellite communications as well (Chan *et al.*, 1983; Chan, 1987).

This chapter focuses mainly on coherent optical fiber systems for transmission of high-speed digital information over distances of 100 km or more. The basic principle, essential technology for development, and recent progress in transmission experiments are described. While coherent modulation–demodulation techniques can be used in analog systems, they are not discussed here because they have their own measures of system performance that are different from those of digital systems, making any discussion needlessly complicated. Similarly, only a brief description will be given of local area applications based on FDM techniques.

Section II describes basic operations and configurations of coherent optical fiber transmission systems compared with those of conventional systems. Section III discusses improvements in both receiver sensitivity and frequency selectivity, the major advantages in optical heterodyne or homodyne detection. From this discussion, Section IV derives three types of system applications: a long-distance point-to-point transmission system, a long-haul optical amplifier repeater system, and an FDM system. Section V discusses the essential technology for realizing such coherent optical fiber transmission: coherent carrier wave generation, coherent modulation and demodulation, single-polarization signal transmission, and optical amplification. Section VI describes the current status of transmission experiments by summarizing significant system experiments in the progress of coherent optical transmission research. Finally, Section VII outlines the present state of technology and summarizes future areas of concern essential to the continuing development of coherent optical fiber transmission systems.

II. System Operations and Configurations

Coherent optical fiber transmission systems have unique system configurations, and therefore unique operating methods, that are very different from those of current optical fiber transmission systems. This is attributed to modulation and demodulation methods, especially to the optical detection methods employed.

The conventional optical fiber transmission systems developed so far use a combination of intensity modulation (IM) and direct detection (DD), and transmit information on the energy of an optical carrier wave. This primitive method allows a simple configuration which, as shown in Fig. 1, appears as an assembly of only three optical components: a semiconductor laser, an optical fiber, and a photodetector. Direct IM feasibility of semiconductor lasers also contributes to system simplicity by enabling the transmitter to do without a modulator. A light-emitting diode (LED) can be used in place of the semiconductor laser. An avalanche photodiode (APD) or a $p-i-n$ photodiode is used as the detector, where receiver sensitivity is improved by adjusting avalanche gain and load resistance. Figure 1 also shows the typical signal waveform launched into the fiber. Because a direct detection scheme is essentially insensitive to the optical signal phase, conventional systems may, and in fact do, use a carrier wave with a rather noisy phase. A launched IM signal generally has a much wider spectrum than the modulation input signal.

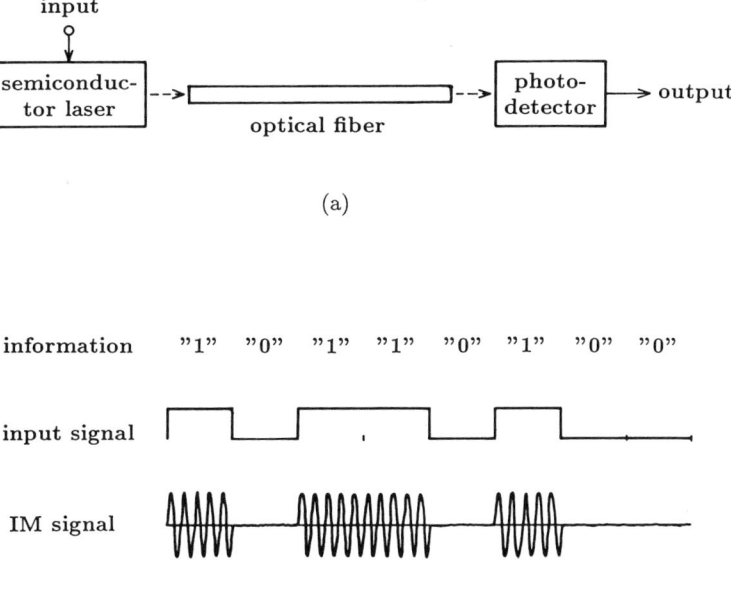

FIG. 1. Conventional optical fiber transmission: (a) basic system configuration; (b) signal waveform.

Coherent optical fiber transmission systems use an optical mixing detection scheme, named optical heterodyne or homodyne detection, to utilize fully the coherent property of a laser light for highly efficient information transmission. Unlike direct detection, which directly converts an optical signal into a demodulated electrical output, the heterodyne receiver first adds a locally generated optical reference wave to the incoming signal, and then detects the combination. The resulting photocurrent contains a beat-note signal—that is, an intermediate-frequency (IF) signal, which carries all the original information but at a lower frequency of several gigahertz. This enables successive electrical signal processing to retrieve the original data completely. As will be discussed in Section III, heterodyne detection achieves significantly better receiver sensitivity and frequency selectivity than direct detection. In a homodyne receiver, the demodulated baseband signal is directly extracted only through the optical mixing process because a local oscillator is phase locked to the carrier wave.

The intensity modulation scheme is not compatible with coherent detection. Therefore, an advanced technique of amplitude-shift keying (ASK), frequency-shift keying (FSK), or phase-shift keying (PSK) is used. Figure 2 illustrates binary optical signals generated using these schemes. A binary ASK signal has amplitudes A_1 or A_0, when the input signal is in the "mark" ("1") or "space" ("0"), respectively. In most cases A_0 is fixed at zero, and this type of ASK is referred to as on–off keying (OOK). A binary FSK signal oscillates at a fixed frequency of either f_1 or f_0 according to the input data, but has a constant amplitude. An important parameter of FSK is the frequency

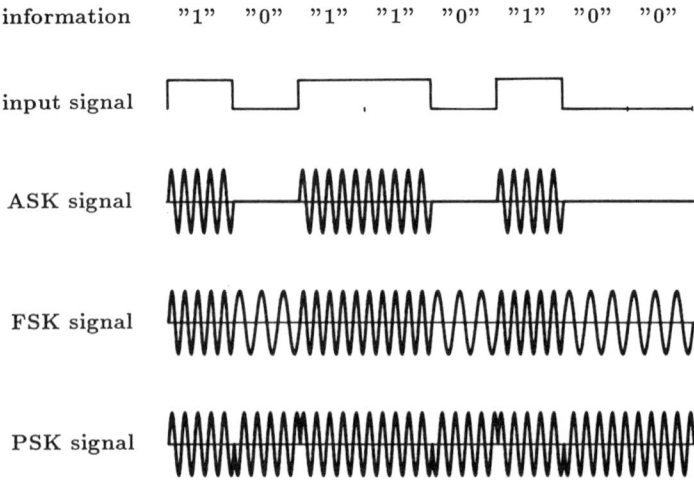

FIG. 2. Signal waveforms in coherent optical fiber transmission.

difference, $|f_1 - f_0|$, which is usually set equal to or at one-half of the transmission data rate to prevent a signal spectrum from spreading excessively. These typical types of FSK are named Sunde's FSK and minimum-shift keying (MSK). Binary PSK assigns a zero phase to the mark and a π phase to the space, or employs differential coding (differential PSK; DPSK), which gives a π phase shift when the input signal changes from the mark to the space, or from the space to the mark. Here, it should be noted that coherent systems use a optical carrier wave stable enough for these coherent modulation signals to have the transform-limited spectra.

Coherent systems can employ various combinations of modulation and demodulation methods. Table I summarizes typical combinations including electrical signal processing in a heterodyne receiver. An optical ASK signal is converted into an IF signal by heterodyning, and is then demodulated to a baseband output by conventional envelope detection because the envelope of an ASK signal corresponds to the original input waveform. Another electrical processing method applicable to the ASK signal is synchronous (coherent) detection, which first multiplies the IF signal by a local electrical wave of the same frequency and phase as the IF carrier wave, and then extracts the low-frequency component carrying the original data. An FSK method combines only with optical heterodyne detection, where a frequency discriminator retrieves the original information through its FM-to-AM conversion process. For a binary FSK signal, a so-called dual-filter detection scheme can also be employed, where the IF signal is first divided by a pair of bandpass filters into two frequency components around f_1 and f_0, which are then envelope

TABLE I

Typical Coherent Modulation–Demodulation Schemes

Modulation	Demodulation	
	Optical Detection	Electrical Detection
ASK	heterodyne	envelope synchronous
	homodyne	—
FSK	heterodyne	frequency discriminator dual filter
PSK	heterodyne	differential synchronous
	homodyne	—

detected and finally compared to decide which state has been transmitted. Synchronous detection can be applied to each path in place of envelope detection, but is rarely used in practice because of the complexity of simultaneous synchronization of two IF reference waves. A differential detection scheme that compares a signal phase to that of the foregoing time slot can cope with a PSK signal, especially a DPSK signal. A synchronous detection scheme can also demodulate a heterodyne-detected PSK signal, since a reference wave is locked to the IF carrier wave with a $\pi/2$ phase difference. Optical homodyne detection responds to optical ASK and PSK signals in a similar manner to electrical synchronous detection and directly obtains the demodulated output.

The basic configuration of a coherent optical fiber transmission system is shown in Fig. 3. The transmitter consists of a laser oscillator, a modulator, and in some cases, an optical post amplifier. The laser oscillator delivers to the system a coherent optical carrier wave of stable frequency and phase. The modulator projects the electrical input data onto the carrier wave and produces an optical ASK, FSK, or PSK signal. Direct modulation of the laser oscillator, if it is practical, may be used in place of the external modulator. The post amplifier is installed optionally to compensate for power loss in the modulator and to boost the optical signal level to the maximum input power for the optical fiber employed. Optical isolators, though not illustrated in

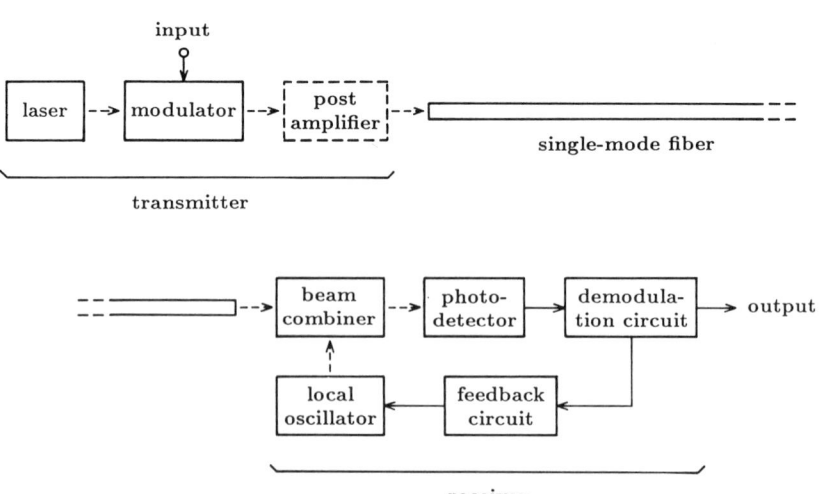

FIG. 3. Basic system configuration for coherent optical fiber transmission.

Fig. 3, are connected after the basic optical devices to secure stable operation against interference caused by reflection.

A heterodyne receiver, the lightwave analogue of a superheterodyne radio set, comprises several basic components: a local oscillator, a beam combiner, a photodetector, an electrical demodulation circuit, and so on. The local oscillator is a laser that provides an optical reference wave of large power and stable phase to improve receiver sensitivity. A beam splitter is a simple and primitive example of a beam combiner. In practice, a single-mode fiber coupler is usually employed instead. The photodetector performs square-law detection to achieve optical mixing. Thus a conventional photodetector, that is, an APD or $p-i-n$ photodiode, can be used. However, it should be noted that optical heterodyne or homodyne detection uses mixing gain rather than avalanche gain to achieve receiver sensitivity improvement. The electrical demodulation circuit, which may have various configurations according to the IF signal processing scheme, can utilize electrical circuitry developed for microwave or millimeter-wave systems. A frequency feedback loop is constructed for IF stabilization. A homodyne receiver has basically the same configuration as the heterodyne receiver but has no electrical demodulation circuit. An optical phase-locked loop (PLL) is installed for the local oscillator to track the carrier wave.

The optical mixing process is sensitive to mode field patterns and polarization states of both incoming signal and local oscillator wave. Thus a single-mode fiber that maintains the polarization state of the propagating signal is a candidate for a transmission medium in a coherent system. A conventional single-mode fiber, in conjunction with a polarization controller or a polarization diversity receiver, can be used instead of the polarization-maintaining fiber.

III. Advantages of Optical Heterodyne or Homodyne Detection

Optical heterodyne or homodyne detection has two principal advantages over direct detection. One is an improvement in receiver sensitivity of 10–20 dB, due to the fact that the optical mixing process is less noisy than the avalanche multiplication commonly used in conventional systems (Yamamoto, 1980; Okoshi *et al.*, 1981). The other is an improvement in frequency selectivity, made possible by the frequency-filtering function of IF or baseband electrical circuits following the frequency conversion process of optical mixing. This section details these two advantages, comparing coherent detection with direct detection.

A. Receiver Sensitivity

1. Signal-to-Noise Ratio in Conventional and Coherent Receivers

When the APD in a conventional receiver is hit by an incoming optical signal, the detector and its associated amplifier generate an output current

$$i(t) = DP_s M + i_{n,\text{APD}}(t), \qquad (1)$$

where the incoming signal field is

$$E_s = \sqrt{2P_s} \cos(2\pi f_s + \phi_s), \qquad (2)$$

and

$$D = e\eta/hf_s,$$

P_s, f_s, and ϕ_s are the optical power, frequency, and phase of the incoming signal, respectively; e is the electron charge; h is Planck's constant; M is the APD multiplication factor; η is the detector quantum efficiency; and $i_{n,\text{APD}}$ is noise current.

The information originally carried by the optical intensity (or power, P_s) is now conveyed by the photocurrent. Under the initial condition of $M = 1$, the photocurrent has an average of DP_s and fluctuates around this average because photoelectrons are generated at random intervals. This results in shot noise that is dominated by Poisson statistics. If the incoming signal photon number in a one-bit time interval is large enough (>100), this distribution is approximately Gaussian. Another shot noise due to dark current is added to the photocurrent. Avalanche multiplication increases the photocurrent by M times, so that the electrical power of a signal component increases in proportion to M^2. The noise power, however, increases by a factor greater than M^2 because the avalanche process generates an extra noise. The resulting noise current, $i_{n,\text{APD}}(t)$, obeying a Gaussian distribution, has a power spectral density

$$N_{\text{APD}} = 2e(DP_s + I_d)M^{2+x} + 4kTF_t/R, \qquad (3)$$

where the thermal noise of the amplifier, the last term, is included. I_d is the dark current, x is the excess noise exponent denoting the extra noise, k is Boltzmann's constant, T is the absolute temperature, F_t is the amplifier noise figure, and R is detector load resistance. The signal-to-noise ratio (SNR) achieved by an APD receiver is

$$(\text{SNR})_{\text{APD}} = \frac{(DP_s M)^2}{N_{\text{APD}} B_0}, \qquad (4)$$

where B_0 is the receiver bandwidth (baseband filter bandwidth). More precisely, SNR should be referred to as carrier-to-noise ratio (CNR) because it is given under no-modulation conditions—that is, only for the mark state.

Figure 4a shows numerical examples of signal and noise power, that is, the numerator and denominator, respectively, of Eq. (4), as a function of the avalanche multiplication factor, M. While signal power is much smaller than noise level under the initial condition of $M = 1$, avalanche multiplication increases SNR enough to retrieve the original information efficiently even if excess noise and dark current exist. The SNR is a maximum at an optimum multiplication factor, which is about 30 for this example. Direct detection, however, cannot achieve the ultimate SNR, $\eta P_s/2hf_s B_0$, which is dominated only by the shot noise arising from the photoelectric conversion process and so is derived from Eq. (1) for the condition where $M = 1$, $I_2 = 0$, and thermal noise $= 0$.

The degradation of SNR from the ultimate limit is more severe in the 1.2–1.6 μm wavelength region than in that below 1 μm. This is because the long-wavelength receiver must use Ge APDs with $x = 0.8$–1.0 and $I_d \sim 1$ μA, or InGaAs APDs with $x = 0.5$–0.7 and $I_d \sim 1$ nA, while the short-wavelength scheme can use Si APDs with $x = 0.2$–0.4 and $I_d \sim 1$ nA.

Heterodyne receivers use photomixing gain rather than avalanche gain to improve the SNR. The incoming signal is converted by a square-law detector into a photocurrent after combining with the local oscillator wave:

$$E_{LO} = \sqrt{2P_{LO}} \cos(2\pi f_{LO} t + \phi_{LO}), \tag{5}$$

where P_{LO}, f_{LO}, and ϕ_{LO} are the power, frequency, and phase of the local wave, respectively. The resulting current is

$$i(t) = D[P_s + P_{LO} + 2\sqrt{P_s P_{LO}} \cos\{2\pi(f_s - f_{LO})t + (\phi_s - \phi_{LO})\}] + i_n(t). \tag{6}$$

The beat component between the signal and local waves, the third term in the brackets of Eq. (6), carries the original information. Here, it should be noted that coherent communications can utilize any of field amplitude ($\sqrt{2P_s}$), frequency, or phase of the optical carrier wave to transmit information, while direct detection can respond only to the information conveyed by the carrier power (P_s). The noise current, $i_n(t)$, in this case, has a power spectral density of

$$N_{coher} = 2e\{D(P_s + P_{LO}) + I_d\} + \frac{4kTF_t}{R}, \tag{7}$$

and a Gaussian distribution. Synchronous detection—which is carried out by multiplying the IF current by the electrical reference wave, $\cos(2\pi f_{IF} t + \phi_{IF})$ ($f_{IF} = f_s - f_{LO}$, ϕ_{IF} is the phase), and then extracting the

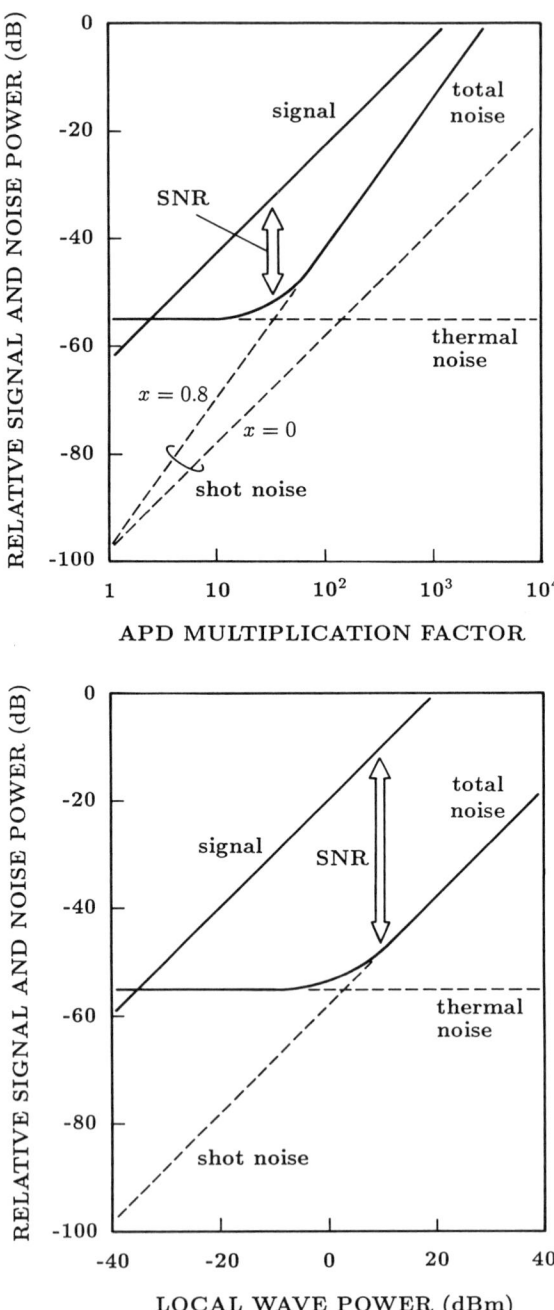

FIG. 4. Signal and noise power levels in receivers. (a) Direct detection. (b) Heterodyne detection. Numerical parameters; signal power $P_s = -40$ dBm, wavelength $\lambda = 1.55$ μm, detector quantum efficiency $\eta = 0.8$, APD excess noise exponent $x = 0.8$, detector dark current $I_d = 1$ nA, receiver bandwidth $B_0 = 100$ MHz, and load resistance $R = 50$ Ω.

low-frequency component—yields the SNR:

$$(\text{SNR})_{\text{hetero}} = \frac{2D^2 P_s P_{\text{LO}}}{N_{\text{coher}} B_0}. \tag{8}$$

Numerical examples of the signal and noise levels are shown in Fig. 4b as a function of the local oscillator power, P_{LO}. Because both signal and noise power increase linearly with local wave power, the heterodyne receiver can achieve its ultimate SNR, $\eta P_s / h f_s B_0$, provided that the local oscillator supplies sufficient power, about 10 dBm for this example. This is one of the most important features of heterodyne detection. A heterodyne receiver can achieve an SNR larger than that of the direct-detection receiver under the same incoming signal power level. Except for the electrical demodulation process, a similar argument is applied to homodyne receivers, directly leading to an SNR of

$$(\text{SNR})_{\text{homo}} = \frac{4D^2 P_s P_{\text{LO}}}{N_{\text{coher}} B_0}, \tag{9}$$

which is 3 dB larger than that of the heterodyne receiver.

2. Minimum Detectable Power

This SNR determines the bit error rate (BER), which is the probability of an error occurring in the decision of a received bit. Figure 5 shows probability density functions of demodulated output signals in the direct detection and heterodyne synchronous detection methods. All the functions consist of two Gaussian distributions, each of which indicates the probability of the output amplitude for either the mark or the space input signal. In direct detection schemes, the average amplitude is $A_1 = DP_s M$ for the mark slot, and $A_0 = 0$ for the space slot, where no input signal is assumed for the space. The distribution variance is different for mark and space slots, as shown in Fig. 5a, since the noise power depends on the input signal power (see Eq. (3)). On the assumption that the mark and the space occur with equal probability and that the threshold of decision, A_T, is adjusted to the optimum point to achieve the smallest BER, we obtained (Personick, 1973)

$$\begin{aligned}(\text{BER})_{\text{IM-DD}} &= \frac{1}{2}\left[\frac{1}{\sqrt{2\pi}\,\sigma_1}\int_{-\infty}^{A_T}\exp\left(-\frac{(A_1-a)^2}{2\sigma_1^2}\right)da \right.\\ &\left.+ \frac{1}{\sqrt{2\pi}\,\sigma_0}\int_{A_T}^{\infty}\exp\left(-\frac{a^2}{2\sigma_0^2}\right)da\right]\\ &= \frac{1}{2}\,\text{erfc}\left(\left\{\frac{1}{8}(\text{SNR})_{\text{DD}}\right\}^{1/2}\right), \end{aligned} \tag{10}$$

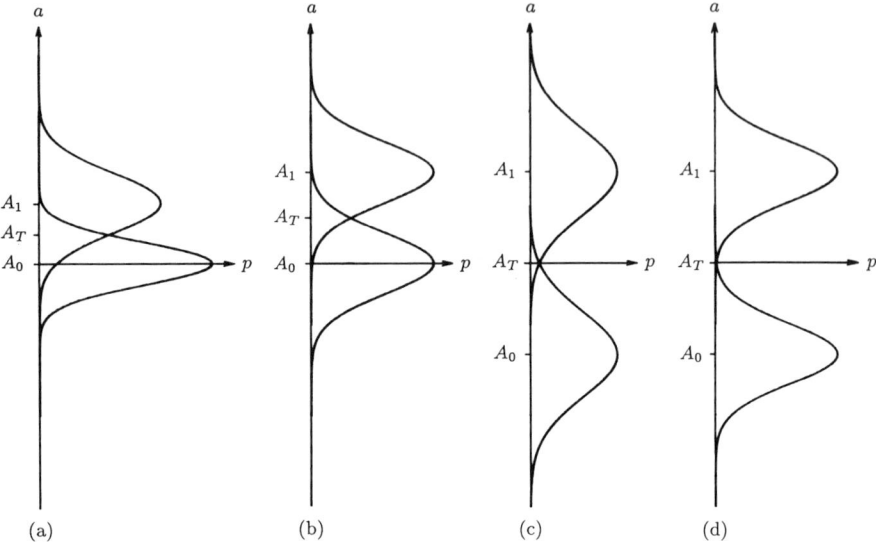

FIG. 5. Probability density functions: (a) IM direct detection; (b) ASK heterodyne synchronous detection; (c) FSK heterodyne synchronous detection; (d) PSK heterodyne synchronous detection.

where

$$\sigma_1^2 = N_{APD} B_0,$$

$$\sigma_2^2 = \left(2eI_d M^{2+x} + \frac{4kTF_t}{R}\right) B_0, \quad (11)$$

$$(SNR)_{DD} = \frac{(DP_s M)^2}{((\sigma_1 + \sigma_2)/2)^2},$$

and erfc(z) is the complementary error function;

$$\text{erfc}(z) = \frac{2}{\sqrt{\pi}} \int_z^\infty \exp(-t^2)\, dt. \quad (12)$$

An $(SNR)_{DD}$ of 21.6 dB results in a 10^{-9} error rate.

An ASK heterodyne synchronous detection method has a probability density function as shown in Fig. 5b and yields

$$A_1 = \sqrt{2DP_s P_{LO}},$$
$$A_0 = 0,$$
$$\sigma_1^2 = \sigma_0^2 = 2N_{coher} B_0, \quad (13)$$

$$A_T = \frac{A_1}{2}, \quad \text{and}$$

$$(BER)_{\text{ASK-hetero}} = \tfrac{1}{2}\text{erfc}(\{\tfrac{1}{8}(SNR)_{\text{hetero}}\}^{1/2}),$$

where the local oscillator power is assumed to be much larger than the input signal power.

For an FSK signal, a heterodyne synchronous detection receiver equips two synchronous-detection circuits, each of which responds to the mark (f_1) or space (f_0) signal. When the f_1 signal is received, one of the twin paths generates an amplitude with an average of $\sqrt{2DP_sP_{LO}}$ and a variance of $N_{\text{coher}}B_0$, and the other yields another amplitude with zero average and $N_{\text{coher}}B_0$ variance. Because the final demodulated output of the receiver is the difference between the two amplitudes, the probability density function for the mark has $A_1 = \sqrt{2DP_sP_{LO}}$ and $\sigma_1^2 = 2N_{\text{coher}}B_0$, as shown in Fig. 5c. Similarly, we get $A_0 = -\sqrt{2DP_sP_{LO}} = -A_1$ and $\sigma_0^2 = 2N_{\text{coher}}B_0 = \sigma_1^2$ for the space. Not only the average amplitude difference, but also the variance, is double that of ASK heterodyne synchronous detection. The threshold level of $A_T = 0$ gives

$$(BER)_{\text{FSK-hetero}} = \tfrac{1}{2}\text{erfc}(\{\tfrac{1}{4}(SNR)_{\text{hetero}}\}^{1/2}), \quad (14)$$

PSK heterodyne synchronous detection has a probability density of $A_1 = \sqrt{2DP_sP_{LO}}$, $A_0 = -A_1$, and $\sigma_1^2 = \sigma_0^2 = N_{\text{coher}}B_0$, and thus yields

$$(BER)_{\text{FSK-hetero}} = \tfrac{1}{2}\text{erfc}(\{\tfrac{1}{2}(SNR)_{\text{hetero}}\}^{1/2}), \quad (15)$$

$(SNR)_{\text{hetero}}$ values of 21.6 dB, 18.6 dB, and 15.6 dB for ASK, FSK, and PSK, respectively, achieve a 10^{-9} error rate. Similarly straightforward analysis can be applied to ASK and PSK homodyne detection. Nonsynchronous detection in a heterodyne receiver, such as envelope and differential detection, complicates the calculation of BER since it does not have a Gaussian probability density function. Without further discussion of the calculation process, the BER performance of various modulation–demodulation schemes is summarized in Fig. 6 (Yamamoto, 1980; Okoshi et al., 1981).

Figure 6 also compares the receiver sensitivity, which is defined as the minimum detectable input power for a receiver to achieve a prescribed BER, say 10^{-9}, and is measured at its peak power for IM and ASK signals. Coherent detection can reduce the required input signal level by 10–20 dB compared with direct detection. This will be confirmed by the numerical calculations described later. In coherent detection schemes, the receiver sensitivity of homodyne detection is 3 dB better than that of heterodyne synchronous detection, as predicted by Eq. (9). Heterodyne nonsynchronous detection schemes have a slightly poor (~ 0.5 dB) sensitivity compared with that of synchronous schemes. With respect to modulation schemes, as indicated by

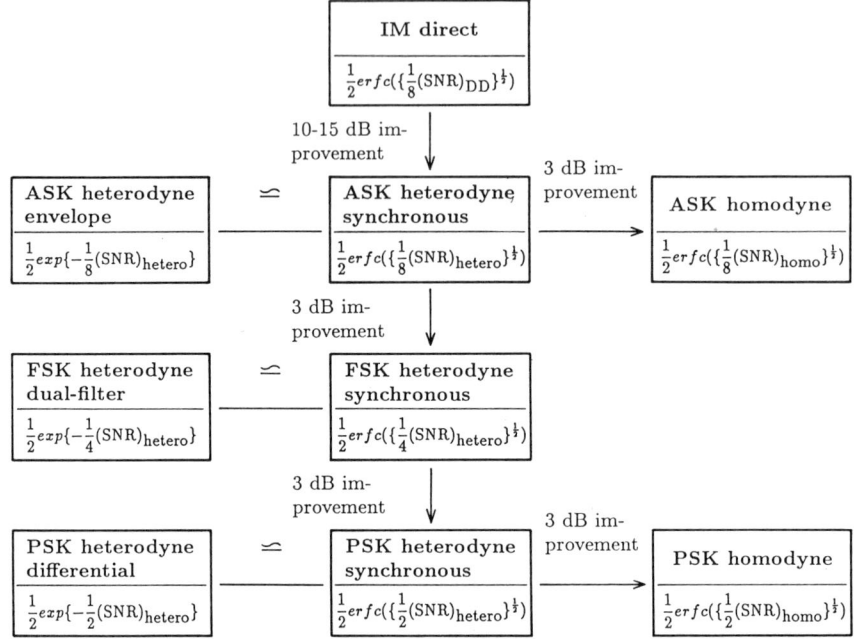

FIG. 6. Error rate performance of various modulation–demodulation schemes, and relative receiver sensitivity.

Eqs. (13)–(15), FSK and PSK schemes can achieve a receiver sensitivity of 3 dB and 6 dB, respectively, better than that of the ASK scheme. Consequently, the most sensitive coherent modulation–demodulation method is PSK homodyne detection.

Minimum detectable power levels of conventional and coherent systems required to achieve a 10^{-9} error rate are shown in Fig. 7 as a function of the data rate, B, where the effect of the capacitance accompanying a photodiode and a load resistance is taken into account (Yamamoto, 1980; Yamamoto and Kimura, 1981). Numerical parameters used for BER calculations are as follows: optical carrier wavelength $\lambda = 1.55$ μm, local oscillator power $P_{LO} = 10$ dBm, photodetector quantum efficiency $\eta = 0.8$, dark current $I_d = 1$ nA, excess noise exponent $x = 0.8$, total capacitance $C = 0.5$ pF, and detector load resistance $R = 1/(BC)$. The equivalent noise input current and voltage of the electronic amplifier are $\sqrt{\langle i_a \rangle^2} = 10$ pA and $\sqrt{\langle e_a \rangle^2} = 0.5$ nV, respectively, and they replace the noise figure. The input signal waveform is assumed to be a nonreturn-to-zero (NRZ) rectangular pulse, and the equalized baseband pulse waveform is a full cosine roll-off. The avalanche gain is optimum for direct detection and unity for coherent detection. The calculated result confirms the previous discussion of receiver sensitivity improvements.

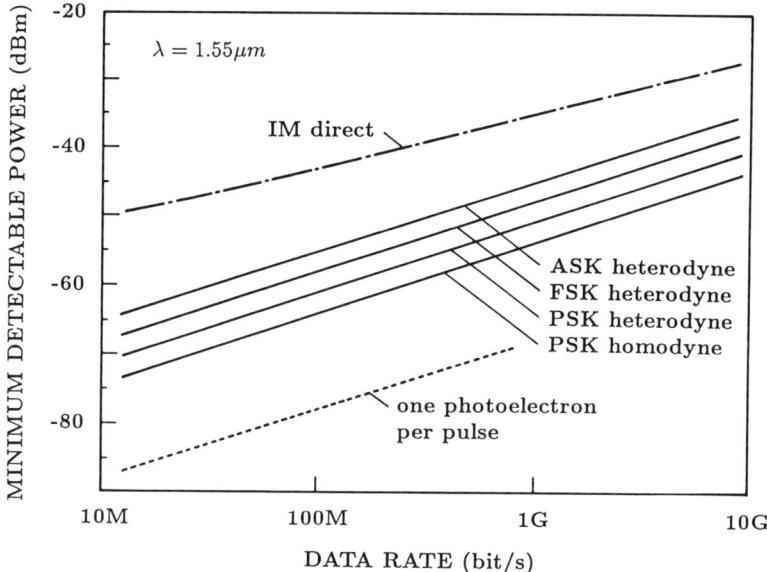

FIG. 7. Minimum detectable power to achieve a 10^{-9} error rate as a function of data rate (Yamamoto, 1980; Yamamoto and Kimura, 1981). (©1980 IEEE. and ©1981 IEEE.)

The advantage of receiver sensitivity improvement becomes more evident in the 2–10 μm wavelength region (Yamamoto, 1980), where the possibility of ultralow-loss fibers has been predicted (Goodman, 1978). Because photodetectors are fabricated from narrow-bandgap semiconductors in this wavelength region, dark current due to diffusion current becomes a dominant noise source. While direct detection performance deteriorates, coherent optical detection can still achieve a shot-noise-limited operation.

3. Ultimate Receiver Sensitivity

The theoretical limit of receiver sensitivity is worth discussion, to make clear the potential of various modulation–demodulation schemes. Here it is assumed that the photodetector quantum efficiency is unity ($\eta = 1$), and that the receiver has the Nyquist bandwidth ($B_0 = B/2$; B is data rate).

For direct detection, let us consider an ideal receiver having no dark current ($I_d = 0$), no excess noise ($x = 0$), no thermal noise, and unit quantum efficiency. The ideal receiver assigns no receiving signal photon to the space state, setting the threshold for decision just above zero. This means that error can occur in the receiver only when the incoming signal is in the mark state. For such a receiver, Poisson statistics must be used instead of the Gaussian approximation to calculate the error rate, since even a small number (<100) of

TABLE II

THEORETICAL LIMIT OF RECEIVER SENSITIVITY FOR A BER OF 10^{-9}

Modulation	Demodulation	Photon Number per Pulse	Signal Power (1.55 μm, 1 Gb/s)
IM	direct	21	−55.7 dBm
ASK	heterodyne	72	−50.4 dBm
	homodyne	36	−53.4 dBm
FSK	heterodyne	36	−53.4 dBm
PSK	heterodyne	18	−56.4 dBm
	homodyne	9	−59.4 dBm

signal photons existing in one mark slot can achieve a BER less than 10^{-9}. Consequently the BER of the ideal direct-detection receiver is given by $(\frac{1}{2}) \exp(-N_s)$, where N_s denotes the photon number per pulse ($N_s = P_s/hf_s B$). This proves that the ultimate receiver sensitivity for a 10^{-9} error rate is 21 photons per pulse.

For coherent schemes, ultimate receiver sensitivity is calculated in a straightforward way: Assuming that the local oscillator wave has sufficient power, we obtain ultimate SNRs of $2P_s/hf_s B$ and $4P_s/hf_s B$ for heterodyne and homodyne detection from Eqs. (8) and (9), respectively, and estimate the minimum detectable power for a particular BER using the formula given in Fig. 6. The theoretical photon number per pulse to achieve a 10^{-9} error rate is shown in Table II. The corresponding signal power at a 1.55-μm wavelength for a 1 Gb/s data rate is also shown. Ideal PSK homodyne detection requires nine signal photons per pulse for a 10^{-9} error rate. In an ideal receiver, there is little difference in sensitivity between direct and coherent detection.

A multivalued system can further improve the receiving performance, as is well known from communication theory. Four-valued, 8-valued, and 16-valued FSK schemes have receiver sensitivities 3 dB, 5 dB, and 6 dB better than that of the binary scheme, respectively. Coded 32-level FSK coherent detection is theoretically shown to achieve a 10^{-9} error rate with about 1.4 photons per bit by using error-correction techniques (Chan et al., 1983). The theoretical channel capacity of optical communication systems derived by quantum mechanical treatment corresponds to a receiving signal level of 0.02 photons per bit (Pierce, 1978).

B. Frequency Selectivity

Let us consider the detection of a multiplexed signal:

$$E_s = \sum_k \sqrt{2P_{s,k}} \cos(2\pi f_{s,k} + \phi_{s,k}), \qquad (16)$$

where $P_{s,k}$, $f_{s,k}$, and $\phi_{s,k}$ ($k = 0, \pm 1, \pm 2, \ldots, \pm m$) are the power, frequency, and phase of each signal, respectively. This incoming signal is shown in Fig. 8a. It is assumed that each signal is modulated independently of other channels at a certain data rate, say 1 Gb/s, and that adjacent channels are separated from each other by a sufficient frequency difference, say 10 GHz. An APD direct-detection receiver responding to the incoming signal generates a current

$$i(t) = DM\left[\sum_k P_{s,k} + \sum_{k>l} 2\sqrt{P_{s,k}P_{s,l}}\cos\{2\pi(f_{s,k} - f_{s,l})t + (\phi_{s,k} - \phi_{s,l})\}\right] + i_n(t), \tag{17}$$

where $i_n(t)$ is the current noise, composed of shot noise and thermal noise. The second term in the brackets denotes beat signals between two channels, appearing outside the signal band, as shown in Fig. 8b. The first term denotes the current carrying all the incoming information. Any receiver in a channel, however, cannot select its own information to be extracted because unwanted information coming from other channels is completely superposed on the object of detection. To detect the desired signal, an optical channel filter or demultiplexer must be installed before the direct-detection receiver.

In response to the multiplexed incoming signal given by Eq. (16), a heterodyne receiver, which provides a sufficiently large local wave given by Eq. (5), generates a current:

$$i(t) = D\left[\sum_k P_{s,k} + P_{LO} + \sum_k 2\sqrt{P_{s,k}P_{LO}}\cos\{2\pi(f_{s,k} - f_{LO}) + (\phi_{s,k} - \phi_{LO})\}\right.$$
$$\left. + \sum_{k>l} 2\sqrt{P_{s,k}P_{s,l}}\cos\{2\pi(f_{s,k} - f_{s,l}) + (\phi_{s,k} - \phi_{s,l})\}\right] + i_n(t). \tag{18}$$

This is shown in Fig. 8c, where f_{LO} is assumed to be $(2f_{s,0} + f_{s,-1})/3$. Beat components between the incoming signal and the local wave, the third term in Eq. (18), convey the information to be demodulated. Because any of them is completely frequency-separated from other components appearing in Eq. (18) except for $i_n(t)$, and IF bandpass filter can distinguish the desired signal satisfactorily. This presents a striking contrast to direct detection, which cannot extract the desired signal. Thus optical heterodyne detection basically does not require an optical channel filter or demultiplexer. In practice, however, these optical devices are installed in advance of heterodyne receivers to suppress branching loss of the incoming signal rather than to improve frequency selectivity. A similar discussion is applied to a homodyne receiver, in which a local oscillator is phase locked to the carrier wave of the signal to be demodulated.

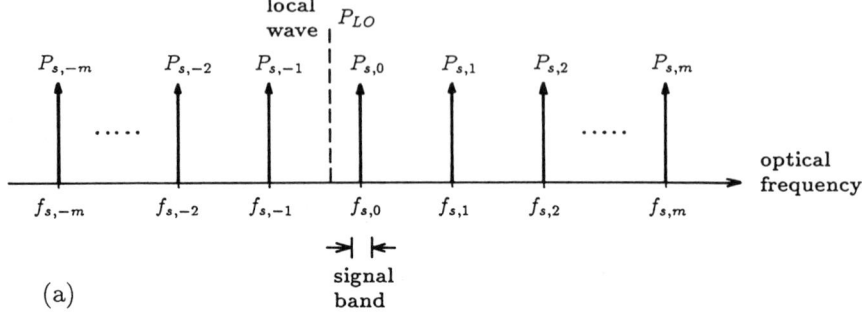

(a)

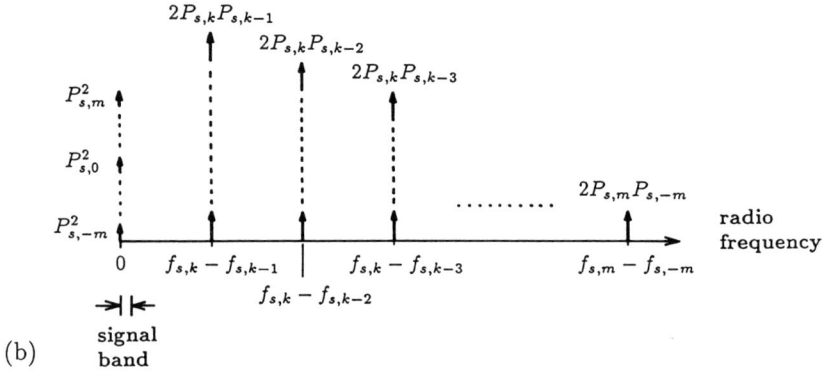

(b)

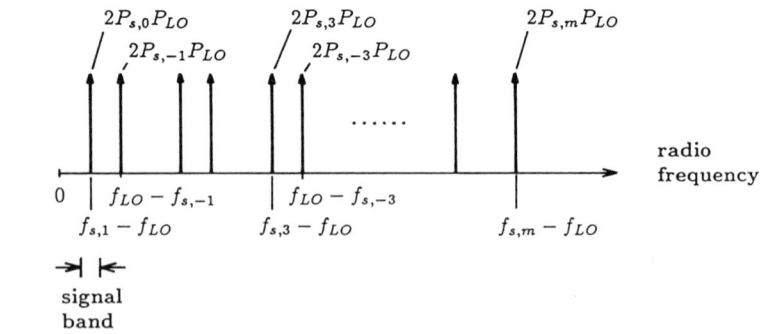

(c)

FIG. 8. Response of both direct and heterodyne detection receivers to an optical frequency-multiplexed signal. (a) Spectrum of an optical incoming signal; (b) spectrum of detector output in a direct detection receiver; (c) spectrum of detector output in a heterodyne detection receiver. A local oscillator wave is indicated in (a).

COHERENT OPTICAL FIBER TRANSMISSION SYSTEMS 223

Now the following conditions are assumed:

(1) only one component in Eq. (16), $\sqrt{2P_{s,0}}\cos(2\pi f_{s,0} + \phi_{s,0})$, denotes the signal carrying information;
(2) other components represent the accompanying noise, so that any two of them have no correlation to each other;
(3) these noise components are smaller than the signal ($P_{s,k} < P_{s,0}, k \neq 0$), but much larger than the shot noise caused by all the incoming waves;
(4) the frequency difference, $f_{s,k} - f_{s,k-1}$, is much smaller than the signal bandwidth, so that some noise components ($P_{s,k}, k = \pm 1, \pm 2, \ldots, \pm k_s$) exist inside the signal bank.

This incoming wave is illustrated in Fig. 9a. The response of an APD receiver to this wave is still given by Eq. (17). $P_{s,0}$ in the first term is the signal to be demodulated. The second term includes two dominant noise sources as shown in Fig. 9b: the beat note between the signal and noise component, and that between two noise components. Dominated by these two types of beat noise, the receiver output yields

$$(\text{SNR})_{\text{APD}} = P_{s,0}^2 \bigg/ \bigg(\sum_{1 \leq |k|} P_{s,k}^2 + \sum_{1 \leq |k| \leq k_s} 2P_{s,0}P_{s,f}$$
$$+ \sum_{1 \leq |k-l| \leq k_s} 2P_{s,k}P_{s,l} + \text{thermal noise} \bigg). \quad (19)$$

Even an incoming noise component existing outside the signal band contributes to the second term of the denominator by making a beat note with other noise components and then falling inside the signal band. Thus the SNR decreases as the noise bandwidth, $f_{s,m} - f_{s,-m}$, increases. Provided that the beat noise between two noise components, as well as the thermal noise, is much smaller than the beat noise between the signal and the noise components, a conventional APD receiver achieves the ultimate SNR of

$$(\text{SNR})_{\text{APD,ultimate}} = P_{s,0} \bigg/ \sum_{1 \leq |k| \leq k_s} 2P_{s,k}. \quad (20)$$

For the noisy incoming wave, a heterodyne receiver generates the current given by Eq. (18). Beat notes between the local wave and noise component, between the signal and noise component, and between two noise components, are the main noise sources. A sufficiency large local wave enables the first beat note to be much larger than the second and third beat notes as well as the thermal noise, even if the noise bandwidth is much larger than the signal bandwidth. This situation is illustrated in Fig. 9c. Besides the noise existing within the optical signal band, the noise appearing in the optical image band, which is symmetrical with the signal band with respect to the local wave

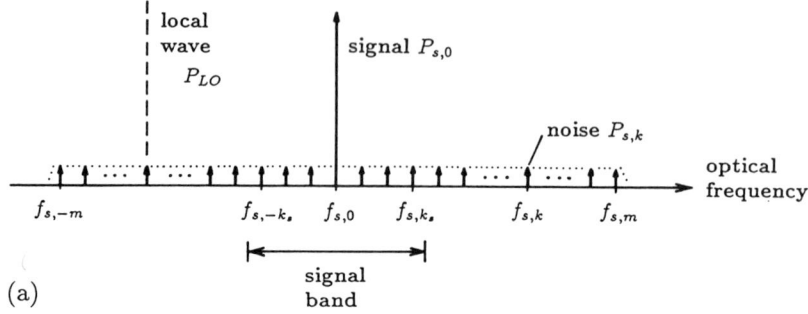

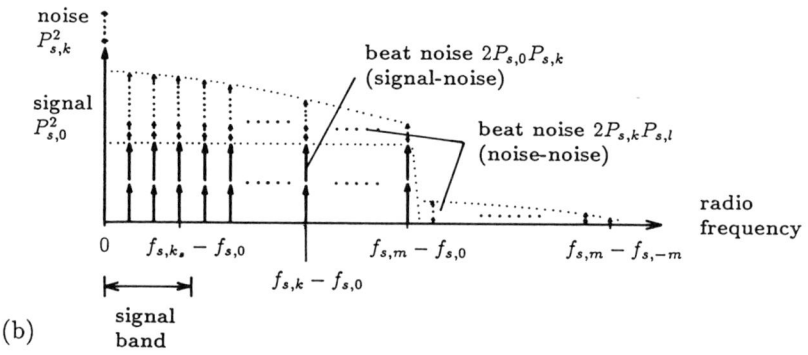

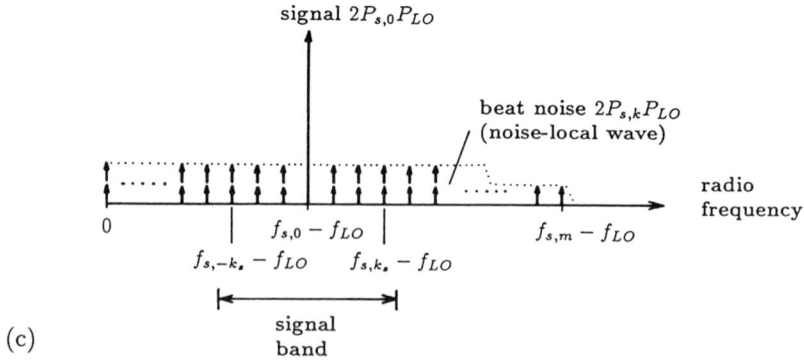

FIG. 9. Response of both direct and heterodyne detection receivers to an optical incoming signal accompanied by noise. (a) Spectrum of an optical incoming signal; (b) spectrum of detector output in a direct detection receiver; (c) spectrum of detector output in a heterodyne detection receiver. A local oscillator wave is indicated in (a).

frequency, falls into the IF signal band through optical heterodyne detection. Synchronous detection of the IF filter output yields

$$(\text{SNR})_{\text{hetero}} = 2P_{s,0} \bigg/ \bigg(\sum_{1 \leq |k| \leq k_s} P_{s,k} + \sum_{k_i} P_{s,ki} \bigg), \qquad (21)$$

where k_i denotes the number of noise components existing in the image band. The obtained SNR is twice the ultimate SNR as given by Eq. (20) for direct detection, provided that all noise components have the same power level. The SNR of homodyne detection is obtained similarly and is 3 dB better than that of heterodyne detection because homodyne detection has no image band. Even in the sense that heterodyne or homodyne detection can suppress SNR degradation, which is caused by the incoming noise components existing outside the signal band, coherent detection schemes offer improved frequency selectivity compared with direct detection schemes.

An optical filter fitted in advance of an APD can improve the frequency selectivity of the direct detection receiver. Even so, coherent detection is still superior to direct detection in frequency selectivity because the frequency filtering function of an optical filter is not so efficient as that of an electrical filter.

IV. System Applications

This section describes three optical fiber communication systems in which coherent lightwave techniques fully display their ability. The first is a long-distance point-to-point transmission system for high data-rate information, which exploits the advantage of improved receiver sensitivity. The second is an optical amplifier repeater system, intended to further improve the performance of the first system toward extremely long regenerative repeater spacing. Improved frequency selectivity, another principal advantage of optical heterodyne or homodyne detection, makes the second system feasible. The third is a multi-terminal local area system using frequency division multiplexing (FDM) techniques. This system is also based on improved frequency selectivity.

A. Long-Distance High-Speed Transmission Systems

The improved receiver sensitivity of coherent detection over direct detection has the advantage that, for a certain transmission data rate, optical fiber transmission systems increase their regenerative repeater spacing. On the other hand, for a fixed receiving signal level, a large transmission data rate is possible. Long-haul high-speed transmission systems, either terrestrial or undersea, are the obvious applications of optical coherent modulation–demodulation techniques. These applications are further promoted by four improvements also resulting from coherent techniques, especially from angle modulation such as FSK and PSK, as follows.

1. Improvement of Power Handling Level in Single-Mode Fibers Using Angle Modulation Schemes

The maximum input power for single-mode fibers is limited by nonlinear interaction occurring in their small-core areas, such as stimulated Brillouin scattering (SBS) or stimulated Raman scattering (SRS) (Stolen, 1979). The critical power, at which the first Stokes power builds up and 3-dB pump-power depletion occurs, is

$$P_C = \frac{21 A \alpha}{G_B(1 - e^{-\alpha L})} \tag{22}$$

for SBS and

$$P_C = \frac{16 A \alpha}{G_R(1 - e^{-\alpha L})} \tag{23}$$

for SRS, where A is the effective core area, which depends on the core diameter and fiber v number; α is the fiber loss; L is the fiber length; G_B is the Brillouin gain; and G_R is the Raman gain. The critical power determined by SBS is only about 2 mW for single-frequency 1.55-μm input light. SRS offers a critical power of 1–2 W. Numerical parameters used for the calculation are: fiber core diameter $d = 10$ μm, $\alpha = 0.2$ dB/km, $L \gg 1/\alpha$, fiber number $v = 2.4$, $G_B = 4.5 \times 10^{-9}$ cm/W, and $G_R = 6.1 \times 10^{-12}$ cm/W.

Critical input power determined by SBS is increased by modulating the single-frequency carrier wave in amplitude or phase (Cotter, 1983), because a Stokes wave, traveling in the opposite direction to the pump wave (carrier wave), is stimulated by the time-dependent pump wave. This shows a striking contrast to SRS, where a Stokes wave travels in the same direction as the pump wave.

Both conventional intensity modulation and ASK for random binary data stream have an effective Brillouin gain of one-fourth of the gain for continuous wave, and thus have a power-handling level of about 8 mW. This shows the fact that the coherent optical carrier wave is not fully suppressed but still maintains at least one-fourth of its initial peak power for nonreturn-to-zero (NRZ) modulation.

On the other hand, the SBS difficulty is drastically alleviated by using constant-amplitude angle modulation signals with a completely suppressed carrier component and a randomly modulated phase (Cotter, 1983). A PSK scheme with a π phase shift completely suppresses the SBS effect. An FSK operation has only a small net Brillouin gain, provided that the frequency shift is sufficiently large. Moreover, these signals can overcome the self-phase modulation effect, which limits fiber input power to only a few hundred milliwatts (Stolen, 1979). Consequently, angle modulation schemes will have a critical fiber input power of a few watts as determined by SRS.

The SBS limit is not evident in current intensity modulation systems because direct intensity modulation of semiconductor lasers causes enormous frequency chirping. Improvement of single-frequency operation under direct modulation, being achieved by distributed feedback (DFB) or distributed Bragg reflector (DBR) lasers (Suematsu et al., 1983), however, must face this limitation.

2. Improvement of Launched Power Using High-Power Post-amplifiers

To boost the transmitted signal level up to the critical fiber input power, a post-amplifier may be installed in the transmitter. Semiconductor-laser linear amplifiers (Mukai et al., 1985), such as the Fabry–Perot type (i.e., resonant type) and traveling-wave type amplifiers, amplify optical angle modulation signals as well as amplitude or intensity modulation signals. These amplifiers have a maximum unsaturation output power of 0–10 dBm at a wavelength of 1.55 μm, and it seems very difficult to increase the output power to a few watts of the fiber input power determined by SRS. On the other hand, injection-locked oscillators (Kobayashi and Kimura, 1981), which amplify only a constant-envelope angle modulation signal, are potentially good post-amplifiers, since they emit an output power larger than that of linear amplifiers. Essentially, an output power of 10–20 dBm is available using the injection-locked amplifier composed of a conventional semiconductor laser. Phase-locked array lasers achieve a high output power above a few watts under the continuous wave (cw) operation condition (Scifres et al., 1983), and thus they can be used as injected-locked oscillators. Since fiber Raman amplifiers (Stolen and Ippen, 1973) are expected to have relatively high saturation power, they may also be used as post-amplifiers, provided that reliable high-power pump lasers are available for 1.55-μm signal amplification. Recently developed erbium-doped fiber amplifiers (Urquhart, 1988), which offer an output power of about 20 dBm, can be used.

3. Dissolution of Bandwidth Limitation in Optical Fibers

Chromatic dispersion of single-mode fibers causes signal waveform distortion, resulting in both modulation-depth degradation and intersymbol interference, the latter of which ends in the former in successive time intervals. As the data rate is increased, the dispersive effect becomes more evident, causing an increased BER. To overcome this additional BER degradation, received power must be increased. The amount of power added is referred to as the power penalty. Under an allowable power penalty, fiber chromatic dispersion finally imposes a bandwidth limitation on optical fiber transmission systems.

In conventional systems, an optical signal generated by direct intensity modulation of semiconductor lasers is accompanied by an unintended spectral spread because of the carrier modulation effect. In this case a 1-dB power penalty gives the dispersion-limited bit rate–distance product:

$$BL \leq \frac{1}{4\delta\sigma\lambda}, \qquad (24)$$

where B is the data rate, L is the fiber length, δ is the fiber chromatic dispersion, and σ_λ is the unintended spectral spread (Henry, 1985). The typical fiber dispersion of $\delta \sim 15$ ps/km/nm at 1.55 μm for a conventional 1.3-μm zero-dispersion fiber, and the typical spectral spread of $\sigma_\lambda \sim 0.3$ nm for DFB lasers, give $BL \leq 55$ Gb/s km.

In coherent systems, the transform-limited spectrum is obtained by modulating the single-frequency carrier wave, so the dispersion limit yields

$$B^2L \leq \frac{c}{2\delta\lambda^2}, \qquad (25)$$

where c is the light velocity in vacuum and λ is the wavelength. At $\lambda = 1.55$ μm, Eq. (25) yields $B^2L \leq 4,200$ (Gb/s)^{2}km. Unlike conventional systems, coherent systems can alleviate the bandwidth limitation. Heterodyne detection schemes have the potential advantage of further relaxing the dispersive effect because they enable electrical delay equalization to compensate for fiber dispersion in the IF stage. A more critical discussion of the bandwidth limitation, though not presented here, reveals a detailed difference among various types of coherent modulation–demodulation schemes (Yamamoto and Kimura, 1981).

The bandwidth limitation is relaxed for both conventional and coherent systems by using single-mode fibers with a zero-dispersion wavelength of 1.55 μm, which is fabricated by modifying the relative refractive index difference and the core diameter (Ainslie and Day, 1986). Residual chromatic dispersion (residual linear chromatic dispersion) and higher-order chromatic dispersion then determine the limitation. Even in this case, coherent systems have a broader bandwidth than conventional systems that use direct intensity modulation.

4. Improvement of Modulation Bandwidth

Modulation speed in direct intensity modulation of semiconductor lasers is limited to the resonance frequency, above which modulation efficiency is inversely proportional to the square of the modulation frequency. The direct frequency modulation of semiconductor lasers (Kobayashi et al., 1982), however, relaxes this limitation. This is because the frequency shift normalized by the unit modulation current, namely the FM efficiency, is inversely proportional to the modulation frequency which is higher than the resonance frequency. Electrooptic waveguide modulators (Alferness, 1981) also achieve a phase modulation whose bandwidth is broader than the direct intensity modulation limit.

According to these four principal improvements, as well as the receiver sensitivity improvement, the potential of coherent optical fiber transmission systems is estimated in terms of transmission distance and data rate. Figure 10 shows the estimated results for PSK homodyne detection, ASK heterodyne detection, and IM direct-detection systems at a 1.55-μm carrier wavelength. In

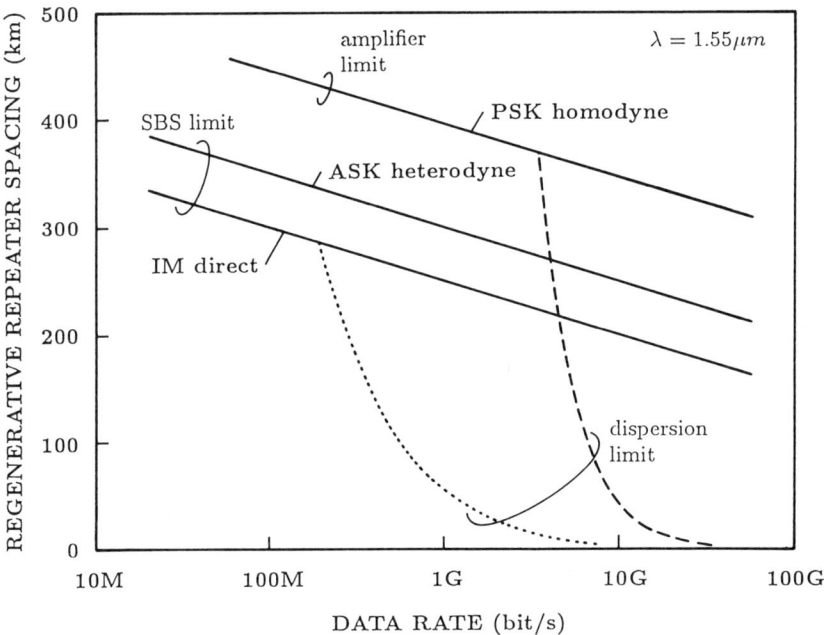

FIG. 10. Regenerative repeater spacing vs. data rate. Dispersion limits are calculated for a conventional single-mode fiber with a 1.3-μm zero-dispersion wavelength. Chromatic dispersion is assumed to be 15 ps/km/nm. The dotted line indicates the dispersion limit for direct intensity modulation, provided that unintended spectral spread is 0.3 nm.

calculations, the ideal receiver sensitivity is used for coherent detection. The sensitivity of 700 photons per pulse (Shikada et al., 1985), which is one of the best so far reported, is used for direct detection. Output power of a post-amplifier is assumed to be 20 dBm for PSK, and 10 dBm for ASK and IM. Fiber attenuation of 0.2 dB/km and dispersion of 15 ps/km/nm are assumed, and no margin is taken into account. Calculated results evidently point out that coherent optical fiber transmission systems have a potential of achieving both long repeater spacing and large information capacity. The maximum bit rate—distance product achieved by PSK homodyne detection described here is about 1,200 Gb/s km, being more than 20 times larger than that of conventional systems using direct intensity modulation of semiconductor lasers. A possible further improvement of the transmission data rate is proposed with the use of single-mode fibers with dispersion-free characteristics at 1.55 μm.

B. Optical Amplifier Repeater Systems

Strict frequency selectivity in optical heterodyne or homodyne detection, described in Section III, enables a coherent optical fiber transmission system to have optical amplifiers installed at intervals, because broadband spontaneous emission, the dominant noise source of the amplifier, is efficiently filtered out (Yamamoto and Kimura, 1981). Transmission distance will be much increased without requiring regenerative repeaters. An optical amplifier is basically a single component that generates a linearly amplified replica of the incoming signal. This forms an obvious contrast to a regenerative repeater, which requires a photodetector and a light source for opto-electronic and electro-optic conversion, as well as electrical circuitry for pulse reshaping, retiming, and regenerating.

Promising approaches to optical amplification are broadly divided into two areas: semiconductor amplifiers (Mukai et al., 1985), and fiber amplifiers (Stolen and Ippen, 1973; Urquhart, 1988). Their device structure, operation, and performance is discussed in Section V. In this section, discussion is concentrated on traveling-wave type semiconductor amplifiers, which seem promising for in-line repeaters and close to practical implementation, because of their simple configuration as well as their excellent amplification performance.

A traveling-wave type semiconductor amplifier has a similar structure to a conventional injection laser, except that both end facets are antireflection-coated. An optical signal incident on the amplifier makes just a single pass through the device, causing the stimulated emission to amplify itself sufficiently. A gain of about 20 dB is feasible, as is described in Section V. This amplifier has the advantage of a broad bandwidth over 1 THz, since a sufficiently small reflectivity of facets suppresses the multiple reflection inside the device that results in cavity resonance. The possibility of low-noise amplification is another desirable feature, which is also attributed to the facet reflectivity reduction. A sufficiently large gain G, where spontaneous emission becomes the dominant noise source, gives the amplifier a simple model consisting of an ideal amplifier of gain G and an equivalent noise source (Henry, 1985). The additive noise source has a power spectral density given by $n_{sp}hf$, where $n_{sp}(\geq 1)$ is the population inversion parameter for the amplifying medium, h is Planck's constant, and f is the optical frequency. n_{sp} is about 2 in practice, but is unity under the ideal case of perfect population inversion.

A simple configuration for an optical amplifier system is illustrated in Fig. 11. A fiber in each section has a length L, and an optical amplifier, which follows each fiber except for the last one, has a gain G just compensating for each fiber loss. As an optical signal through the cascade of amplifiers, it accumulates spontaneous emission generated by each amplifier and gets noisier. This effect is also shown in Fig. 11 in the form of level diagrams for both signal and spontaneous emission. The signal power first is reduced by $1/G$, and then undergoes repetitive amplification and attenuation. On the other hand, the spontaneous emission generated by each amplifier undergoes similar repetition but finally keeps its initial power level. The total power of spontaneous emission components at the receiver input end is m times each noise-source power, where m is the number of amplifiers. At the end of the chain, a

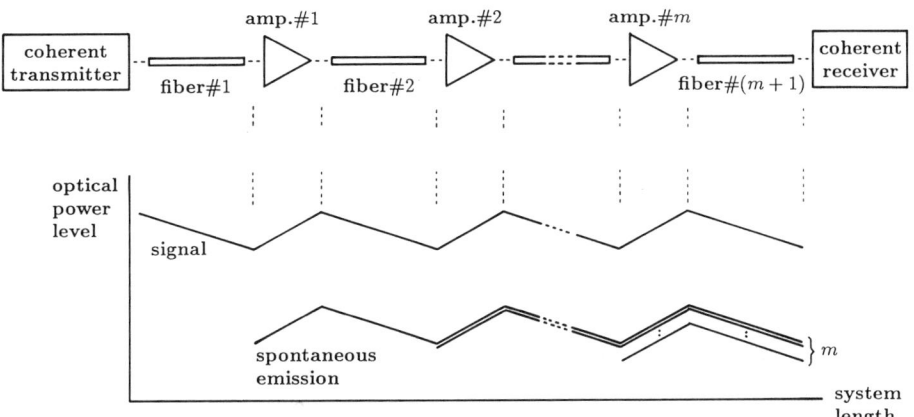

FIG. 11. Amplifier repeater system, and level diagram of signal and spontaneous emission.

heterodyne synchronous detection receiver yields

$$(\text{SNR})_{\text{hetero}} = \frac{P_0/G}{2mn_{\text{sp}}hf_sB_0}, \tag{26}$$

provided that traveling-wave amplifiers are used. Here, P_0 is the signal power launched by the transmitter into the first fiber, f_s is the carrier frequency, and B_0 is the receiver bandwidth (baseband filter bandwidth).

Where $P_0 = 0$ dBm, $G = 20$ dB, $n_{\text{sp}} = 2$, $B_0 = 500$ MHz (data rate $B = 1$ Gb/s), and wavelength $\lambda = 1.55$ μm, it is found that $m \sim 1070$ for SNR $= 15.6$ dB, which achieves 10^{-9} error rate in PSK heterodyne detection. A fiber attenuation of 0.2 dB/km yields $L = 100$ km and a total system length of about 1.07×10^5 km. This is a little more than 2.5 times the circumference of the earth. A critical discussion taking into account coupling loss, gain margin, etc., gives a total system length of about 10^4 km at 1 Gbit/s, which is mainly determined by the fiber dispersion (Yamamoto and Kimura, 1981). This is still long enough for a transoceanic undersea optical-fiber transmission system without using regenerative repeaters.

The dominant noise determining the denominator in Eq. (26) is the beat noise between the local oscillator and spontaneous emission. As discussed in Section III, coherent detection schemes successfully remove the beat noise between spontaneous emission noise components. An optical filter attached to the detector enables a direct detection system to have optical amplifiers installed to expand the transmission distance. However, even if an ideal optical filter is used, a coherent detection system maintains superiority over the conventional system because of the improved receiver sensitivity.

C. FDM Systems

Improved frequency selectivity over direct detection, together with the use of FDM techniques, enables coherent systems to have improved information capacity. This allows collective communication of numerous signals of densely allocated channels through the same transmission medium. Thus, the vast silica-fiber bandwidth of 60 THz in a 1.2–1.6 μm low-loss window can be exploited.

Recently, coherent FDM techniques have received increasing attention for application to local networks (Nosu and Iwashita, 1988; Stanley et al., 1987; Wagner et al., 1987) rather than long-distance point-to-point transmission. A multiterminal system capable of bidirectional transmission, where each optical carrier frequency is assigned to the terminal of destination, is a possible configuration. A distribution network, the fiber network analog of a broadcasting system, in which wideband multi-channel signals are delivered through an optical fiber and then distributed so that each end terminal can selectively receive the desired information, is also a possible form of the system.

In these systems, a large signal gain achieved by coherent techniques, that is, a high receiver sensitivity, plays an important role in compensating for both coupling and branching losses, which are the dominant loss factors in passive networks.

V. Essential Technology for Developing Coherent Systems

Single-frequency carrier generation, angle modulation, heterodyne or homodyne detection, single-polarization signal transmission, and optical amplification are essential to development of the coherent optical fiber communication systems described in Section IV. Table III summarizes this technology and related key subjects to be investigated.

With respect to coherent optical carrier generation, frequency stabilization and spectral linewidth reduction of semiconductor lasers are important because coherent detection methods are very sensitive to the spectral coherence of optical carrier waves. A frequency-stabilized master oscillator is required to synchronize an entire system, especially a large-scale composite network consisting of a number of coherent systems. To achieve highly

TABLE III

Essential Technology and Key Subjects of Coherent Optical Fiber Transmission

Technology	Subjects
Single-frequency optical carrier generation	Absolute frequency stabilization and linewidth reduction of semiconductor lasers
Angle modulation	High-speed and high-efficiency phase/frequency modulation in semiconductor lasers High-speed and low-loss external modulators
Heterodyne/homodyne detection	Suppression of local oscillator AM and FM noise Optical phase-locked loop
Single-polarization signal transmission	Low-loss single-polarization fibers Polarization state control Polarization diversity receiver
Optical amplification	Low-noise optical amplifier having stable gain and frequency

efficient angle modulation of the optical carrier wave, frequency or phase modulation schemes using semiconductor lasers need to be extensively studied. It is important to maintain a narrow carrier-wave linewidth, even in the modulation condition. High-speed and low-loss external modulators also need to be developed.

In coherent detection schemes, suppression of SNR degradation caused by the AM and FM noise of a local oscillator is one of the main items of concern. Since homodyne receivers must track the carrier phase, an optical phase-locked loop (PLL) technique is indispensable. To accomplish polarization-plane matching between the transmitted signal and the local oscillator waves in coherent receivers, it is important to develop low-loss single-polarization fibers that maintain the polarization state of the transmitted signal light. Development of polarization control devices attached to the fiber output end, or of polarization diversity receivers, also becomes important if polarization fluctuation in conventional single-mode fibers is very small and slow. With respect to optical amplification, low-noise optical amplifiers with stable gain and frequency are important for post-amplifier as well as optical repeater application.

A. Coherent Laser Sources

Light sources used in optical fiber communication systems must satisfy several basic requirements, such as long life, high efficiency, compact size, good reproducibility, and so on. Also of great importance is the oscillation wavelength, which must be in close harmony with the low-loss wavelength of optical fibers. A stable oscillation frequency and a narrow spectral linewidth are also indispensable, especially in coherent systems. Although the required frequency stability has not been quantified, it generally must be much smaller than the data rate so that the local oscillator can be successfully frequency offset-locked or phase-locked to the carrier wave. When a number of coherent systems are synchronized, the master oscillator of the entire network has to deliver an absolutely stable frequency standard.

Required spectral linewidths have been evaluated for various coherent modulation–demodulation schemes (Yamamoto and Kimura, 1981; Kikuchi et al., 1984; Garrett and Jacobsen, 1986; Kazovsky, 1985, 1986; Hodgkinson, 1986) through critical evaluation of the laser FM noise. Spontaneous emission causes a laser output wave to have a randomly fluctuating phase in the form of a Lorentzian lineshape or a flat FM noise spectrum (Manes and Siegman, 1971; Yamamoto, 1983). This affects receiver performance in two ways. First, signal power bulging outside the signal band is removed by the IF or baseband filter, so the spectral spread gives rise to a power penalty. Second, FM and PM noise causes an instantaneous frequency or phase slip at a sampling time for decision of mark or space state, resulting in bit errors irrespective of the receiving signal power. A power penalty of less than 1 dB requires the spectral linewidth (full width at half maximum) to be less than 0.16 times the data rate B for ASK heterodyne envelope detection, which suffers only from the first effect. Allowable maximum linewidths are $1.1 \times 10^{-2} B$, $2.7 \times 10^{-3} B$, $4 \times 10^{-3} B$, and $5 \times 10^{-4} B$, for Sunde's FSK heterodyne discriminator detection, MSK heterodyne discriminator detection, PSK heterodyne differential detection, and PSK heterodyne synchronous detection or homodyne detection, respectively. These are mainly due to the second effect, calculated for a 10^{-9} error rate, provided that the transmitter laser and local oscillator have the same spectral linewidth.

In the low-loss wavelength region of optical fibers, there are several candidates for coherent light sources: $1.3-1.6$-μm InGaAsP semiconductor lasers, 1.32-μm Nd:YAG lasers, 1.32-μm lithium neodymium tetraphosphate (LNP) lasers, and 1.52-μm He–Ne lasers. Semiconductor lasers are the most promising because of their long life and high efficiency. However, they have three principal drawbacks when used in coherent systems. First, the oscillation longitudinal mode jumps because of temperature change and long-term aging degradation. Second, the oscillation frequency is sensitive to both temperature and injection current changes, even if laser operation is continued in the single longitudinal mode. The frequency changes by 10–20 GHz/degree with respect to temperature and by 1–5 GHz/mA with respect to injection current. Third, the FM noise, which determines the spectral linewidth, is large, since semiconductor lasers have small cavity lengths and hence small cavity-Q values. Both Fabry–Perot and distributed feedback (DFB) InGaAsP lasers oscillating at 1.5 μm exhibit spectral linewidths of 10–100 MHz (Olesen *et al.*, 1983; Henning *et al.*, 1984).

DFB lasers, phase-shift DFB lasers, and distributed Bragg reflector (DBR) lasers show promise in overcoming the first problem. Their frequency-selective structures control oscillation behavior and achieve single-longitudinal-mode operation.

To solve the second problem, the oscillation frequency of a semiconductor laser can be locked to a frequency reference such as a Fabry–Perot interferometer or to an absorption spectral line of gaseous atoms or molecules (Tako *et al.*, 1983; Okoshi and Kikuchi, 1980; Yanagawa *et al.*, 1984). Figure 12 shows a basic configuration of this scheme. Frequency stabilization is carried out by extracting an error signal and feeding it back to countermodulate the laser temperature or injection current. A frequency stability of $10^{-12}-10^{-11}$ for an average time of 100 s has thus far been achieved (Tako *et al.*, 1983), where this characteristic actually means the frequency traceability of a semiconductor laser to the frequency reference. An absorption line

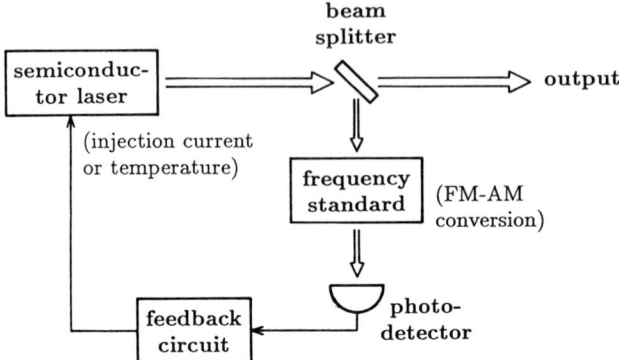

FIG. 12. Basic configuration for frequency stabilization of a semiconductor laser.

exhibits long-term highly stable frequency, enabling development of a semiconductor laser with an absolute standard frequency. For example, an absorption line of NH_3, which is found in the 1.5-μm wavelength region, exhibits a center frequency shift toward a temperature change $(\Delta f/f)/\Delta T$ of 10^{-14} deg^{-1} through the second-order Doppler effect (Shimoda, 1973), while a Fabry–Perot cavity whose housing is made of a low-expansion material, such as super-invar or quartz glass, shows a frequency shift of 10^{-7}–10^{-6} deg^{-1}. An InGaAsP DFB laser stabilized to an NH_3 absorption line at 1.52 μm has been reported (Yanagawa et al., 1984). Besides NH_3, such gaseous molecules as CO_2, H_2O, C_2H_2, and CH_3Cl offer absorption spectra in the 1.3–1.7-μm wavelength region. Important for future systems, then, are the detailed assignment of absorption lines and the search for new absorption media.

Enlarging the laser cavity Q value by employing a long cavity structure is one effective way of solving the third problem (Yamamoto, 1983). Therefore, a DFB laser having a number of coupled units of quarter-wave shifted corrugation has been proposed (Kimura and Sugimura, 1987). An external cavity structure having a mirror or a diffraction grating, as shown in Fig. 13, also increases the effective cavity Q value, and thus reduces the spectral linewidth (Saito et al., 1982). An external-cavity 1.5-μm InGaAsP laser, in which the laser facet facing an external grating is antireflection-coated, achieves a spectral linewidth of less than 1 kHz (Wyatt, 1985). Furthermore, the linewidth reduction factor (Patzak et al., 1983) and operation stability (Tromborg et al., 1984) of external cavity lasers have been extensively discussed as promising solutions. A spectral linewidth of 135 kHz at 5 mW output power has been achieved using a 1.5-μm semiconductor laser butt-coupled to an external resonant reflector, which is made from dielectric waveguides on a silicon substrate (Olsson et al., 1987).

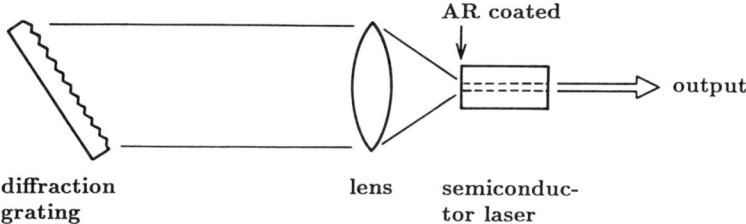

FIG. 13. Basic configuration for external cavity laser.

Negative frequency feedback control of semiconductor lasers reduces FM noise and spectral linewidth (Saito et al., 1985; Ohtsu and Kotajima, 1985). Except for a broader feedback bandwidth, this scheme basically involves the same operation principle as the frequency stabilization scheme described above. Theoretically, negative frequency feedback control offers the possibility of spectral linewidth decreasing below the modified Schawlow–Townes limit (Yamamoto et al., 1985). Reported experiments confirm, however, that time delay around a feedback loop dominates the FM noise reduction factor. Reduction of loop time delay, which can be achieved by utilizing optoelectronic integration circuit (OEIC) technology (Forrest, 1985), is indispensable to the success of this method.

It is important for semiconductor lasers to achieve stable frequency and narrow linewidth simultaneously. One way of achieving this is to combine the frequency stabilization and linewidth reduction schemes. The frequency stabilization of a narrow-linewidth external-cavity laser may be achieved automatically by inserting a dispersive medium into the cavity.

The study of the individual noise characteristics of semiconductor lasers is also important. Laser FM noise basically stems from spontaneous emission coupled to a lasing mode. In semiconductor lasers, AM noise, that is, photon number fluctuation, generated through the same mechanism as FM noise, competes with carrier number fluctuation and then enhances FM noise through the dispersion characteristics of the laser medium (Yamamoto, 1983). This FM noise enhancement factor is the so-called α parameter (Henry, 1982; Vahara and Yariv, 1983). Although FM noise possessing the f^{-1} power spectrum has been observed in a low-frequency region (Walther and Kaufman, 1983), its cause is still being discussed. The theory of noise in semiconductor lasers has been well established using the quantum mechanical Langevin equation (Haug, 1969). This method has also been expanded for application to the density-matrix master equation and the quantum mechanical Fokker–Planck equation (Yamamoto, 1983). Moreover, it is essential to take a close look at the electron system present in semiconductor lasers by means of quantum statistical treatment.

An optical isolator is one important associated device for maintaining stable operation of coherent laser sources. This is because the stability of semiconductor lasers is adversely affected by incident waves reflected from the following device. At a feedback power ratio as low as -80 dB, narrowing or broadening of the oscillation spectrum was observed for a 1.5-μm DFB laser, depending on the phase of the feedback (Tkach and Chraplyvy, 1986). Even if a long-length or an external-cavity configuration can enlarge cavity Q value, reflection can still affect laser stability. Conventional isolators, consisting of a Faraday rotator and two polarizers situated before and after the rotator, have an isolation of about 30 dB as well as an insertion loss of about 1 dB, so tandem installation may be required. Optical isolators, preferably with a guided-wave configuration, having a larger isolation and a lower insertion loss should be developed.

B. Modulation–Demodulation Technology

1. Modulation

Angle modulation is of great interest in coherent transmission application because it has the potential to improve both transmitting and receiving signal levels, as described in Sections III and IV. There are two promising approaches to angle modulation: direct modulation of semiconductor lasers (Kobayashi et al., 1982), mainly for FSK application, and modulation using external modulators (Alferness, 1981), for PSK application. The latter can also be applied to ASK signal generation.

An optical FM signal is obtained by directly modulating the injection current of a conventional single-longitudinal-mode semiconductor laser (Kobayashi et al., 1982). This is illustrated schematically in Fig. 14a. The carrier modulation effect dominates the direct FM modulation in the high-modulation-frequency region, and the temperature modulation effect dominates in the low-frequency region. Amplitude and phase characteristics of the FM response have been previously measured in detail (Kobayashi et al., 1982; Jacobsen et al., 1982). Although FM efficiency is not uniform with respect to modulation frequency, a frequency shift of 100 MHz to 1 GHz is easily obtained without serious intensity modulation. This has the advantage of simple transmitter configuration. Nonuniform FM characteristics can be compensated for by electrically equalizing the modulation input current (Saito et al., 1983). The inherent spectral linewidth of a semiconductor laser, which may be one drawback of this method, can also be suppressed, for example, by an external cavity configuration at the expense of FM efficiency degradation (Saito et al., 1982).

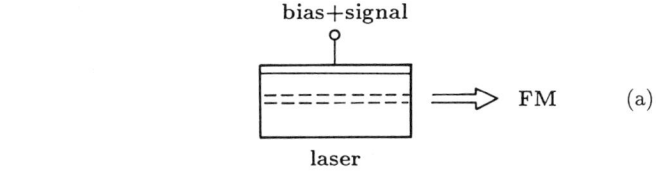
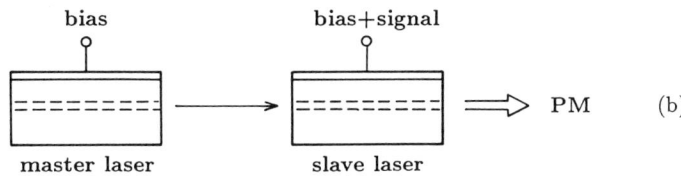
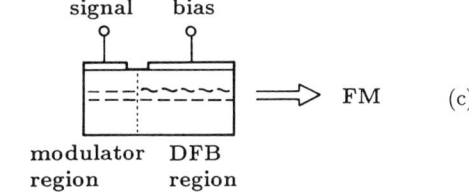
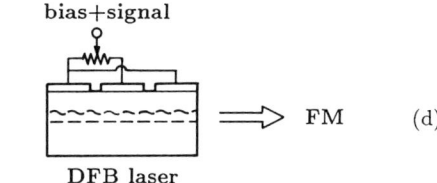
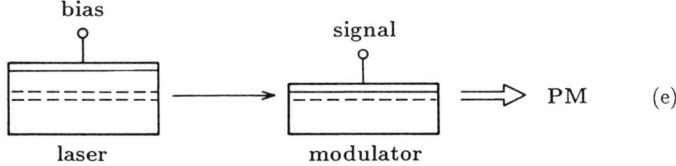

FIG. 14. Optical FM or PM signal generation. (a) FM signal generation by direct modulation of a semiconductor laser. (b) PM signal generation by direct modulation of an injection-locked semiconductor laser. (c) FM signal generation with a DFB laser monolithically integrating a modulation region. (d) FM signal generation with a multi-electrode DFB laser. (e) PM signal generation using a guided-wave modulator.

Direct current modulation of a semiconductor laser can be applied to optical phase modulation, where external coherent laser light is injected into the laser (Kobayashi and Kimura, 1982), as shown in Fig. 14b. The output signal phase relative to the input light phase is zero when the injected laser frequency is tuned exactly to the input signal frequency. It changes by $\pi/2$ radians when the injected laser frequency is shifted away from the input light frequency to the locking bandwidth limit. The cutoff modulation frequency in this method is determined by the injection locking bandwidth. Even if the inherent spectral linewidth of the injection locked laser is rather broad, it is reduced to the level of the injected signal linewidth (Kobayashi et al., 1981).

The above modulation methods using semiconductor lasers need to have a flat FM response, a high FM efficiency, and a suppressed spurious intensity modulation. A composite semiconductor laser having a DFB laser region and a modulator region, as shown in Fig. 14c, has been developed to satisfy such requirements (Yamazaki et al., 1985). Supplying a modulation input signal to the modulator region effectively demonstrated flat frequency modulation caused mainly by the carrier modulation effect. A modulation efficiency of 1–2 GHz/mA and a cutoff frequency of several hundred megahertz were achieved. A two-section laser with different α parameters was also proposed after a critical look at the direct frequency modulation mechanism in semiconductor lasers (Nilsson and Yamamoto, 1985). Push–pull operation was indicated as a possible means for achieving these requirements. Furthermore, a DFB laser with electrically separated multielectrodes has been developed (Yoshikuni and Motosugi, 1987). Figure 14d schematically illustrates this laser. By controlling the modulation current to each electrode, frequency modulation exhibiting a frequency shift of up to 12 GHz was demonstrated at a 1-GHz modulation frequency under constant output power operation. Quite recently, a three-electrode 1.55-μm DFB laser module having an 8-GHz flat FM response, a 1.5-MHz linewidth, and 10 dBm output power has been reported (Onaka et al., 1990).

A conventional guided-wave electro-optic modulation (Alferness, 1981) can be used for generating PSK signals as well as ASK signals, being installed just behind a frequency stabilized cw laser, as shown in Fig. 14e. A LiNbO$_3$ traveling-wave optical modulator with a 20-GHz bandwidth has been demonstrated (Jungerman et al., 1987). Besides such dielectric substances as LiNbO$_3$, semiconductor materials, for example III-V compounds, can be used for guided-wave modulators. Phase modulation due to the carrier modulation effect was demonstrated in a semiconductor guided-wave modulator with a multiple quantum well (MQW) structure (Wakita et al., 1987). Intensity modulation was also achieved by the large electroabsorption effect resulting from exciton resonance (Wood et al., 1987).

2. Demodulation

To fulfill their function adequately, optical heterodyne or homodyne receivers must overcome the AM and FM noise of the local oscillator as well as the finite power transmittance of the beam combiner, all of which degrade performance. Phase locking of the local oscillator is a subject of technological significance for homodyne receivers. It is also important to develop such constituent optical devices as a tunable local oscillator and a high-speed photodetector.

Owing to its resonance characteristics, a semiconductor laser has relatively large AM noise (Yamamoto, 1983), and this noise appears in a photodetector current as excess AM noise, which is often larger than shot noise. AM noise of a semiconductor-laser local oscillator causes SNR degradation in coherent receivers (Saito et al., 1983) and finally deteriorates the minimum detectable power. Since excess AM noise decreases with an increase in laser driving current, high-bias current operation is promising for AM noise suppression. If a semiconductor laser can provide sufficient power, it is also effective to introduce suitable attenuation for the local oscillator wave. This is because excess AM noise is inversely proportional to the square of the attenuation, while shot noise is inversely proportional to the attenuation. SNR degradation due to excess AM noise can be completely removed by the so-called balanced mixer receiver (Yuen and Chan, 1983) as well, which will be discussed later.

FM noise of the local oscillator, like the transmitter laser, degrades SNR, and then determines a lower limit of error rate performance. This effect, however, can be removed by a semiconductor laser whose spectral linewidth is suppressed by, for example, an external cavity configuration, as described in Section V.A.

Signal power attenuation in the beam combiner also causes SNR degration, although this is neglected in the discussion in Section III, where the combiner is assumed to transmit both signal and local waves completely. In practice, signal power P_s and local wave power P_{LO} should be replaced by εP_s and $(1 - \varepsilon)P_{LO}$, respectively. Here, ε is the power transmittance for the signal wave in the combiner, so $(1 - \varepsilon)$ denotes that for the local wave. Even if transmittance approaches unity and can thus relax degradation, it creates another difficulty in that the local oscillator wave suffers from large attenuation and then fails to achieve shot-noise-limited detection.

A balanced-mixer receiver can overcome this problem, as well as that caused by the local oscillator's excess AM noise (Yuen and Chan, 1983). This receiver consists basically of a 50–50% beam splitter, two balanced photodetectors, and a hybrid junction with a π phase shift, as shown in Fig. 15. The twin detectors generate signal components of equal amplitude but reversed

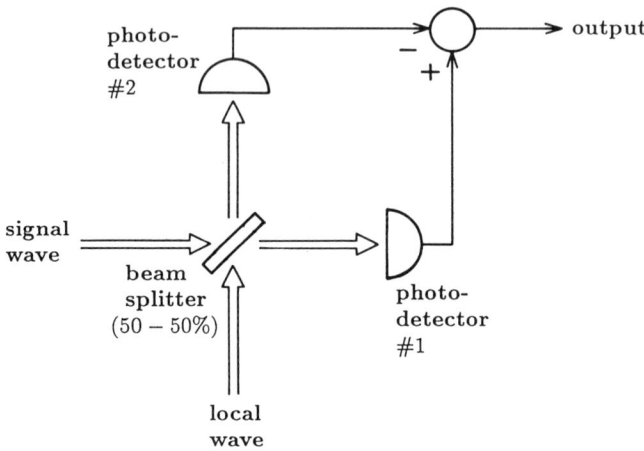

FIG. 15. Basic configuration of a balanced-mixer receiver.

phase. According to the notation in Section III, signal currents in heterodyne detection are given by $\pm D\sqrt{P_s P_{LO}}\cos\{2\pi(f_s - f_{LO})t + (\phi_s - \phi_{LO})\}$. On the other hand, noise components caused by excess AM noise from the local oscillator have equal amplitude and phase—that is, they are both $D\Delta P_{LO}/2$, where $\Delta P_{LO}(t)$ denotes the power fluctuation of the local wave. The differential composition of detector currents doubles the signal amplitude and then yields an output of $2D\sqrt{P_s P_{LO}}\cos\{2\pi(f_s - f_{LO})t + (\phi_s - \phi_{LO})\}$, but completely cancels the excess AM noise components. Here, shot noise and thermal noise are simply added in power because both of them display no correlation between twin-photodiode outputs. Thus the resulting SNR is still given by Eq. (7), except that the thermal noise is doubled. Provided that local oscillator power is large enough to overcome the doubled thermal noise, a balanced-mixer receiver can achieve the ultimate SNR described in Section III. The same argument is applied to homodyne detection. Since a balanced receiver automatically cancels the dc output component, homodyne detection enjoys the advantage that an extra circuit for dc compensation becomes unnecessary.

In homodyne detection, optical PLL techniques are essential to achieve homodyne receiver sensitivity 3 dB better than heterodyne performance. An optical PLL consists basically of a phase detector, a loop filter, and a voltage-controlled oscillator (VCO). A photodetector and a local oscillator act as phase detector and VCO, respectively. The optical phase difference between the incoming signal and the local oscillator wave, that is, the phase error signal, is detected by the phase detector and then fed back to the VCO to track the carrier phase effectively. A PLL configuration, however, is very complicated for a binary PSK signal with a π phase shift because the optical carrier component is completely suppressed. Incomplete modulation with a phase shift slightly less than π can be employed to maintain a simple PLL configuration. Otherwise, the optical carrier must be retrieved from a demodulated signal through, for example, nonlinear signal processing. An optical PLL was demonstrated using 1.52-μm He–Ne lasers in a homodyne detection experiment, where a 140-Mb/s PSK signal with incomplete phase shift was transmitted over a 30-km cabled fiber (Malyon, 1984). A complicated PLL like the Costas-type loop has also been demonstrated, using 10.6-μm CO_2 lasers (Philipp et al., 1983). In heterodyne synchronous detection, similar techniques are required to retrieve an IF carrier wave.

An important parameter of optical PLLs is the stationary phase error, which needs to be minimized since it causes SNR degradation. Also important is phase error noise caused by FM noise of both signal and local waves, as well as by shot noise and excess AM noise of the local oscillator. A broad feedback bandwidth can suppress phase error noise due to laser FM noise, but increases phase error noise due to shot noise and excess AM noise. With this in mind, the optimum system parameters and the required spectral linewidth have been evaluated for a balanced PLL (Kazovsky, 1986), decision-driven PLL (Kazovsky, 1985), and Costas PLL (Hodgkinson, 1986).

Since an optical PLL is presently rather difficult to construct, efforts to demonstrate coherent optical fiber transmission have mainly concentrated on heterodyne detection systems. Such systems, however, are inferior to homodyne systems in their attainable transmission data rate for two reasons. First, heterodyne detection requires a receiver wider bandwidth than homodyne detection, and second, the response speed of photodetectors limits the maximum data rate. To overcome these problems, phase-diversity receivers (Davis et al., 1987; Hodgkinson et al., 1985), such as the three-phase detection receiver and the in-phase and quadrature (IQ) detection receiver, have been demonstrated. Phase diversity techniques achieve heterodyne detection performance and allow the homodyne receiver bandwidth to be used without the need for optical PLLs.

Local oscillators, as well as photodetectors, are important optical devices in heterodyne or homodyne receivers. Generally, the local oscillator is a semiconductor laser whose characteristics are similar to those of the transmitter laser. An important property peculiar to the local oscillator is frequency tunability, which is necessary for frequency offset locking or phase locking to the carrier wave. FDM systems in particular may require local lasers to tune their frequencies at considerable speed over a wide range when selecting the desired channel. Oscillation frequency tuning in a semiconductor laser is accomplished by changing the device temperature. A change in reflection angle of an external grating is also applicable to laser frequency tuning, although it is often accompanied by mode jumping. Recently a 1.3-μm

InGaAsP laser monolithically integrating a thermoelectrically controlled DBR achieved 5 Å tunability (Cella et al., 1987). A 1.5-μm fiber-extended-cavity laser, consisting of an InGaAsP superluminescent diode, a single-mode fiber with an integral microlens, and a variable-period diffraction grating coupled to the fiber, demonstrated a tuning range of 450 Å without mode jumping (Whalen et al., 1987).

Important parameters of photodetectors used in coherent systems are quantum efficiency and bandwidth. Heterodyne detection systems, in particular, require detectors to have a bandwidth more than three times the data rate because the intermediate frequency must generally be at least twice the data rate for efficient separation between the IF band and baseband. Recent transmission experiments commonly use InGaAs $p-i-n$ photodiodes of typically 70-80% quantum efficiency. A recently developed 67 GHz bandwidth InGaAs $p-i-n$ photodiode demonstrates the possibility of high-speed response (Tucker et al., 1986). In coherent systems, photodetectors must constantly receive incident light of at least a few milliwatts, so their reliability must be confirmed under this condition.

Optical devices essential to FDM systems are multiplexer/demultiplexers, which couple into a fiber several signals of different carrier frequencies, or distribute frequency-multiplexed signals to their objective channels. These functions are accomplished by utilizing one of two basic effects: chromatic dispersion in such optical devices as prisms or diffraction gratings, and optical interference resulting, for example, from multiple reflection in multilayered films. A multiplexer/demultiplexer has been designed in the 1.5-μm wavelength range using a diffraction grating (McMahon et al., 1987). A tunable fiber Fabry–Perot (Stone and Stulz, 1987), which utilizes the latter effect, can be used as a demultiplexer in FDM networks. A waveguide device having a Mach–Zehnder interferometer configuration has been fabricated to demonstrate 5-GHz-spaced eight-channel multiplexing/demultiplexing (Toba et al., 1987). Although not frequency-selective, optical power couplers or dividers, for example star couplers, are also important devices to be developed for multiterminal systems.

C. Optical Fibers

Optical heterodyne or homodyne efficiency depends greatly on polarization-state matching between the transmitted signal and local oscillator waves. Since the local oscillator generally emits linearly polarized light, the polarization state of the optical receiving signal must be made linear for efficient detection. To achieve this, two approaches have thus far been attempted. One is to develop optical fibers that maintain the polarization plane of the incident linearly polarized light during propagation (Kaminow, 1981). The other is to install polarization control devices at the output end of conventional single-mode fibers (Ulrich, 1979). In place of polarization control devices, polarization diversity receivers can be used, which do not necessarily recover the linear polarization, but detect two orthogonal polarization modes individually, then combine the detected signals (Okoshi et al., 1983).

1. Polarization-Maintaining Fibers

With respect to the first approach, an optical fiber with axial asymmetry in its cross-section is used, where degeneracy of two possible polarization modes is removed. Two eigenpolarization modes (Sakai et al., 1981), orthogonal to each other, can then propagate through such deformed fibers. To transmit one of the two eigenpolarization modes, only the polarization mode should be launched into the fiber. Moreover, mode conversion to another eigenmode must be suppressed during propagation.

Mode coupling occurs between the two orthogonal polarization modes when spatial perturbations exist along an optical fiber (Kaminow, 1981; Marcuse, 1974). The mode coupling coefficient, the parameter describing the degree of mode conversion, is in proportion to

$$\Psi(\varphi, \delta\beta) = \left[\frac{\sin\{(\varphi - \delta\beta)L/2\}}{(\varphi - \delta\beta)L/2}\right]^2, \qquad (27)$$

where L is the fiber length, φ is the spatial frequency of perturbations, and $\delta\beta$ is the modal birefringence ($\delta\beta = \beta_1 - \beta_2$; β_1 and β_2 are the propagation constants for the two eigenmodes.). This equation indicates that mode conversion is likely to occur when the spatial frequency agrees with the modal birefringence. However, fiber perturbations do not have only one spatial frequency, but rather a power spectrum existing mostly in the low-frequency region, as shown in Fig. 16. Contribution of the spread power spectrum must be taken into account when the actual mode coupling coefficient is evaluated. The mode coupling coefficient determines the extinction ratio at the fiber output end, an important measure of the result of mode conversion, which is defined as the power ratio of the undesirable mode to the launched mode. To achieve a small extinction ratio, basically the modal birefringence, $\delta\beta$, should be well separated from the spatial frequency of perturbation. Thus $\delta\beta$ should be made large, so that the so-called beat length, $\Lambda \equiv 2\pi/\delta\beta$, should be made small. It is also important to suppress the power spectrum of fiber perturbations and confine it to the low-spatial-frequency region.

Several types of polarization-maintaining optical fibers have been proposed to increase modal birefringence. Early proposals included fibers with

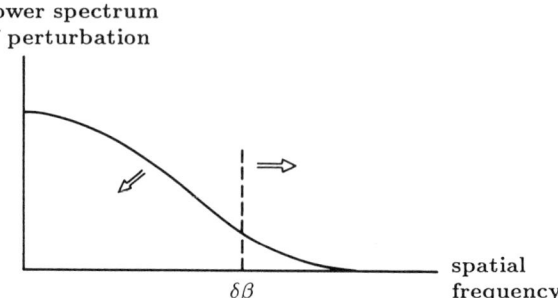

FIG. 16. Power spectrum of spatial perturbation existing along an optical fiber. The broken line indicates the modal birefringence in terms of the spatial frequency. In order to suppress mode conversion, the power spectrum should be confined to a low-frequency region and the modal birefringence should be made large. This is indicated by arrows.

geometrically asymmetric refractive-index profiles, such as elliptical-core fibers (Ramaswamy et al., 1979). The thermal-stress-induced anisotropy due to the asymmetric thermal expansion coefficient was subsequently pointed out as being more effective for enhancing modal birefringence (Ramaswamy et al., 1979; Kaminow and Ramaswamy, 1979). The thermally induced modal birefringence has been theoretically treated for specific profiles (Sakai and Kimura, 1982; Varnham et al., 1983a). Thermally stressed polarization-maintaining fibers, such as elliptical jacket (Katsuyama et al., 1983), PANDA (Sasaki et al., 1984), and bow-tie fibers (Birch et al., 1982), have been fabricated. Of these, the shortest beat length thus far achieved is $\Lambda = 0.8$ mm (Katsuyama et al., 1983). These polarization-maintaining fibers can convey either of the two eigenpolarization modes.

To restrict propagation through such fibers to only one polarization mode, an attempt to impose a cutoff condition on the other mode has been made by introducing a loss difference between the two eigenpolarization modes (Simpson et al., 1983; Varnham et al., 1983b). Figure 17 shows the extinction ratio as a function of the mode coupling coefficient, with and without a loss difference between the two polarization modes (Marrone, 1985). Introduction of the loss difference, leading to saturation in the extinction ratio for a long fiber, is effective in achieving a small extinction ratio, although it may cause an increase in propagation mode loss.

Fiber perturbations are attributed to both internal and external origins. Internal origins are, for example, scattering center inhomogeneity and core-cladding interface irregularity, which depend on the fabrication technology employed. Improvement of uniformity along the fiber axis effectively suppresses mode conversion and simultaneously decreases fiber loss. To obtain information about internal perturbations, the coupling coefficient should

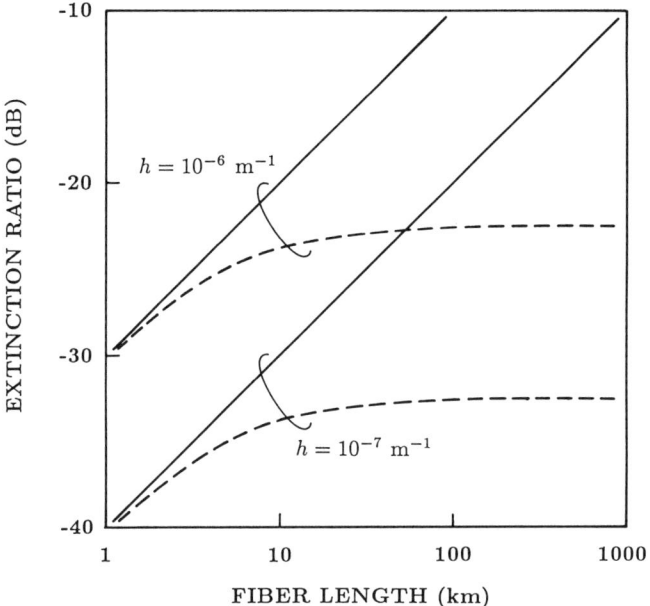

FIG. 17. Extinction ratio as a function of fiber length (Kimura, 1987). The parameter h is the mode coupling coefficient. Solid lines indicate no loss difference between two orthogonal modes. Broken lines indicate a loss difference (0.2 dB/km for the launched mode and 1 dB/km for another mode). (©1987 IEEE.)

be directly measured under suppressed external fluctuation conditions (Nakazawa et al., 1984). It is also important to evaluate the power spectra of external perturbations, such as ambient temperature change and bending along the fiber axis.

It is essential that polarization-maintaining fibers have a small loss comparable to that of conventional single-mode fibers. Significantly, a 2.4-km polarization-maintaining fiber whose loss approached that of the single-mode fiber has been fabricated (Hosaka et al., 1985). The minimal fiber loss was 0.25 dB/km at a 1.57-μm wavelength. Moreover, an extinction ratio of -29 dB was achieved for a 2.4-km fiber length, which corresponds to a mode coupling coefficient of 5.2×10^{-7} m^{-1}. These results show that polarization-maintaining fibers can be used in coherent optical fiber transmission systems provided that mass production problems are overcome. It is thus important to continue development of thermally stressed polarization-maintaining fibers possessing both low loss and a small extinction ratio.

By introducing polarization-maintaining fibers instead of conventional single-mode fibers, polarization-state matching between signal and local oscillator waves can be carried out easily, without the necessity for any additional polarization control scheme. Polarization-maintaining fibers are of great advantage in the development of sophisticated coherent systems using optical amplifiers and optical integrated circuits, which are often sensitive to the polarization state of input signals.

A twisted fiber (Jeunhomme and Monerie, 1980), in which circularly polarized light can keep its polarization state during propagation, may be classified as a polarization-maintaining fiber. It utilizes the fact that clockwise and counterclockwise circular polarizations become eigenmodes when shearing stress is induced by twisting a nearly round core fiber (Machida et al., 1982). Polarization dispersion of the twisted fiber approaches zero as the twist rate increases. If local oscillator light can be circularly polarized, a polarization controller may be unnecessary. Twisted fibers have the advantage of easy connection or splicing free from precise polarization-plane matching.

2. Polarization Control and Diversity Techniques

Since polarization-maintaining fibers are still under development, coherent optical transmission experiments have thus far mainly used only conventional single-mode fibers. High heterodyne or homodyne detection efficiency can be achieved by controlling the polarization state of the transmitted signal at the fiber output end, unless conventional single-mode fibers give rise to crucial polarization-state fluctuation. In fact, it has been reported that the fluctuation after 30-km single-mode fiber transmission is small and slow enough to be easily suppressed by automatic polarization control devices (Harmon, 1982). This method presents the obvious advantage of using well-developed low-loss single-mode fibers, without the need for any structural modification.

The basic function of polarization control schemes is to compensate for phase difference between the two orthogonal polarization modes of the transmitted signal light (Ulrich, 1979). Phase compensation, basically simulating the operation of half- and quarter-wave plates, adjusts the polarization state of the transmitted signal to that of the local oscillator light. After extracting a fraction of the transmitted light and detecting its polarization state, phase compensation is accomplished by the feedback control of newly introduced birefringence. The additional birefringence is induced by, for example, applying stress to optical fiber, or controlling birefringence in a dedicated electro-optic crystal. Important parameters in polarization control schemes are response speed, insertion loss, output stability, and device size. Continuous tracking of the input signal polarization state is also an important requirement (Okoshi, 1985).

Besides polarization control schemes, polarization diversity techniques can be applied at an output end of a conventional single-mode fiber (Okoshi et al., 1983). In the polarization diversity receiver, the two orthogonal polarization modes of the signal light are heterodyne- or homodyne-detected individually and then combined electrically to obtain a polarization-independent signal. Bit error rate performance was evaluated for polarization diversity receivers as well as phase diversity receivers (Siuzdak and Etten, 1989). A practical receiver configuration that achieves a high sensitivity is an important issue to be discussed.

D. Optical Amplifiers

Optical amplifiers may be used as post-amplifiers and intermediate line repeaters. There are two promising types of optical amplifiers: semiconductor amplifiers (Mukai et al., 1985) and fiber amplifiers (Stolen and Ippen, 1973; Urquhart, 1988). Semiconductor amplifiers are classified into two categories: linear amplifiers and injection-locked oscillators. Fabry–Perot cavity type (Mukai et al., 1982, 1983) and traveling-wave type (Simon, 1982; Eisenstein et al., 1985; Saitoh and Mukai, 1987) semiconductors amplifiers belong to the first category. The Fabry–Perot cavity type amplifier is basically a semiconductor laser biased just below the threshold. This laser achieves linear amplification by utilizing stimulated emission. The traveling-wave type amplifier, made by antireflection-coating the facets of a Fabry–Perot type amplifier, uses a single-pass gain through its active layer, as described in Section IV. The injection-locked oscillator, on the other hand, is a semiconductor laser whose phase or frequency is controlled by weak optical injected signals (Kobayashi and Kimura, 1981; Kobayashi et al., 1981).

Fiber-Raman amplifiers (Stolen and Ippen, 1973; Washio et al., 1985; Hegarty et al., 1985) and fiber-Brillouin amplifiers (Olsson and van der Ziel, 1986), which utilize stimulated Raman scattering and stimulated Brillouin scattering in an optical fiber, may be applied to coherent optical fiber transmission systems. Erbium (Er)-doped fiber amplifiers (Urquhart, 1988; Mears et al., 1987) utilize stimulated emission, where population inversion is generated by optical pumping. These fiber amplifiers require a pump laser with sufficient output power and a suitable operation wavelength.

Important device characteristics of optical amplifiers as applied to coherent optical fiber transmission are small-signal gain, saturation output power, frequency bandwidth, and noise figure (Mukai et al., 1985). These characteristics, as well as features and problems, are summarized in Table IV for the semiconductor, fiber-Raman, and Er-doped fiber amplifiers.

TABLE IV
Characteristics, Principal Features, and Problems of Optical Amplifiers

	Fabry–Perot Type Semiconductor Amplifier	Traveling-wave Type Semiconductor Amplifier	Injection-locked Semiconductor Laser	Fiber Raman Amplifier	Er-Doped Fiber Amplifier
Small-signal gain	25–35 dB	~20 dB	20–30 dB	20–30 dB	20–40 dB
Saturation output level	−10–0 dBm	0–10 dBm	10–20 dBm	~20 dBm	~20 dBm
Bandwidth	1–3 GHz	~1 THz	0.5–1 GHz	~1 THz	0.1–1 THz
Noise figure (experiment)	3 dB (6–9 dB)	3 dB (5.2–9 dB)	3 dB (6–9 dB)	3 dB	3 dB (3.2–6 dB)
Excess noise bandwidth	Medium	Large	Small	Medium	Medium
Input/output coupling loss	Medium	Medium	Medium	Low	Low
Gain stability against frequency deviation	Medium	High	Low	Medium	Medium
Features	Spontaneous emission filtering function	Broad bandwidth Common amplification	Suppression of spontaneous emission by gain saturation and filtering function	Broad bandwidth Small coupling loss	Polarization-insensitive gain Small coupling loss
Problems	Small output power	Need for narrow-band optical filter	Need for frequency stabilization	Need for reliable pump laser in 1.5-μm region	Need for reliable pump laser
References	Mukai et al. (1982, 1983)	Simon, (1982), Eisenstein et al. (1985)	Kobayashi and Kimura (1981), Kobayashi et al. (1981)	Washio et al. (1985), Hegarty et al. (1985)	Mears et al. (1987), Miniscalco et al. (1990), Yamada et al. (1990)

Since a Fabry–Perot cavity-type amplifier inherently has an optical frequency filtering function, its noise figure and excess noise bandwidth can be decreased by employing a low-reflectivity input facet and a high-reflectivity output facet (Mukai et al., 1982). Here, the excess noise bandwidth is defined by the spontaneous emission power of the amplifier normalized by the small-signal gain.

Traveling-wave type amplifiers and fiber Raman amplifiers have a broad bandwidth of over 1 THz, making them advantageous for the common amplification of frequency-division multiplexing signals. Fiber-Brillouin amplifiers have narrow bandwidths, which are inherently about 15 MHz for silica fibers at a 1.5-μm wavelength, prompting attempts to broaden the Brillouin bandwidth through frequency modulation of the pump light (Olsson and van der Ziel, 1986). Injection-locked oscillators, which have a small excess noise bandwidth and a high power output, can be applied to the power amplification of angle modulation signals.

Er-doped fiber amplifiers allow 1.53–1.56 μm input signal light to be amplified. A small-signal gain of more than 40 dB has been achieved (Miniscalco et al., 1990). They have the advantages of polarization-insensitive gain and small coupling loss. 1.48-μm and 0.98-μm semiconductor lasers are available for pumping the amplifier. Owing to these advantages, Er-doped fiber amplifiers have already demonstrated their potential in system experiments (Iaubal et al., 1989; Nakagawa et al., 1990), even though they were developed quite recently.

All optical amplifiers described here have a theoretical noise-figure limit of 3 dB. A small noise figure of 5.2 dB has been achieved for InGaAsP traveling-wave amplifiers (Saitoh and Mukai, 1987). Er-doped fiber amplifiers have also demonstrated a small noise figure of 3.2 dB (Yamada et al., 1990).

VI. System Experiments

A large number of system experiments in laboratories using coherent techniques have been reported. Generally, however, these experiments have not yet exhibited the ultimate in coherent system performance, since they are based mainly on conventional optical device technology. The principal points in the development stage continue to be verifying system operation, confirming feasibility, and elucidating the technological problems to be solved. This section reviews significant experiments in coherent optical transmission technology. The latest transmission experiments are also introduced.

Table V summarizes four of the most significant reports on coherent optical system experiments (Kimura, 1987). The first coherent optical

TABLE V
DISTINCTIVE COHERENT OPTICAL SYSTEM EXPERIMENTS

	(a)	(b)	(c)	(d)
Light sources	He–Ne lasers	AlGaAs lasers	He–Ne laser (TR) External-cavity InGaAsP laser (LO)	InGaAsP DFB lasers
Wavelength	3.39 μm	0.83 μm	1.52 μm	1.57 μm
Modulation	FM (50 kHz) Direct modulation (cavity length)	FSK (100 Mb/s) Direct modulation (injection current)	PSK (140 Mb/s) Phase modulator	FSK (100 Mb/s) Direct modulation (injection current)
Demodulation	Heterodyne frequency-discrimination detection	Heterodyne frequency-discrimination detection	Heterodyne synchronous detection	Heterodyne single-filter detection
Transmission medium	Space (150 m)	Space (Face-to-face)	Single-mode fiber (109 km)	Single-mode fiber (105 km)
Features	Quantitative evaluation of laser FM noise and atmospheric turbulence	Usefulness of direct frequency modulation in semiconductor laser Broadband FM signal	Small polarization-state fluctuation 1.5-μm narrow-linewidth lasers	Suppression of nonflat FM response using Manchester coding Single-filter detection suppressing S/N degradation due to FM noise
Year	1967	1980	1983	1984
References	Goodwin (1967)	Saito et al. (1980)	Wyatt et al. (1983)	Emura et al. (1984)

transmission experiment was made in 1967 using two 3.39-μm He–Ne lasers (Goodwin, 1967). A 50-kHz FM signal transmitted across 150 m of space was received by heterodyne frequency-discrimination detection. Better sensitivity than that of a conventional photodetector receiver was also demonstrated. Laser FM noise and atmospheric turbulence, both of which determined the system performance, were quantitatively evaluated. The causes of laser FM noise were discussed in detail, and stable He–Ne lasers were specially designed accordingly.

Two independent semiconductor lasers were used in the heterodyne detection system experiment reported in 1980 (Saito et al., 1980). An AlGaAs laser operating in a single longitudinal mode was directly frequency-modulated through injection current modulation. A 100-Mb/s FSK signal was heterodyne-detected using another AlGaAs laser as a local oscillator. Direct frequency modulation of a semiconductor laser, whose characteristics were measured in the modulation frequency region of 150–900 MHz, demonstrated its usefulness for coherent optical transmission.

Long-fiber transmission over 100 km was demonstrated by Wyatt et al. in 1983, using a 1.52-μm He–Ne laser and an external cavity InGaAsP laser as a transmitter laser and a local oscillator, respectively. 140-Mb/s PSK heterodyne coherent detection was carried out, and a receiver sensitivity improvement of 14 dB above direct detection was achieved. This experiment, using conventional single-mode fibers, indicated that the polarization state fluctuation following propagation over 100 km was small and slow enough to be compensated for by manual adjustment using a fiber polarization controller.

In 1984, two 1.57-μm InGaAsP DFB lasers were used in a coherent transmission system of 105-km line length (Emura et al., 1984). Direct frequency modulation generated a 100-Mb/s FSK signal in which nonuniform FM response was suppressed by utilizing Manchester coding. SNR degradation due to laser FM noise was relaxed by using a single-filter detection scheme in the electronic demodulation stage after optical heterodyne detection. Thereafter, 140-Mbit/s, 243-km and 280-Mbit/s, 204-km FSK heterodyne single-filter detection transmission experiments were demonstrated using a phase-tunable DFB laser in the transmitter. Moreover, a 34-Mb/s FSK heterodyne dual-filter detection experiment achieved signal transmission over 300 km (Emura et al., 1986).

Recently 400-Mb/s, 270-km (Iwashita et al., 1986a) and 1-Gb/s, 150-km (Linke et al., 1986) heterodyne detection experiments have been reported. The first experiment, whose system configuration is shown in Fig. 18a, used two external-cavity DFB lasers oscillating at 1.546 μm for its transmitter and local oscillator. The beat spectrum linewidth between the two lasers was less than 200 kHz. Direct modulation of the transmitter laser, in which an LC circuit was used to achieve a flat FM response, created a 400-Mb/s continuous-phase

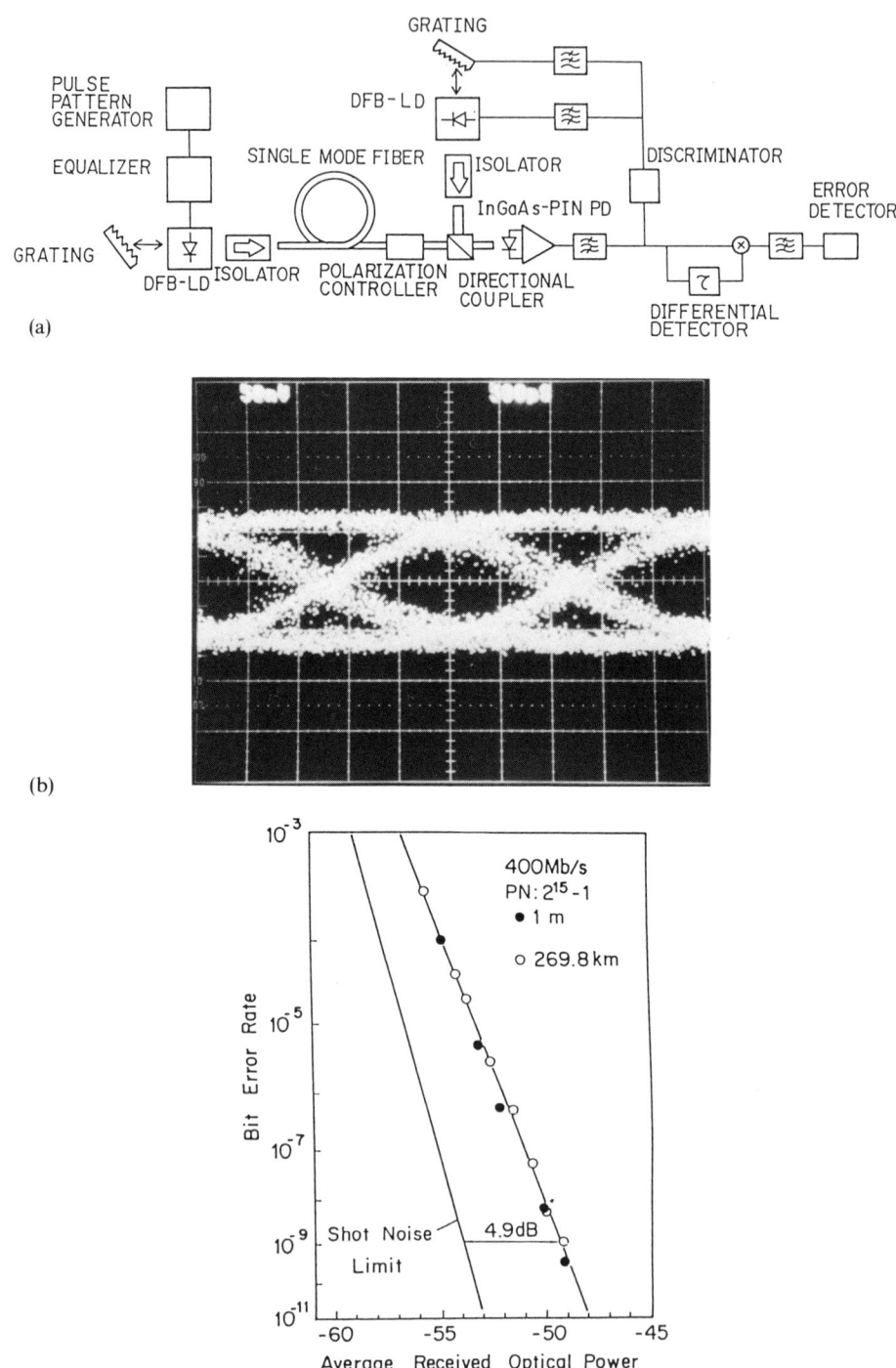

FIG. 18. 400-Mbit/s · 270-km FSK transmission experiment (Iwashita et al., 1986a): (a) experimental setup; (b) eye pattern of a received signal; (c) error-rate performance.

FSK signal with a frequency shift of nearly 250 MHz, where the residual amplitude modulation was less than 0.5 dB. The transmitter laser output power and fiber input power were 10.6 dBm and 5.5 dBm, respectively. A conventional single-mode fiber was used, and transmitted signal polarization was manually adjusted with a fiber-optic polarization controller. The total fiber loss, including splicing, was 53 dB at 1.546 μm. After heterodyne detection with an InGaAs $p-i-n$ photodiode, the continuous-phase FSK (CPFSK) signal was demodulated by differential demodulation. The received eye pattern and the error-rate performance for a $2^{15} - 1$ pseudorandom binary sequence are shown in Figs. 18b and c, respectively. A receiver sensitivity of -49 dBm was achieved for an error rate of 10^{-9}. In the course of this experiment, a 400-Mb/s continuous-phase FSK signal with a 0.5 modulation index, that is, a 400-Mb/s minimum-shift-keying (MSK) signal, was transmitted over a 289-km pure-silica-core single-mode fiber (Iwashita et al., 1986b). Moreover, 2.5-Gb/s, 308-km (Imai et al., 1990a) and 8-Gb/s, 202-km (Takachio et al., 1990) CPFSK heterodyne experiments were clearly demonstrated. Here, the former achieved a receiver sensitivity of 67 photons/bit at a 10^{-9} error rate by using a three-electrode DFB laser module and a low-noise balanced receiver. At a wavelength of 1.55 μm, the latter successfully compensated for the dispersive effect of a 1.3-μm zero-dispersion single-mode fiber by using a microstrip delay equalizer in the IF stage.

The second experimental system was a DPSK system using separate 1.5-μm buried heterostructure (BH) lasers with external grating cavities of 5-kHz spectral linewidth (Linke et al., 1986). PSK modulation was performed with a titanium-diffused lithium niobate waveguide phase modulator with 1-cm-long traveling-wave electrodes. A signal greater than -3 dBm was coupled into a single-mode fiber. The total fiber loss, including connector and splice losses, was 39.6 dB. The polarization state of local oscillator light was adjusted to that of the transmitted signal. A balanced-mixer receiver composed of InGaAs $p-i-n$ photodetectors was employed. A receiver sensitivity of -44.5 dBm achieved at 1 Gb/s corresponded to an improvement of 7.5 dB over the best direct detection result. Recently, this experiment has been followed by a 2-Gb/s, 170-km DPSK heterodyne detection experiment (Gnauck et al., 1987). In addition to these experiments, a 1.2-Gb/s DPSK heterodyne detection system using an external-cavity DFB laser module has achieved a transmission distance of 201 km (Chikama et al., 1990).

For intensity-modulation direct-detection schemes, 565-Mb/s, 204-km (Shikada et al., 1985), 2-Gb/s, 141-km (Shikada et al., 1985), 5-Gb/s, 111-km (Heidemann et al., 1987), and 10-Gb/s 80-km (Fujita et al., 1988) system experiments utilizing direct intensity modulation of 1.5-μm DFB lasers have been reported. All of the experiments commonly employed single-mode fibers with a shifted zero-dispersion wavelength to suppress the dispersive effect.

System experiments, using external waveguide modulators and conventional 1.3-μm zero-dispersion single-mode fibers, also achieved 4-Gb/s, 117-km (Korotky et al., 1985) and 8-Gb/s, 68-km (Gnauck et al., 1986) signal transmission.

Transmission distances and data rates achieved by both coherent and conventional system experiments are shown in Fig. 19. Laboratory experiments have already produced coherent transmission systems of comparable or longer transmission distance than conventional systems.

All of the above coherent transmission experiments used heterodyne detection schemes. Moreover, they used, not polarization-maintaining fibers, but conventional single-mode fibers in conjunction with polarization-control devices. However, other types of coherent transmission experiments have also been reported. For homodyne detection in the 1.3–1.6-μm wavelength range, two separate He–Ne lasers demonstrated a 140-Mb/s PSK signal transmission over a 30-km single-mode fiber (Malyon, 1984). An impressive experiment of PSK homodyne detection using a phase-locked semiconductor laser has been done in a 1-Gb/s, 209-km system (Kahn, 1989). Recently, this experiment has been extended to a 4-Gb/s, 167-km system (Kahn, 1990).

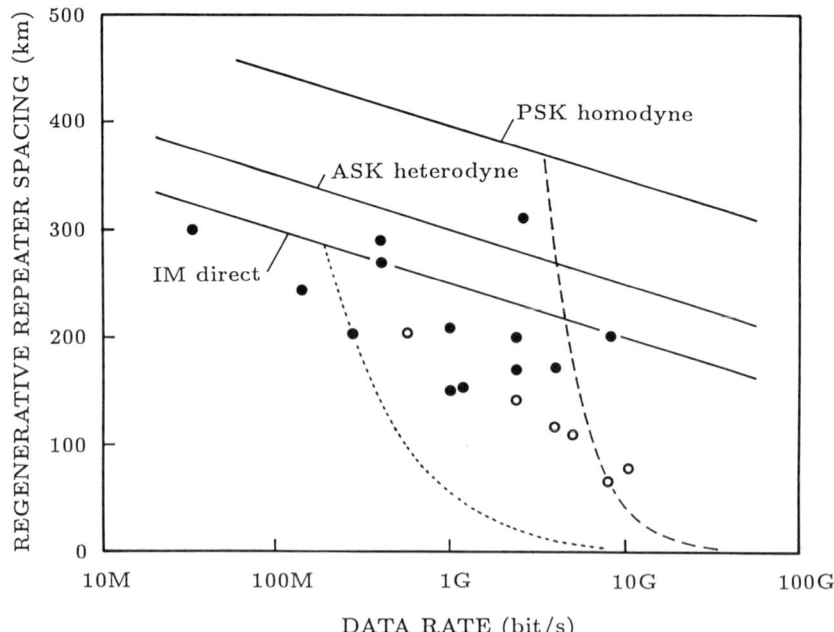

FIG. 19. Transmission distance versus data rate for coherent and conventional system experiments.

A three-port phase-diversity receiver displayed its ability in a 140-Mb/s, 80-km FSK transmission experiment using conventional 1.5-μm DFB lasers (Pettit et al., 1987). A 10.5-km polarization-maintaining fiber was employed in a 400-Mb/s FSK heterodyne detection experiment (Iwashita et al., 1986c). A polarization diversity technique was also applied to 560-Mb/s MSK signal transmission over a 150-km single-mode fiber (Ryu et al., 1987) and 1.2-Gb/s DPSK signal transmission over a 100-km fiber (Chikama et al., 1990).

With respect to amplifier repeaters, two Fabry–Perot-type 1.5-μm amplifiers were used as in-line repeaters in a 100-Mb/s ASK heterodyne detection experiment (Olsson, 1985), which followed 0.8-μm tandem amplifier experiments (Mukai et al., 1982). Four in-line amplifiers with a 0.5–3% reflectivity were installed to demonstrate 400-Mb/s FSK signal transmission over a total distance of 372 km (Olsson et al., 1988). Together with this experiment, a 1-Gb/s, 313-km IM direct detection experiment using three in-line amplifiers was reported (Oberg et al., 1988). Quite recently, a 2.5-Gb/s CPFSK signal has been successfully transmitted over 2,200 km of single-mode fiber, in which 25 Er-doped-fiber in-line amplifiers were installed at 80-km intervals (Saito et al., 1990). Moreover, an Er-doped-fiber post amplifier with 18 dBm of output power has enabled a 2.5-Gb/s signal to be transmitted over 350 km of single-mode fiber in a heterodyne detection system, where the stimulated Brillouin scattering limit was alleviated with FSK modulation (Sugie et al., 1990).

Recent progress in coherent technology allows transmission system experiments to be conducted in a field environment (Brain et al., 1990; Imai et al., 1990). One of these experiments (Imai et al., 1990) achieved error-free transmission over 17 hours with a 2.5-Gb/s CPFSK heterodyne detection system consisting of three cascaded sections. Installed submarine fiber cable was used, and the total system length was 431 km. Three-electrode DFB laser modules having a flat FM response and about 2 MHz linewidth were used in the transmitters. Polarization diversity receivers enabled the system to be unaffected by polarization fluctuation occurring in the installed fiber cables.

VII. CONCLUSION

Research on coherent optical transmission has produced many fruitful results since it was revived in the late 1970s. A demonstration of direct frequency modulation capability in semiconductor lasers was one of the most encouraging studies leading to steady improvements in coherent transmission technology. The absolute frequency stabilization of 1.5-μm semiconductor lasers also made available an essential element in providing stable carrier waves to systems in the low-loss wavelength region of fibers. A double-balanced optical heterodyne or homodyne receiver additionally proved to be a unique method of removing the excess AM noise of a local oscillator and achieving quantum-limited performance. In addition, an injection-locked semiconductor laser demonstrated its own potential as a post-amplifier and as a PM signal transmitter. Also, an evaluation of fiber birefringence showing the usefulness of thermal-stress-induced anisotropy opened the way for single-polarization fiber development. In a further important development, transmission experiments have successfully demonstrated the principal system features leading to increased transmission lengths and data rates.

The ultimate goal of coherent optical fiber transmission is to utilize fully the coherent properties of laser light and the low-loss and broad-bandwidth characteristics of the optical fibers. Single-frequency semiconductor laser technology, opto-electronic integrated circuit (OEIC) technology, and new technologies based on these fundamental device technologies will solidify the foundation necessary to develop coherent optical fiber transmission systems.

Definite research targets for the future include the following:

(1) A narrow-linewidth semiconductor laser whose oscillation frequency is stabilized to an absolute frequency reference during a long period of time, for example, over 10,000 hours or more.

(2) A high-speed FM semiconductor laser with suppressed temperature modulation effect as well as suppressed intensity modulation.

(3) An OEIC negative-feedback-controlled semiconductor laser configuration potentially applicable to wideband optical PLLs.

(4) A single-polarization fiber capable of simultaneously decreasing fiber loss and mode coupling.

(5) A low-noise optical amplifier with stable frequency and gain.

(6) Optical passive devices such as low-loss narrow-band optical filters and low-loss optical isolators.

References

Ainslie, B. J., and Day, C. R. (1986). "A review of single-mode fibers with modified dispersion characteristics," *IEEE J. Technol.* **LT-4**(8), 967–979.

Alferness, R. C. (1981). "Guided-wave devices for optical communication," *IEEE J. Quantum Electron.* **QE-17**(6), 946–959.

Birch, R. D., Payne, D. N., and Varnham, M. P. (1982). "Fabrication of polarization-maintaining fibers using gas-phase etching," *Electron. Lett.* **18**(24), 1036–1038.

Brain, M. C., Creaner, M. J., Steele, R. C., Walker, N. G., Walker, G. R., Mellis, J., Al-Chalabi, S., Davidson, J., Rutherford, M., and Sturgess, I. C. (1990). "Progress towards the field deployment of coherent optical fiber systems. "*IEEE J. Lightwave Technol.* **L-8**(3), 423–437.

Cella, T., Dutta, N. K., Piccirilli, A. B., and Brown, R. L. (1987). "Monolithically integrated thermoelectrically tunable distributed Bragg reflector laser," *Electron. Lett.* **23**(19), 1031–1032.

Chan, V. W. S. (1987). "Space coherent optical communications—An introduction," *IEEE J. Lightwave Tech.* **LT-5**(4), 633–637.

Chan, V. W. S., Jeromin, L. L., and Kaufmann, J. E. (1983). "Heterodyne lasercom systems using GaAs lasers for ISL applications," *IEEE Int. Conf. Commun.* E1-5, Boston, Massachusetts, USA, Jun. 19–22.

Chikama, T., Watanabe, S., Naito, T., Onaka, H., Kiyonaga, T., Onoda, Y., Miyata, H., Suyama, M., Seino, M., and Kuwahara, H. (1990). "Modulation and demodulation techniques in optical heterodyne PSK transmission systems," *IEEE J. Lightwave Technol.* **LT-8**(3), 309–322.

Cotter, D. (1983). "Stimulated Brillouin scattering in monomode optical fiber," *J. Opt. Commun.* **4**(1), 10–19.

Davis, A. W., Pettitt, M. J., King, J. P., and Wright, S. (1987). "Phase diversity techniques for coherent optical receivers," *IEEE J. Lightwave Technol.* **LT-5**(4), 561–572.

DeLange, O. E. (1970). "Wide-band optical communication systems: Part II—Frequency-division multiplexing," *Proc. IEEE* **58**(10), 1683–1690.

Eisenstein, G., Jopson, R. M., Linke, R. A., Burrus, C. A., Koren, U., Whalen, M. S., and Hall, K. L. (1985). "Gain measurements of InGaAsP 1.5 μm optical amplifiers," *Electron. Lett.* **21**(23), 1076–1077.

Emura, K., Shikada, M., Fujita, S., Mito, I., Honmou, H., and Minemura, K. (1984). "Novel optical FSK heterodyne single filter detection system using a directly modulated DFB-laser diode," *Electron. Lett.* **20**(24), 1022–1023.

Emura, K., Yamazaki, S., Fujita, S., Shikada, M., Mito, I., and Minemura, K. (1986). "Over 300 km transmission experiment on an optical FSK dual heterodyne dual filter detection system," *Electron. Lett.* **22**(21), 1096–1097.

Emura, E., Yamazaki, S., Shikada, M., Fujita, S., Yamaguchi, M., Mito, I., and Minemura, K. (1987). "System design and long-span transmission experiments on an optical FSK heterodyne single filter detection system," *IEEE J. Lightwave Technol.* **LT-5**(4). 469–477.

Favre, F., Jeunhomme, L., Joindot, I., Monerie, M., and Simon, J. C. (1981). "Progress towards heterodyne-type single-mode fiber communication systems," *IEEE J. Quantum Electron.* **QE-17**(6), 897–906.

Forrest, S. R. (1985). "Monolithic optoelectronic integration: A new component technology for lightwave communications," *IEEE J. Lightwave Technol.* **LT-3**(6), 1248–1263.

Fujita, S., Henmi, N., Takano, I., Yamaguchi, M., Torikai, T., Suzaki, T., Takano, S., Ishihara, H. and Shikada, M. (1988). "A 10-Gb/s-80 km optical fiber transmission experiment using a directly modulated DFB-LD and a high speed InGaAs-APD," *Opt Fiber Commun. Conf.*, PD16, New Orleans, Louisiana, USA, Jan. 25–28.

Garrett, I., and Jacobsen, G. (1986). "Theoretical analysis of heterodyne optical receivers for transmission systems using (semiconductor) lasers with nonnegligible linewidth," *IEEE J. Lightwave Technol.*, **LT-4**(3), 323–334.

Gimlett, J. L., Vodhanel, R. S., Choy, M. M., Elrefaie, A. F., Cheung, N. K., and Wagner, R. E. (1987). "A 2-Gbit/s optical heterodyne transmission experiment using a 1520-nm DFB laser transmitter," *IEEE J. Lightwave Technol.* **LT-5**(9), 1315–1324.

Gnauck, A. H., Korotky, S. K., Kasper, B. L., Campbell, J. C., Talman, J. R., Veselka, J. J., and McCormick, A. R. (1986). "Information-bandwidth-limited transmission at 8 Gb/s over 68.3 km of optical fiber," *9th Conf. Opt. Fiber Commun.*, PD9, Atlanta, Georgia, USA, Feb. 24–26.

Gnauck, A. H., Linke, R. A., Kasper, B. L., Pollock, K. J., Reichmann, K. C., Valenzuela, R., and Alferness, R. C. (1987). "Coherent lightwave transmission at 2 Gbit/s over 170 km of optical fiber using phase modulation," *Electron. Lett.* **23**(6), 286–287.

Goodman, C. H. L. (1978). "Devices and materials for 4 μm-band fiber-optical communication," *Solid-State and Electron Devices* **2**(5), 129–137.

Goodwin, F. E. (1967). "A 3.39-micron infrared optical heterodyne communication system," *IEEE J. Quantum Electron.* **QE-3**(11), 524–531.

Gordon, J. P. (1962). "Quantum effects in communications system," *Proc. IRE* **50**(9), 1898–1908.

Harmon, R. A. (1982). "Polarization stability in long lengths of monomode fiber," *Electron. Lett.* **18**(24), 1058–1061.

Haug, H. (1969). "Quantum-mechanical rate equations for semiconductor lasers," *Phys. Rev.* **184**(2), 338–348.

Haus, H. A., and Townes, C. H. (1962). 'Comments on Noise in photoelectric mixing,'" *Proc. IRE* **50**(6), 1544–1546.

Hegarty, J., Olsson, N. A., and Goldner, L. (1985). "CW pumped Raman preamplifier in a 45 km-long fiber transmission system operating at 1.5 μm and 1 Gbit/s," *Electron. Lett.* **21**(7), 290–292.

Heidemann, R., Scholz, U., and Wedding, B. (1987). "5 Gbit/s transmission system experiment over 111 km of optical fiber," *Electron. Lett.* **23**(19), 1030–1031.

Henning, I. D., Westbrook, L. D., Nelson, A. W., and Fiddyment, P. J. (1984). "Measurements of the linewidth of ridge-guide DFB lasers," *Electron. Lett.* **20**(21), 885–887.

Henry, C. H. (1982). "Theory of the linewidth of semiconductor lasers," *IEEE J. Quantum Electron.* **QE-18**(2), 259–264.

Henry, P. S. (1985). "Lightwave primer," *IEEE J. Quantum Electron.* **QE-21**(12), 1862–1879.

Hodgkinson, T. G. (1986) "Costas loop analysis for coherent optical receivers," *Electron. Lett.* **22**(7), 394–396.

Hodgkinson, T. G., Harmon, R. A., and Smith, D. W. (1985). "Demodulation of optical DPSK using in-phase and quadrature detection," *Electron. Lett.* **21**(19), 867–868.

Hooper, R. C., Midwinter, J. E., Smith, D. W., and Stanley, I. W. (1983). "Progress in monomode transmission techniques in the United Kingdom," *IEEE J. Lightwave Technol.* **LT-1**(4), 596–611.

Hosaka, T., Sasaki, Y., Noda, J., and Horiguchi, M. (1985). "Low-loss and low-crosstalk polarization-maintaining optical fibers," *Electron. Lett.* **21**(20), 920–921.

Imai, T., Okhawa, N., Ichihashi, Y., Sugie, T., and Ito, T. (1990a). "Over 300 km CPFSK heterodyne transmission experiment using 67 photon/bit sensitivity receiver at 2.5 Gbit/s," *Electron Lett.* **26**(6), 357–358.

Imai, T., Hayashi, Y., Okhawa, N., Sugie, T., Ichihashi, Y., and Ito, T. (1990b). "Field demonstration of 2.5 Gbit/s coherent optical transmission through installed submarine fiber cables," *Electron. Lett.* **26**(17), 1407–1408.

Iqubal, M. Z., Gimlett, J. L., Choy, M. M., Yi-Yan, A., Andrejco, M. J., Curtis, L., Saifi, M. A., Lin, C., and Cheung, N. K. (1989). "An 11 Gb/s, 151 km transmission experiment employing a 1480 nm pumped erbium-doped in-line Fiber amplifier," *7th Int. Conf. Integrated Opt. and Optical Fiber Commun.*, 20PDA-7, Kobe, Japan, Jul. 18–21.

Iwashita, K., Imai, T., Matsumoto, T., and Motosugi, G. (1986a). "400 Mbit/s optical FSK transmission experiment over 270 km of single-mode fiber," *Electron. Lett.* **22**(3), 164–165.

Iwashita, K., Matsumoto, T., Tanaka, C., and Motosugi, G. (1986b). "Linewidth requirement evaluation and 290 km transmission experiment for optical CPFSK differential detection," *Electron. Lett.* **22**(15), 791–792.

Iwashita, K., Kano, H., Matsumoto, T., and Sasaki, Y. (1986c). "FSK transmission experiment using 10.5 km polarization-maintaining fiber," *Electron. Lett.* **22**(4), 214–215.

Jacobsen, G., Olesen, H., and Birkedahl, F. (1982). "Current/frequency-modulation characteristics for directly optical frequency-modulated injection lasers at 830 nm and 1.3 μm," *Electron. Lett.* **18**(20), 874–876.

Jeunhomme, L., and Monerie, M. (1980). "Polarization-maintaining single-mode fiber cable design," *Electron. Lett.* **16**(24), 921–922.

Jungerman, R. L., Johnsen, C. A., Dolfi, D. W., and Nazarathy, M. (1987). "Coded phase-reversal $LiNbO_3$ modulator with bandwidth greater than 20 GHz at 1.3 μm wavelength," *Electron. Lett.* **23**(4), 172–174.

Kahn, J. M. (1989). "1 Gbit/s PSK homodyne transmission system using phase-locked semiconductor lasers," *IEEE Photonics Technol. Lett.* **1**(10), 340–342.

Kahn, J. M. (1990). "4-Gb/s PSK homodyne transmission system using phase-locked semiconductor lasers," *IEEE Photonics Technol. Lett.* **2**(4), 285–287.

Kaminow, I. P. (1981). "Polarization in optical fibers," *IEEE J. Quantum Electron.* **QE-17**(1), 15–22.

Kaminow, I. P., and Ramaswamy, V. (1979). "Single-polarization optical fibers: Slab model," *Appl. Phys. Lett.* **34**(4), 268–270.

Katsuyama, T., Matsumura H., and Suganuma, T. (1983). "Low-loss single polarization fibers," *Appl. Opt.* **22**(11), 1741–1747.

Kazovsky, L. G. (1985). "Decision-driven phase-locked loop for optical homodyne receivers: Performance analysis and laser linewidth requirements," *IEEE J. Lightwave Technol.* **LT-3**(6), 1238–1247.

Kazovsky, L. G. (1986). "Balanced phase-locked loops for optical homodyne receivers: Performance analysis, design considerations, and laser linewidth requirements," *IEEE J. Lightwave Technol.* **LT-4**(2), 182–195.

Kikuchi, K., Okoshi, T., Nagamatsu, M., Henmi, N. (1984). "Degradation of bit-error rate in coherent optical communications due to spectral spread of the transmitter and the local oscillator," *IEEE J. Lightwave Technol.* **LT-2**(6), 1024–1033.

Kimura, T. (1987). "Coherent optical fiber transmission," *IEEE J. Lightwave Technol.* **LT-5**(4), 414–428.

Kimura, T., and Sugimura, A. (1987). "Linewidth reduction by coupled phase-shift distributed-feedback lasers," *Electron Lett.* **23**(19), 1014–1015.

Kobayashi S., and Kimura, T. (1981). "Injection locking in AlGaAs semiconductor laser," *IEEE J. Quantum Electron.* **QE-17**(5), 681–689.

Kobayashi, S., and Kimura, T. (1982). "Optical phase modulation in an injection locked AlGaAs semiconductor laser," *IEEE J. Quantum Electron.* **QE-18**(10), 1662–1669.

Kobayashi, S., Yamamoto, Y., and Kimura, T. (1981). "Optical FM signal amplification and FM noise reduction in an injection locked AlGaAs semiconductor laser," *Electron. Lett.* **17**(22), 849–851.

Kobayashi, S., Yamamoto, Y., Ito, M., and Kimura, T. (1982). "Direct frequency modulation in AlGaAs semiconductor lasers," *IEEE J. Quantum Electron.* **QE-18**(4), 582–595.

Korotky, S. K., Eisenstein, G., Gnauck, A. H., Kasper, B. L., Veselka, J. J., Alferness, R. C., Buhl, L. L., Burrus, C. A., Huo, T. C. D., Stulz. L. W., Nelson, K. C., Cohen, L. G., Dawson, R. W., and Campbell, J. C. (1985). "4-Gb/s transmission experiment over 117 km of optical fiber using a Ti:$LiNbO_3$ external modulator," *IEEE J. Lightwave Technol.* **L-3**(5), 1027–1031.

Linke, R. A., Kasper, B. L., Olsson, N. A., and Alferness, R. C., (1986). "Coherent lightwave transmission over 150 km fiber lengths at 400 Mbit/s and 1 Gbit/s data rates using phase modulation," *Electron. Lett.* **22**(1), 30–31.

Machida, S., Kawana, A., Ishihara, K., and Tsuchiya, H. (1979). "Interference of an AlGaAs laser diode using a 4.15 km single mode fiber cable," *IEEE J. Quantum Electron.* **QE-15**(7), 535–537.

Machida, S., Sakai, J., and Kimura, T. (1982). "Polarization preservation in long-length twisted single-mode optical fibers," *Trans. IECE Japan* **E65**(11), 642–648.

Malyon, D. J. (1984). "Digital fiber transmission using optical homodyne detection," *Electron. Lett.* **20**(7), 281—283.

Manes, K. R., and Siegman, A. E. (1971). "Observation of quantum phase fluctuations in infrared gas lasers," *Phys. Rev. A* **4**(1), 373–386.

Marcuse, D. (1974). "Theory of Dielectric Optical Waveguides." Academic Press, New York.

Marrone, M. J. (1985). "Polarization holding in long-length polarizing fibers," *Electron. Lett.* **21**(6), 244–245.

McMahon, D. H., Dyes, W. A., Cooper, R. F., Robinson, W. C., and Mahapatra, A. (1987). "Echelon grating multiplexers for hierarchically multiplexed fiber-optic communication networks," *Appl. Opt.* **26**(11), 2188–2196.

Mears, R. J., Reelie, L., Jauncey, I. M., and Payne, D. N. (1987). "Low-noise erbium-doped fiber amplifier operating at 1.54 μm," *Electron. Lett.* **23**(19), 1026–1028.

Miniscalco, W. J., Thompson, B. A., Eichen, E., and Wei, T. (1990). "Very high gain Er^{3+} fiber amplifier pumped at 980 nm," *Opt. Fiber Commun. Conf.*, FA2, San Francisco, California, USA, Jan. 22–26.

Mukai, T., Yamamoto, Y., and Kimura, T. (1982). "S/N and error rate performance in AlGaAs semiconductor laser preamplifier and linear repeater systems," *IEEE J. Quantum Electron.* **QE-18**(10), 1560–1568.

Mukai, T., Saitoh, T., Mikami, O., and Kimura, T. (1983). "Fabry–Perot cavity type 1.5 μm InGaAsP BH-laser amplifier with small optical-mode confinement," *Electron. Lett.* **19**(15), 582–583.

Mukai, T., Yamamoto, Y., and Kimura, T. (1985). "Optical amplification by semiconductor lasers," *in* "Semiconductors and Semimetals, Lightwave Communication Technology" (W. T. Tsang, ed.). Academic Press, New York.

Nakagawa, K., Hagimoto, K., and Nishi, S. (1990). "Optical bit rate flexible transmission experiment over 10 Gbit/s 375 km employing in-line fiber amplifiers," *Opt. Fiber Commun. Conf.*, WC2, San Francisco, California, USA, Jan. 22–26.

Nakazawa, M., Shibata, N., Tokuda, M., and Negishi, Y. (1984). "Measurement of polarization mode couplings along polarization-maintaining single-mode optical fibers," *J. Opt. Soc. Am. A.* **1**(3) 285–292.

Nilsson, O., and Yamamoto, Y. (1985). "Small-signal response of a semiconductor laser with inhomogeneous linewidth enhancement factor: Possibilities of a flat carrier-induced FM response," *Appl. Phys. Lett.* **46**(3), 223–225.

Nosu, K., and Iwashita, K. (1988). "A consideration of factors affecting future coherent lightwave communication systems," *IEEE J. Lightwave Technol.* **LT-6**(5), 686–694.

Oberg, M. G., Olsson, N. A., Koszi, L. A., and Przybylek, G. J. (1988). "313 km transmission experiment at 1 Gbit/s using optical amplifiers and a low chirp laser," *Electron. Lett.* **24**(1), 38–39.

Ohtsu, M., and Kotajima, S. (1985). "Linewidth reduction of a semiconductor laser by electrical feedback," *IEEE J. Quantum Electron.* **QE-21**(12), 1905–1912.

Okoshi, T. (1985). "Polarization-state control schemes for heterodyne or homodyne optical fiber communications," *IEEE J. Lightwave Technol.* **LT-3**(6), 1232–1237.

Okoshi, T. (1987). "Recent advances in coherent optical fiber communication systems," *IEEE J. Lightwave Technol.* **LT-5**(1) 44–52.

Okoshi, T., and Kikuchi, K. (1980). "Frequency stabilization of semiconductor lasers for heterodyne-type optical communication systems," *Electron. Lett.* **16**(5), 179–181.

Okoshi, T., and Kikuchi, K. (1981). "Heterodyne-type optical fiber communications," *J. Opt. Commun.* **2**(3), 82–88.

Okoshi, T., Emura, K., Kikuchi, K., and Kersten, R. Th. (1981). "Computation of bit-error rate of various heterodyne and coherent-type optical communication schemes," *J. Opt. Commun.* **2**(3), 89–96.

Okoshi, T., Ryu, S., and Kikuchi, K. (1983). "Polarization-diversity receiver for heterodyne/coherent optical fiber communications," *4th Int. Conf. Integrated Opt. and Optical Fiber Commun.*, 30C3-2, Tokyo, Japan, Jun. 27–30.

Olesen, H., Saito, S., Mukai, T., Saitoh, T., and Mikami, O. (1983). "Solitary spectral linewidth and its reduction with external grating feedback for a 1.55 μm InGaAsP BH laser," *Jpn. J. Appl. Phys.* **22**(10), L664–L666.

Oliver, B. M. (1961). "Signal-to-noise ratios in photoelectric mixing," *Proc. IRE* **49**(12), 1960–1961.

Olsson, N. A. (1985). "ASK heterodyne receiver sensitivity measurements with two in-line 1.5 μm optical amplifiers," *Electron. Lett.* **21**(23), 1085–1087.

Olsson, N.A., van der Ziel, J. P. (1986). "Fiber Brillouin amplifier with electronically controlled bandwidth," *9th Conf. Opt. Fiber Commun.*, PD6, Atlanta, Georgia, USA, Feb. 24–26.

Olsson, N. A., Henry, C. H., Kazarinov, R. F., Lee, H. J., Johnson, B. H., and Orlowsky, K. J. (1987). "Narrow linewidth 1.5 μm semiconductor laser with a resonant optical reflector," *Appl. Phys. Lett.* **51**(15), 1141–1142.

Olsson, N. A., Oberg, M. G., Koszi, L. A., and Przybylek, G. J. (1988). "400 Mbit/s, 372 km coherent transmission experiment using in-line optical amplifiers," *Electron. Lett.* **24**(1), 36–38.

Onaka, H., Miyata, H., Kotaki, Y., Kuwahara, H., Takada, T., and Takeuchi, Y. (1990). "Distributed feedback laser diode module with 8-GHz flat FM response and +10-dBm output for coherent lightwave transmission," *Opt. Fiber Commun. Conf.*, FD4, San Francisco, California, USA, Jan. 22–26.

Patzak, E., Sugimura, A., Saito, S., Mukai, T., and Olesen, H. (1983). "Semiconductor laser linewidth in optical feedback configurations," *Electron. Lett.* **19**(24), 1026–1027.

Personick, S. D. (1973). "Receiver design for digital fiber optic communication systems, I," *Bells Syst. Tech. J.* **52**(6), 843–874.

Pettitt, M. J., Remedios, D., Davis, A. W., Hadjifotiou, A., and Wright, S. (1987). "Optical FSK transmission system using a phase-diversity receiver," *Electron. Lett.* **23**(20), 1075–1076.

Philipp, H. K., Scholtz, A. L., Bonek, E., and Leeb, W. R. (1983). "Costas loop experiments for a 10.6 μm communications receiver," *IEEE Trans. Commun.* **COM-31**(8), 1000–1002.

Pierce, J. R. (1978). "Optical channels: Practical limits with photon counting," *IEEE Trans. Commun.* **COM-26**(12), 1819–1821.

Ramaswamy, V., Stolen, R. H., Divino, M. D., and Pleibel, W. (1979). "Birefringence in elliptically clad borosilicate single-mode fibers," *Appl. Opt.* **18**(24), 4080–4084.

Ryu, S., Yamamoto, S., and Mochizuki, K. (1987). "Polarization-insensitive operation of coherent FSK transmission system using polarization diversity," *Electron. Lett.* **23**(25), 1382–1384.

Saito, S., Yamamoto, Y., and Kimura, T. (1980). "Optical heterodyne detection of directly frequency modulated semiconductor laser signals," *Electron. Lett.* **16**(22), 826–827.

Saito, S., Nilsson, O., and Yamamoto, Y. (1982). "Oscillation center frequency tuning, quantum FM noise and direct frequency modulation characteristics in external grating loaded semiconductor lasers," *IEEE J. Quantum Electron.* **QE-18**(6), 961–970.

Saito, S., Yamamoto, Y., and Kimura, T. (1983). "S/N and error rate evaluation for an optical FSK-heterodyne detection system using semiconductor lasers," *IEEE J. Quantum Electron.* **QE-19**(2), 180–193.

Saito, S., Nilsson, O., and Yamamoto, Y. (1985). "Frequency modulation noise and linewidth reduction in a semiconductor laser by means of negative frequency feedback technique," *Appl. Phys. Lett.* **46**(1), 3–5.

Saito, S., Imai, T., Sugie, T., Ohkawa, N., Ichihashi, Y., and Ito, T. (1990). "Coherent transmission experiment over 2,223 km at 2.5 Gbit/s using erbium-doped fiber amplifiers," *Electron. Lett.* **26**(10), 669–671.

Saitoh, T., and Mukai, T. (1987). "1.5 μm GaInAsP traveling-wave semiconductor laser amplifier," *IEEE J. Quantum. Electron.* **QE-23**(6), 1010–1020.

Sakai, J., and Kimura, T. (1982). "Birefringence caused by thermal stress in elliptically deformed core optical fibers," *IEEE J. Quantum Electron.* **QE-18**(11), 1899–1909.

Sakai, J., Machida, S., and Kimura, T. (1981). "Existence of eigen polarization modes in anisotropic single-mode optical fibers," *Opt. Lett.* **6**(10), 496–498.

Sasaki, Y., Hosaka, T., and Noda, J. (1984). "Low crosstalk polarization-maintaining optical fiber with an 11 km length," *Electron. Lett.* **20**(19), 784–785.

Scifres, D. R., Lindström, C., Burnham, R. D., Streifer, W., and Paoli, T. L. (1983). "Phase-locked (GaAl) As laser diode emitting 2.6 W CW from a single mirror," *Electron. Lett.* **19**(5), 169–171.

Shikada, M., Fujita, S., Takano, I., Henmi, N., Mito, I., Taguchi, K., and Minemura, K. (1985). "1.5-μm high bit rate long span transmission experiments employing a high power DFB-DC-PBH laser diode," *5th Int. Conf. Integrated Opt. and Optical Fiber Commun./11th European Conf. Optical Commun.*, postdeadline paper, Venice, Italy, Oct. 1–4.

Shimoda, K. (1973). "Frequency shifts in methane-stabilized lasers," *Jpn. J. Appl. Phys.* **12**(9), 1393–1402.

Simon, J. C. (1982). "Polarization characteristics of a traveling-wave-type semiconductor laser amplifier," *Electron. Lett.* **18**(11), 438–439.

Simpson, J. R., Stolen, R. H., Sears, F. M., Pleibel, W., MacChesney, J. B., and Howard, R. E. (1983). "A single-polarization fiber," *IEEE J. Lightwave Technol.* **LT-1**(2), 370–374.

Siuzdak, J., and Etten, W. V. (1989). "BER evaluation for phase and polarization diversity optical homodyne receivers using noncoherent ASK and DPSK demodulation," *IEEE J. Lightwave Technol.* **LT-7**(4), 584–599.

Stanley, I. W., Hill, G. R., and Smith, D. W. (1987). "The application of coherent optical techniques to wide-band networks," *IEEE J. Lightwave Technol.* **LT-5**(4), 439–451.

Stolen, R. H. (1979). "Nonlinear properties of optical fibers," in "Optical Fiber Telecommunications" (S. E. Miller and A. G. Chynoweth, eds.). Academic Press, New York.

Stolen, R. H., and Ippen, E. P. (1973). "Raman gains in glass optical waveguides," *Appl. Phys. Lett.* **22**(6), 276–278.

Stone, J., and Stulz, L. W. (1987). "Pigtailed high-finesse tunable fiber Fabry–Perot interferometers with large, medium and small free spectral ranges," *Electron. Lett.* **23**(15), 781–783.

Suematsu, Y., Arai, S., and Kishino, K. (1983). "Dynamic single-mode semiconductor lasers with a distributed reflector," *IEEE J. Lightwave Technol.* **LT-1**(1), 161–176.

Sugie, T., Imai, T., and Ito, T. (1990). "Over 350 km CPFSK repeaterless transmission at 2.5 Gb/s employing high-output power erbium-doped fiber amplifiers." *Electron. Lett.* **26**(19), 1577–1578.

Takachio, N., Iwashita, K., Nakanishi, K., and Koike, S. (1990). "8 Gbit/s 202 km optical CPFSK transmission experiment using 1.3 μm zero dispersion fiber," *Electron. Lett.* **26**(8), 506–508.

Tako, T., Ohtsu, M., and Tsuchida, H. (1983). "Frequency control of semiconductor lasers," *3rd Conf. Lasers and Electro-Optics*, WB5, Baltimore, Maryland, USA, May 17–20.

Tkach, R. W., and Chraplyvy, A. R. (1986). "Regimes of feedback effects in 1.5 μm distributed feedback lasers," *IEEE J. Lightwave Technol.* **LT-4**(11), 1655–1661.

Toba, H., Oda, K., Takato, N., and Nosu, K. (1987). "5 GHz-spaced, eight-channel, guided-wave tunable multi/demultiplexer for optical FDM transmission systems," *Electron. Lett.* **23**(15), 788–789.

Tromborg, B., Osmundsen, J. H., and Olesen, H. (1984). "Stability analysis for a semiconductor laser in an external cavity," *IEEE J. Quantum Electron.* **QE-20**(9), 1023–1032.

Tucker, R. S., Taylor, A. J., Burrus, C. A., Eisenstein, G., and Wiesenfeld, J. M. (1986). "Co-axially mounted 67 GHz bandwidth InGaAs PIN photodiode," *Electron. Lett.* **22**(17), 917–918.

Ulrich, R. (1979). "Polarization stabilization on single-mode fiber," *Appl. Phys. Lett.* **35**(11), 840–842.

Urquhart, P. (1988). "A review of rare earth doped fiber lasers and amplifiers," *IEEE Proc.* **135J**(6), 385–407.

Vahara, K., and Yariv, A. (1983). "Semiclassical theory of noise in semiconductor lasers—Parts I and II," *IEEE J. Quantum Electron.* **QE-19**(6), 1096–1109.

Varnham, M. P., Payne, D. N., Barlow, A. J., and Birch, R. D. (1983a). "Analytic solution for the birefringence produced by thermal stress in polarization-maintaining optical fibers," *IEEE J. Lightwave Technol.* **LT-1**(2), 332–339.

Varnham, M. P., Payne, D. N., Birch, R. D., and Tarbox, E. J. (1983b). "Single-polarization operation of highly birefringent bow-tie optical fibers," *Electron. Lett.* **19**(7), 246–247.

Wagner, R. E., Cheung, N. K., and Kaiser, P. (1987). "Coherent lightwave systems for interoffice and loop-feeder applications," *IEEE J. Lightwave Technol.* **LT-5**(4), 429–438.

Wakita, K. Yoshikuni, Y., and Kawamura, Y. (1987). "Highly efficient InGaAs/InAlAs MQW waveguide phase shifter," *Electron. Lett.* **23**(6), 303–304.

Walther, F. G., and Kaufman, J. E. (1983). "Characterization of GaAlAs laser diode frequency noise," *6th Conf. Opt. Fiber Commun.*, TuJ5, New Orleans, Louisiana, USA, Feb. 28–Mar. 2.

Washio, K., Aoki, Y., and Kishida, S. (1985). "Raman amplification in optical fibers," *in* "Japan Annual Reviews in Electronics, Components & Telecommunications: Vol. 17, Optical Devices & Fibers (Y. Suematsu ed.)." Ohmsha, Tokyo.

Whalen, M. S., Hall, K. L., Tennant, D. M., Koren, U., and Raybon, G. (1987). "Tunable fiber-extended-cavity laser," *Electron. Lett.* **23**(7), 313–314.

Wood, T. H., Carr, E. C., Burrus, C. A., Tucker, R. S., Chiu, T. H., and Tsang, W. T. (1987). "High-speed waveguide optical modulator made from GaSb/AlGaSb multiple quantum wells (MQWs)," *Electron. Lett.* **23**(10), 540–542.

Wyatt, R. (1985). "Spectral linewidth of external cavity semiconductor lasers with strong, frequency-selective feedback," *Electron Lett.* **21**(14), 658–659.

Wyatt, R., Hodgkinson, T. G., and Smith, D. W. (1983). "1.52 μm PSK heterodyne experiment featuring an external cavity diode laser local oscillator," *Electron. Lett.* **19**(14), 550–552.

Yamada, M., Shimizu, M., Okayasu, M., Takeshita, T., Horiguchi, M., Tachikawa, Y., and Sugita, E. (1990). "Noise characteristics of Er^{3+}-doped fiber amplifiers pumped by 0.98 and 1.48 μm laser diodes," *IEEE Photonics Technol. Lett.* **2**(3), 205–207.

Yamamoto, Y. (1980). "Receiver performance evaluation of various digital optical modulation–demodulation systems in the 0.5–10 μm wavelength region," *IEEE J. Quantum Electron.* **QE-16**(11), 1251–1259.

Yamamoto, Y. (1983). "AM and FM quantum noise in semiconductor lasers—Parts I and II," *IEEE J. Quantum Electron.* **QE-19**(1), 34–58.

Yamamoto, Y., and Kimura, T. (1981). "Coherent optical fiber transmission systems," *IEEE J. Quantum Electron.* **QE-17**(6), 919–935.

Yamamoto, Y., Nilsson, O., and Saito, S. (1985). "Theory of a negative frequency feedback semiconductor laser," *IEEE J. Quantum Electron.* **QE-21**(12), 1919–1928.

Yamazaki, S., Emura, K., Shikada, M., Yamaguchi, M., and Mito, I. (1985). "Realization of flat FM response by directly modulating a phase tunable DFB laser diode," *Electron. Lett.* **21**(7), 283–285.

Yanagawa, T., Saito, S., and Yamamoto, Y. (1984). "Frequency stabilization of 1.5-μm InGaAsP distributed feedback laser to NH_3 absorption lines," *Appl. Phys. Lett.* **45**(8), 826–828.

Yoshikuni, Y., and Motosugi, G. (1987). "Multielectrode distributed feedback laser for pure frequency modulation and chirping suppressed amplitude modulation," *IEEE J. Lightwave Technol.* **LT-5**(4), 516–522.

Yuen, H. P., and Chan, V. W. S. (1983). "Noise in homodyne and heterodyne detection," *Opt. Let.* **8**(3), 177–179.

Nonlinear Effects in Optical Fibers

ANDREW R. CHRAPLYVY

AT&T Bell Laboratories
Crawford Hill Laboratory
Holmdel, New Jersey

I. Introduction 267
II. Nonlinear Gain and System Parameters 268
III. Stimulated Raman Scattering 269
IV. Carrier-Induced Phase Modulation 278
V. Stimulated Brillouin Scattering 281
VI. Four-Photon Mixing 285
VII. Multiplexing Effects 288
VIII. Scaling 290
IX. Conclusion 292
References 293

I. Introduction

The attractiveness of lightwave communications is the ability of silica optical fibers to carry large amounts of information over long repeaterless spans. To utilize the available bandwidth, numerous channels at different wavelengths can be multiplexed on the same fiber. To increase system margins, higher transmitter powers or lower fiber losses are required. All these attempts to utilize fully the capabilities of silica fibers will ultimately be limited by nonlinear interactions between the information-bearing lightwaves and the transmission medium. These optical nonlinearities can lead to interference, distortion, and excess attenuation of the lightwaves, resulting in system degradations. On the other hand, these same nonlinearities, if cleverly exploited, can be utilized to enhance the performance of communication systems.

There exists a rich collection of nonlinear optical effects in fused-silica fibers, each of which manifests itself in a unique way. Stimulated Raman scattering, an interaction between light and vibrations of silica molecules, causes frequency conversion of light and results in excess attenuation of short-wavelength channels in wavelength-multiplexed systems. Stimulated Brillouin scattering, an interaction between light and sound waves in the fiber, causes frequency conversion and reversal of the propagation direction of light.

Cross-phase modulation is an interaction between the intensity of one light wave and the optical phase of other light waves. Four-photon mixing is analogous to third-order intermodulation distortion, whereby two or more optical waves at different wavelengths mix to produce new optical waves at other wavelengths.

Each of these nonlinearities will affect specific lightwave systems in different ways. In general, however, stimulated Raman scattering, stimulated Brillouin scattering, and four-photon mixing will deplete certain optical waves and, by means of frequency conversion, will generate interfering signals for other channels. These will degrade both direct detection and heterodyne systems. Cross-phase modulation, on the other hand, affects only the phase of optical signals. Consequently, only angle-modulated systems will be affected by this nonlinearity.

This chapter describes optical nonlinearities in the context of lightwave system limitations. The four nonlinearities mentioned above are discussed, and the nature and severity of system degradation caused by each nonlinearity are described. In particular, the system power limitations are plotted as a function of number of optical channels. Methods for scaling these results with changes in system parameters such as fiber loss, core diameter, and length are discussed. Some nonlinearities can be exploited to enhance lightwave system performance. Although most such applications are only in the research stage, they are promising techniques and are briefly mentioned.

II. Nonlinear Gain and System Parameters

Most nonlinear optical interactions involving two overlapping optical waves propagating in a medium can be characterized generally by

$$P_1(L) = P_1(O)\exp(gP_2L/A), \tag{1}$$

where $P_1(O)$ and $P_1(L)$ are the power of one wave entering and exiting, respectively, a medium of length L. This amplified wave is commonly called the probe wave. P_2 is the injected power of the other wave, called the pump, which generates the gain for the first wave. The cross-sectional area common to the light beams is A; the gain coefficient g (expressed in cm/W) is a direct measure of the strength of the nonlinearity. Equation (1) assumes that P_2 is constant throughout the nonlinear medium; that is, there is no pump depletion due to the nonlinearity and no intrinsic loss. Furthermore, Eq. (1) assumes that the polarization states of the pump and probe waves are the same. Neither of these assumptions typically holds in fibers. Attenuation in long fibers is non-negligible, and the polarization states of the pump and probe waves can evolve differently in the fiber. Consequently, Eq. (1) must be modified to be applicable to single-mode optical fibers (Smith, 1972). The correct expression is

$$P_1(L) = P_1(O)\exp(gL_e P_2/bA_e), \qquad (2)$$

where P_2, $P_1(O)$, $P_1(L)$, and g are defined as before. The effective area of the propagating waves, A_e, is evaluated by calculating the average modal overlap between the pump and probe waves (Hill et al., 1978; Stolen, 1979; Stolen and Bjorkholm, 1982). However, in general, if the pump and probe wavelengths are comparable and both are slightly longer than the fiber cutoff wavelength, then $A_e \approx A$, where A is the core area of the fiber (Johnson et al., 1977). The effective fiber length, L_e, replaces the actual length L in order to account for the exponential decay with length of the pump power due to fiber loss. A simple integration shows that

$$L_e = (1 - e^{-\alpha L})/\alpha, \qquad (3)$$

where α is the loss coefficient of the fiber. For $\alpha L \ll 1$, $L_e \approx L$; for $\alpha L \gg 1$, $L \approx 1/\alpha$. The factor b accounts for the relative polarizations of pump and probe waves and the polarization properties of the fiber. In a polarization-maintaining fiber with identical pump and probe polarization states, $b = 1$. In a conventional fiber that does not maintain polarization, $b = 2$, which will be assumed in what follows.

Equation (2) describes the strength of optical nonlinearities as a function of system parameters. As a starting point, a long-haul ($L > 30$ km) single-mode system operating at 1.55 µm is assumed. The fiber is assumed to have core area of 5×10^{-7} cm^2 (core diameter = 8 µm), a fiber loss of 0.2 dB/km, and a chromatic dispersion of 16 ps/nm–km at $\lambda = 1.55$ µm. For operation in a densely packed frequency-multiplexed mode, channel spacing of 10 GHz is assumed. System limitations using these parameters will be discussed for the various nonlinearities. Also, the dependence of nonlinear effects on changes in system parameters will be described.

III. Stimulated Raman Scattering

Stimulated Raman scattering (SRS) is an intriguing nonlinearity because of its potential applications. SRS can be a source of signal degradation in some lightwave systems, while in other configurations it can provide optical amplification of weak signals. SRS can play a key role in megameter soliton transmission without signal regeneration. In addition, SRS in silica optical fibers can be exploited to generate wideband infrared radiation useful for fiber diagnostics. In this section, the deleterious effects of SRS are discussed first. A summary of the various amplification experiments relevant to lightwave systems then follows. Finally, several examples of infrared sources based on SRS in optical fibers and their applications are presented.

There are several ways that SRS can be described. Classically, Raman scattering can be viewed as modulation of light by the molecular vibrations in the silica matrix. This modulation produces sidebands spaced by a frequency equal to that of the vibrating molecules. The lower-frequency side-band is called the Stokes line, and the upper-frequency sideband is the anti-Stokes line. The Stokes line is typically significantly stronger than the anti-Stokes line. Quantum mechanically, in Raman scattering a pump photon is annihilated and a Stokes photon is created, along with a quantum of vibrational energy in the scattering molecule. The details of Raman scattering are not important in this paper and can be found in numerous references (Bloembergen, 1967; Kaiser and Maier, 1972; Yariv, 1975; Stoicheff, 1963). The important point is that if two optical waves separated by the Stokes frequency are co-injected into a Raman-active medium, the lower-frequency (probe) wave will experience optical gain generated by, and at the expense of, the higher-frequency (pump) wave. This gain process is called stimulated Raman scattering (SRS) and can be described by Eq. (2). Because fused silica is a glass there is, in fact, a continuum of Stokes frequencies (Shuker and Gammon, 1970) corresponding to a spectral dependence of the gain coefficient g (in Eq. (2)), as shown in Fig. 1 (Stolen and Ippen, 1973). (One cm^{-1} equals 30 GHz.) Note that the gain coefficient increases approximately linearly with pump-probe frequency separation up to a separation of about 500 cm^{-1}. This means that any channels separated by up to 15,000 GHz will be coupled via SRS. The magnitude of the gain coefficient shown in Fig. 1 is for a pump wavelength of 1 μm. The gain coefficient scales inversely with pump wavelength (Stolen and Ippen, 1973), so that at 1.55 μm, which is the wavelength region under consideration, the peak Raman gain coefficient is about 7×10^{-12} cm/W.

In a single-channel lightwave system, only one wavelength of light is injected into the fiber. However, this signal generates spontaneous Raman-scattered light that can then be amplified. It has been shown both theoretically (Smith, 1972; Auyeung and Yariv, 1978; Stolen, 1980) and experimentally (Ohmori et al., 1981; Ikeda, 1981a; Stolen et al., 1984) that amplification of the Raman-scattered light will cause severe degradation (50% signal depletion) when

$$gL_e P/bA_e = 16. \tag{4}$$

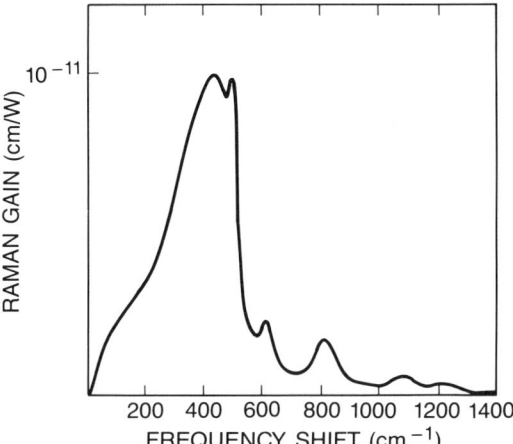

FIG. 1. Raman gain coefficient g vs. frequency shift for fused silica at a pump wavelength of 1.0 μm. The gain coefficient scales inversely with pump wavelength (Stolen, 1980). $1 \text{ cm}^{-1} = 30 \text{ GHz}$.

For the assumed system parameters, the injected signal power required to produce system degradation is about 1 W. It is clear that SRS will not be a factor in single-channel silica-fiber-based lightwave systems.

In wavelength-multiplexed systems, the situation is quite different because channels at numerous wavelengths are injected into the fiber, and the signals at longer wavelengths will be amplified by the shorter-wavelength signals. In other words, the probe photons no longer build up from spontaneous Raman noise, but are injected in macroscopic quantities as signal channels. This leads to system degradation at lower optical powers than in the single-channel case.

The degradation due to SRS for a two-channel system is schematically shown in Fig. 2 (Chraplyvy and Henry, 1983). Suppose channel 1 and channel 2 are spaced such that SRS couples the two channels. This assumption will usually be satisfied in the 1.5-μm region because the broad stimulated-Raman gain profile of silica (Fig. 1) will couple channels that are separated in wavelength by up to 100 nm.

Let channel 1 (pump) operate at a wavelength λ_1, which is shorter than λ_2, the wavelength of channel 2 (probe). Assume initially that both channels have equal optical power injected into the fiber. Suppose that in a return-to-zero (RZ) modulation format the bit pattern of the two channels is shown in Figure 2a. Schematically, the effect of SRS is to produce bit patterns as shown in Figure 2b. Thus far we have ignored the effects of dispersion. Note that whenever there is a mark in both channels, the pump channel (λ_1) is depleted and the probe channel (λ_2) is amplified. If a space (zero light intensity) appears

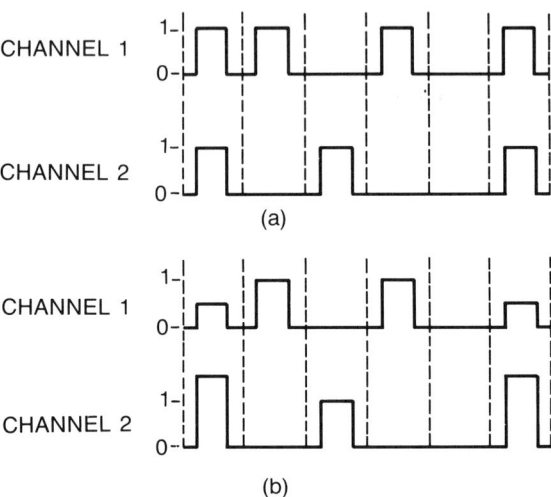

FIG. 2. (a) Bit pattern in two-channel wavelength-multiplexed system with no stimulated Raman interaction between channels. (b) Bit pattern with SRS ($\lambda_1 < \lambda_2$) (Chraplyvy and Henry, 1983).

in either channel, no intensity change occurs. (In conventional crosstalk, a mark in channel 1 can produce a signal in channel 2 even if there is no mark in channel 2.) Furthermore, the effects of SRS on the two channels are not symmetric. Channel 1 experiences a partial closing of the eye pattern because of the depletion of individual bits, and therefore a degradation in signal-to-noise ratio. The opening of the eye in channel 2 is, in principle, unaffected because in the worst case some of the bits are amplified while the rest of the bits are unaltered. However, in practice this can also lead to degradations, especially in receivers with automatic gain control.

For multiple-channel systems, the interactions are more complicated but qualitatively similar. In general, the longer-wavelength channels will be amplified at the expense of the shorter-wavelength channels. The degradation can be estimated (Chraplyvy, 1984) by assuming that the Raman gain profile (Fig. 1) between 0 and 500 cm^{-1} is triangular. The result is that in a system of N channels with channel spacing Δf and power P per channel, no channel will experience a 1-dB penalty provided

$$[NP][(N-1)\Delta f] < 500 \text{ GHz-W}. \tag{5}$$

Note that NP is the total optical power injected into the fiber, and $(N-1)\Delta f$ is the total occupied optical bandwidth. Therefore, Eq. (5) is a very general result: The product of total power and total optical bandwidth must be smaller than 500 GHz-W to reduce degradation due to SRS to acceptable levels.

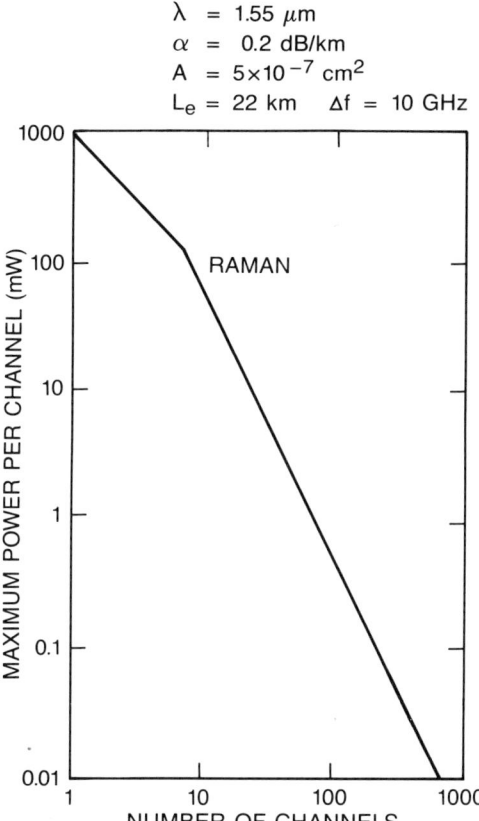

FIG. 3. Maximum power per channel, vs. number of channels, that ensures SRS degradation below 1 dB for all channels.

All the SRS results can be summarized assuming the system parameters in Section II. Figure 3 shows the maximum allowable power per channel as a function of number of channels. For several channels, the power limit decreases as $1/N$ because the Raman gain profile is extremely broad and the powers in all N channels contribute to the SRS process (Eq. (4)). As more channels are added, the occupied optical bandwidth increases, the interchannel interactions become more significant, and the maximum power per channel decrease as $1/N^2$ (Eq. (5)). These results have been derived assuming equal group velocities. It has been shown (Cotter and Hill, 1984) that the effect of group velocity dispersion on the nonlinear Raman interaction decreases the effect by a factor between one and two. For high bit rates and nonzero group velocity dispersion, the effects of SRS are reduced by a factor of two. Consequently, the curve in Fig. 3 will be 3 dB higher in power.

These results were derived assuming there is optical power in every channel. In wavelength-multiplexed amplitude-shift-keyed (ASK) systems with many channels, the probability of marks occurring in all channels is small and statistical considerations must be employed. The overall Raman degradation will be reduced by occurrence of spaces in some channels. In frequency-shift-keyed (FSK) and phase-shift-keyed (PSK) systems, the optical power in each channel is nominally constant and statistical treatment is not needed.

The effects of SRS in two-channel configurations have been measured in several ways (Ikeda, 1981b; Tomita, 1983; Hegarty et al., 1985). The crosstalk due to SRS was measured (Tomita, 1983) using light from two injection lasers multiplexed on a fiber 21 km long. The power in the long-wavelength channel (1.34 μm) was deliberately reduced to 0.05 mW to avoid depletion of the short-wavelength channel (1.26 μm). Thus, the effects of SRS were monitored not by measuring the degradation of the short-wavelength channel, but by measuring the amplification of the long-wavelength channel as a function of power in the short-wavelength channel. Both continuous wave (cw) and modulated (230-MHz square wave) signals were employed. For 1 mW of power at 1.26 μm, a crosstalk of -25 dB was measured, in good agreement with theoretical prediction.

The degradation of a short-wavelength channel due to depletion by a long-wavelength channel in a two-channel configuration was measured directly by determining power penalties from bit-error-rate (BER) curves (Hegarty et al., 1985). Light from a DFB injection laser operating at 1.5 μm was transmitted through 43 km of fiber. The laser was modulated with a $2^{15} - 1$ pseudorandom bit stream at 1 Gb/s, and the BER was measured as a function of received power. Light from a color-center laser (FCL) emitting up to 150 mW at 1.57 μm was than also injected into the fiber, and the BER curves were measured for several FCL powers. The BER measurements displayed power penalties up to 2.5 dB. The measured power penalties corresponded to the observed depletion levels in the short-wavelength channels and agreed with the predicted degradations.

In a wavelength-multiplexing experiment (Olsson et al., 1985), 10 channels occupying a total optical bandwidth of 30 nm were multiplexed on a 68-km long fiber. The average total injected power in all channels was about 5 mW. Therefore, total power × optical bandwidth = 20 GHz W, well below the figure needed to produce an observable penalty (Eq. (5)). Indeed, no power penalty due to SRS was observed in the BER measurements.

There is considerable interest in exploiting SRS to amplify optical signals in fibers. Fiber-optic Raman amplification offers the possibility of replacing conventional regenerators with simple amplifiers consisting of a pump laser and a wavelength-selective coupler. This would lead to extremely large repeater spacings (Mochizuki, 1985). The transmission medium also serves as the amplification medium (this is referred to as active transmission lines), thereby eliminating various optical and electrical components normally required to regenerate or amplify a signal. Furthermore, fiber-optic Raman amplifiers have rather high saturation powers, which is important in amplifier chain configurations. A schematic diagram of a Raman amplifier is shown in Fig. 4. The information-bearing signal is transmitted in the usual fashion. However, in addition to the light from the signal laser, shorter-wavelength pump light is injected into the fiber either copropagating with or counterpropagating to the signal wave. (In some applications, such as soliton amplification, it is desirable to inject both copropagating and counterpropagating pump light.) The pump and signal light is combined or separated by some wavelength-selective element such as a dichroic mirror, grating, or wavelength-selective coupler. The counterpropagating configuration probably is preferred because such a system is much less sensitive to pump power fluctuations. In effect, the signal channel sees an average pump intensity over the effective length of the fiber (~ 22 km) that equivalently corresponds to averaging the pump power for about 100 μs.

There are two major requirements on the pump used in an active transmission line. It must provide several hundred milliwatts of cw power, and it must have spectral properties that are not conducive to creation of stimulated Brillouin scattering (SBS). The latter point will be discussed in a subsequent section. The cw power requirements initially limited demonstrations of active transmission lines to those regions of the spectrum near

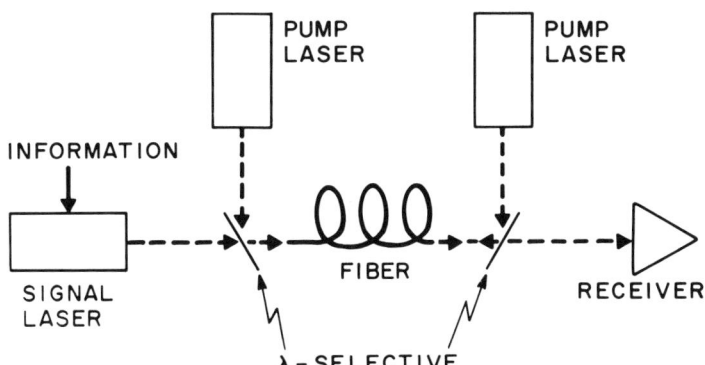

FIG. 4. Schematic diagram of Raman amplification of propagating signal. Both forward and backward Raman amplification are shown.

Nd:YAG laser lines, 1.06 μm (Ikeda, 1981b; Koepf et al., 1982; Desurvire et al., 1983; Nakazawa et al., 1984) and 1.32 μm (Aoki et al., 1983, 1985; Nakazawa, 1985). The highest gains (up to 45 dB) were achieved with pulsed or Q-switched pump lasers (Desurvire et al., 1983). However, pulsed amplification is not applicable in communication systems. Active transmission lines using cw Nd:YAG pump sources at 1.06 μm (Koepf et al., 1982) and 1.32 μm (Aoki et al., 1983) have provided optical gain of 3 dB at 1.118 μm and over 20 dB at 1.4 μm, respectively.

Recently the emphasis in amplification experiments has shifted to measuring not only optical gains, but also bit error rates, using a pseudorandom bit stream in the signal channel. A gain of 14 dB at 1.4 μm was achieved (Aoki et al., 1985) using a mode-locked Nd: YAG laser at 1.32 μm to amplify a 100-Mb/s NRZ signal. A small (1-dB) penalty was measured and attributed to fluctuations of the pump power. An active transmission line in the low-loss region of silica was demonstrated (Hegarty et al., 1986; Olsson and Hegarty, 1986) using a tunable FCL pump laser operating on two longitudinal modes at 1.48 μm. The 1.57-μm signal laser was modulated at 1 Gb/s with a pseudorandom bit stream. An optical gain of 5–8 dB was achieved, but a 1.4-dB power penalty attributed to backscattered pump light reduced the net increase in receiver sensitivity to 4.4 dB.

The noise properties of Raman amplifiers have been measured (Cohen and Lin, 1978) using an FCL pump laser and a 1.55-μm DFB laser. The Raman amplifier was shown to have no excess noise for signal powers between −30 and −50 dBm using an APD receiver with an excess noise factor of 10. However, when signal levels are low, the Raman amplifier does add shot noise due to spontaneous Raman scattered light. For low-bit-rate systems (< 500 Mb/s), optical filtering of the light from spontaneous Raman scattering is required (Cohen and Lin, 1978). The performance of Raman amplifiers, with respect to signal-to-noise ratio, is comparable to that of semiconductor amplifiers. The advantage of Raman amplifiers over semiconductor laser amplifiers is their high saturation powers. Gain saturation in Raman amplifiers occurs when the amplified signal power becomes comparable to the pump power, which is typically a few hundred milliwatts. Semiconductor amplifiers saturate at powers between 0.1 and 1 mW (Cohen and Lin, 1978). The biggest problem confronting implementation of active transmission lines is the lack of suitable pump sources. Until a compact pump source (preferably an injection laser) is developed, Raman amplifiers are not practical for communications use.

Besides being potentially useful in active transmission lines, SRS in single-mode fibers has provided a convenient source of infrared light that can be used for diagnostic measurements of optical fibers. A few hundred meters of conventional single-mode fiber pumped by a Q-switched and/or mode-locked

1.06-μm Nd: YAG laser will emit a wavelength continuum of light (Cohen and Lin, 1978) between 1.1 and 1.7 μm (Fig. 5). After spectral filtering, this light can serve as a tunable source of infrared light. Infrared pulses as short as 100 ps or as long as hundreds of nanoseconds are available from such a source. The short pulses can be used for measurements of group velocity dispersion (Stone *et al.*, 1982), whereas the long pulses can be used as pump pulses in SRS amplification measurements (Desurvire *et al.*, 1983; Nakazawa *et al.*, 1984; Nakazawa, 1985). Single-mode silica fibers can also be used as host matrices for molecular species such as H_2 and D_2 (Chraplyvy *et al.*, 1983; Stone *et al.*, 1982). SRS from these molecules can provide extremely large single-step Raman shifts (3000–4000 cm^{-1}). For example, D_2 in silica-fiber Raman lasers converts 1.06-μm Nd: YAG light efficiently in a single-step

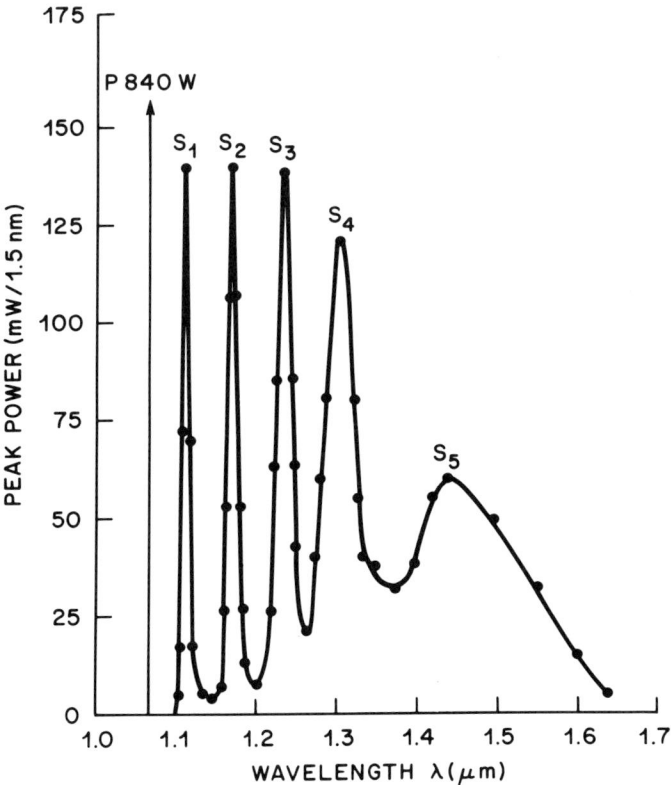

FIG. 5. Spectral emission curve for a single-mode fiber Raman laser (Cohen and Lin, 1978). Power was measured through a monochromator with 1.5-μm resolution. The pump was a 1.064-nm wavelength Nd:YAG laser. Five orders of stimulated Stokes emission S_1 to S_5 are generated.

process to 1.56-μm light (Chraplyvy and Stone, 1984; Chraplyvy and Stone, 1985), which can be used for various measurement purposes. Pulses as short as 15 ps (Chraplyvy and Stone, 1984; Chraplyvy and Stone, 1985) can be used to study nonlinear propagation effects, whereas pulses several hundred nanoseconds long (Chraplyvy and Stone, 1985) have been used for extremely long-range optical time-domain reflectometry (Stone et al., 1984). Single-pass fiber Raman lasers all share the common quality that efficient conversion of pump light to Raman can be realized at relatively low powers (hundreds of watts), orders of magnitude smaller than those required in bulk samples.

IV. Carrier-Induced Phase Modulation

In phase-shift-keyed systems, information is digitally impressed on the phase of the wave, typically toggling between $+\pi/2$ and $-\pi/2$ to represent a logic "1" and a logic "0." Any source of phase noise in such systems will degrade system performance. An example of an optical nonlinearity that affects only the phase of the propagating signal is the nonlinear refractive index of the fiber material, which gives rise to carrier-induced phase modulation (CIP) (Chraplyvy et al., 1984). In single-channel configurations, CIP is called self-phase modulation and converts optical power fluctuations in a lightwave to phase fluctuations in the same wave. In wavelength-multiplexed systems, cross-phase modulation converts power fluctuations in a particular channel to phase fluctuations in the other channels.

CIP in silica fibers exists because of an intensity-dependent refractive index. The refractive index of most transparent solids, including silica, has the form

$$n = n_0 + n_2 I \tag{6}$$

where n_0 is the ordinary refractive index associated with the material, n_2 is the intensity-dependent refractive index, and I is the optical intensity (P/A). Consequently, the phase of light after propagating through a fiber with length L (relative to the phase of the injected light) is

$$\phi(L) = \frac{2\pi n_0 L}{\lambda} + \frac{2\pi n_2 I L_e}{\lambda}. \tag{7}$$

Clearly, any changes in optical intensity I will produce corresponding changes in the phase and can potentially affect PSK systems. In silica for self-modulation $n_2 = 3 \times 10^{-16}$ cm^2/W, and for cross-phase modulation, $n_2 = 6 \times 10^{-16}$ cm^2/W (Stolen and Bjorkholm, 1982). Although these are very small refractive indices, the long interaction lengths in optical fibers magnify these effects.

Using Eq. (7) and accounting for random polarization (Stolen and Lin, 1978), it can be shown that in single-channel systems the phase change in the received signal due to the nonlinear refractive index is given by

$$\sigma_\phi = 0.035\sigma_p, \qquad (8)$$

where σ_ϕ is the rms phase fluctuation in radians and σ_p is the rms power fluctuation in milliwatts.

Power fluctuations in InGaAsP injection lasers at power levels of a few milliwatts are $\sigma_p \approx 0.1$ mW (Liu *et al.*, 1983). These fluctuations increase roughly as the square root of the optical power (Yamamoto *et al.*, 1983), so that for transmitter powers up to 100 mW we expect the power fluctuations σ_p to be less than 1 mW. (We assume that the bandwidth of the power fluctuations is comparable to or less than the information bandwidth of the transmission system. This is a reasonable approximation for data rates in the gigabit-per-second range.) The resultant phase noise is less than 0.04 radians, which is negligibly small in angle-modulated systems (Prabhu, 1976) (0.15 rad of phase noise corresponds to a power penalty of roughly 0.5 dB).

In wavelength-multiplexed systems, in addition to self-phase modulation, there are cross-phase modulation effects due to power fluctuations in other optical channels. In a system with N channels, the rms phase fluctuations in a particular channel due to power fluctuations in the other channel is

$$\sigma_\phi = 0.07\sqrt{N}\,\sigma_p \qquad (9)$$

where σ_ϕ is in radians and σ_p is in milliwatts. The power fluctuations σ_p in all the channels have been assumed to be the same. Assuming the laser noise characteristics just described, the limitations due to CIP will be negligible even for large numbers of channels.

Much greater CIP can be generated from residual AM present when semiconductor lasers are directly phase modulated (Vodhanel, 1989). Residual AM as large as 20% of the laser output power is typical (Vodhanel, 1989). Furthermore, the degradation in this case grows linearly with N rather than as $\sqrt{N}$ (Chraplyvy *et al.*, 1984). In oder to limit power penalties due to CIP to less than 1 dB, the power per channel (assuming 20% residual AM) must satisfy

$$P < 21/N. \qquad (10)$$

This requirement is plotted in Fig. 6. For comparison, the Raman results are included.

As in the case of SRS, a third-order electric susceptibility gives rise to CIP. Consequently, this is essentially an instantaneous effect and the results apply to both cw and modulated lightwaves. Similarly, the CIP results in Fig. 6 assume marks or spaces in all channels. For large number of channels,

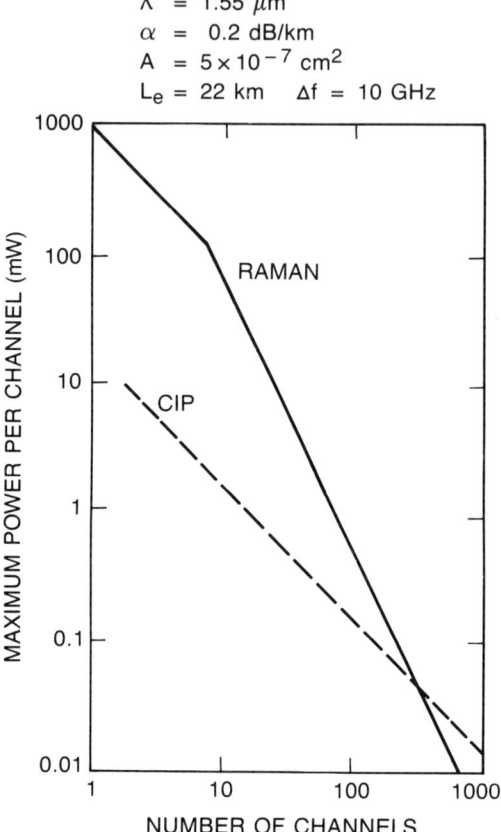

FIG. 6. Maximum power per channel, vs. number of channels, that ensures degradations due to SRS and CIP below 1 dB for all channels. The CIP curve is for directly phase-modulated lasers with 20% residual AM.

statistical occurrences of marks and spaces will change the results by a small factor.

Cross-phase modulation has been experimentally observed using light from two conventional InGaAsP injection lasers multiplexed on a 15-km-long single-mode fiber (Chraplyvy and Stone, 1984b). In the experiment to measure the effects of cross-phase modulation, a novel self-reflexive interferometer was employed. The channel 1 source was a cw 1.5-μm InGaAsP single-frequency distributed-feedback laser with a coherence length of several meters, and channel 2 used a 1.3-μm InGaAsP V-groove buried-crescent multifrequency laser. The two beams were combined by a dichroic mirror and coupled into a 15-km-long depressed-step-index single-mode fiber. The effects of on-off modulation in channel 2 on the phase of the cw light in channel 1 were measured. A 1-mW change in the power of channel 2 produced at 0.024-rad phase shift in channel 1. The predicted value is 0.022 rad.

A factor that limits the performance of high-speed, long-haul lightwave systems is chromatic dispersion. This leads to pulse broadening. The nonlinear refractive index of silica can be exploited to combat the effects of chromatic dispersion. If the intensity of the signal pulses is carefully adjusted, the effects of the nonlinear refractive index can be made to cancel exactly the effects of chromatic dispersion (Hasegawa and Tappert, 1973; Mollenauer et al., 1980)—i.e., chromatic dispersion broadens the pulse, but the nonlinear refractive index narrows the pulse. As a result, the light pulses propagate with no change in shape. Such pulses are called solitons. In addition, if the fiber loss is exactly compensated by stimulated Raman gain, the solitons can propagate over enormous distances with no change in shape or amplitude. This was recently demonstrated in an experiment in which solitons were transmitted through more than 600 km of fiber with no distortion or attenuation of the injected pulses (Mollenauer and Smith, 1988).

V. Stimulated Brillouin Scattering

Superficially, stimulated Brillouin scattering (SBS) (Ippen and Stolen, 1972) is similar to SRS except that SBS involves acoustic phonons rather than molecular vibrations. In this respect, both scattering processes are three-wave processes in which the incident (pump) light is converted into (Stokes) light of longer wavelength with a concomitant excitation of a molecular vibration (SRS) or an acoustic phonon (SBS). However, there are a number of significant differences between SBS and SRS that lead to markedly different systems consequences.

First, the peak SBS gain coefficient in single-mode fibers is over two orders of magnitude larger ($g_B \approx 4 \times 10^{-9}$ cm/W) than the gain coefficient for SRS. Consequently, under the proper conditions SBS will be the dominant nonlinear process. Second, the optical gain bandwidth Δv_R for SRS is on the order of 200 cm^{-1} FWHM (6000 GHz). Therefore, there is essentially no reduction in Raman gain for pump lasers with large linewidths. The optical bandwidth Δv_B for SBS, on the other hand, is about 20 MHz at 1.55 μm and varies as λ^{-2} (Hamilton and Hellwarth, 1979). Maximum SBS gain will occur for pump lasers with linewidths less than 20 MHz. For lasers with linewidths Δv_L larger than 20 MHz, SBS gain decreases as the ratio $\Delta v_B/\Delta v_L$, that is, $g = g_B \Delta v_B/\Delta v_L$ (Stolen, 1979), where g_B is the maximum steady-state Brillouin gain. Unlike SRS, which can occur in copropagating or counterpropagating geometries, SBS (because of phase-matching considerations) occurs only in the backward direction in single-mode fibers. This process obviously depletes the incident wave, and, in addition, generates a potentially strong scattered beam propagating back toward the transmitter (Stolen, 1979, 1980). The scattered light is shifted to a lower frequency by an amount $f_B = (2nV_s)/\lambda$, where n is the refractive index, nd V_s is the velocity of sound in the fiber. At 1.55 μm, $f_B \approx 11$ GHz for silica glasses.

For a single channel, the critical power level at which SBS degrades system performance (the scattered wave in the backward direction depletes the power of the signal propagating in the forward direction) is (Smith, 1972)

$$P_c = 21bA_e/(g_B L_e), \tag{11}$$

where the various symbols are defined following Eq. (2). For the previously assumed system parameters, $P_c = 2.4$ mW. In multichannel systems it can be shown (Aoki and Tajima, 1987; Lichtman and Friesem, 1987) that each channel interacts with the fiber independent of other channels. Consequently the critical power is constant with increasing number of channels (Fig. 7).

The above results have been derived assuming cw signal waves. Unlike SRS and CIP, SBS is very sensitive to signal modulation because the origin of SBS involves a process that, unlike electronic susceptibilities, is not instantaneous on the time scale of the information rate. The acoustic phonons that scatter light have long lifetimes, as evidenced by the narrow Brillouin linewidths (20 MHz). High modulation rates produce broad optical spectra, and a reduction in stimulated Brillouin amplification can be expected. The analysis developed for SBS generated by a narrow-linewidth source (Tang, 1966) can be extended (Aoki and Tajima, 1987; Lichtman and Friesem, 1987) to multimode sources, and finally to sources with pseudorandom modulation (Lichtman et al., 1989) used in communications. The results depend on the particular encoding scheme used (ASK, FSK, or PSK), and on the ratio $B/\Delta v_B$, where B is the bit rate.

For ASK the launched field amplitude can be described by

$$E(t) = E_0(1 - [1 - m(t)][1 - (1 - k_a)^{1/2}]) \tag{12}$$

where the binary data stream is represented by the function $m(t)$ that can take values of 0 and 1 with equal probability, and k_a is the depth of intensity modulation ($0 < k_a \leq 1$). In this case the SBS gain is (Lichtman et al., 1989)

$$g = g_B\left[\left(1 - \frac{a}{2}\right)^2 + \frac{a^2}{4}\left(1 - \frac{B}{\Delta v_B}(1 - e^{-\Delta v_B/B})\right)\right], \tag{13}$$

where $a = 1 - (1 - k_a)^{1/2}$.

The SBS gain for ASK can be minimized by using a 100% modulation depth ($k_a = 1$). For bit rates much smaller than the Brillouin linewidth, g

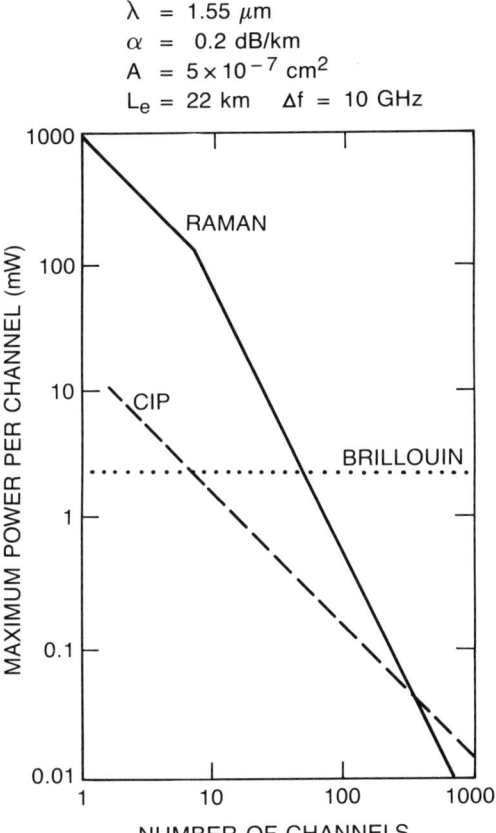

FIG. 7. Maximum power per channel, vs. number of channels, that ensures SRS, CIP, and SBS degradations below 1 dB for all channels. The SBS curve assumes cw power.

approaches $g_B/2$. For high bit rates, g approaches $g_B/4$. The dependence of g on $B/\Delta v_B$ is summarized in Fig. 8.

In PSK, the information is impressed on the phase of the electric field, as given by

$$E(t) = E_0 e^{i\phi(t)}, \tag{14}$$

where $\phi(t) = k_p m(t)$, and k_p is the keyed phase shift. The SBS gain for PSK is (Lichtman et al., 1989)

$$g = g_B \left[\frac{1}{2}(1 + \cos k_p) + \frac{1}{2}(1 - \cos k_p)\left[1 - \frac{B}{\Delta v_B}(1 - e^{-\Delta v_B/B})\right]\right]. \tag{15}$$

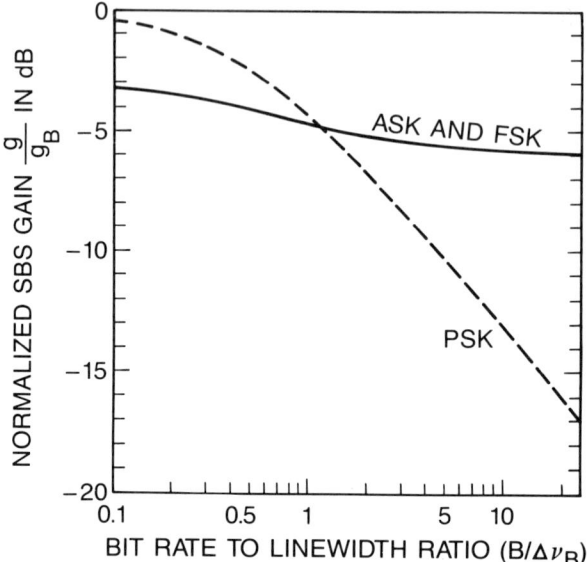

FIG. 8. Normalized SBS gain as a function of the ratio of bit rate to Brillouin linewidth.

The gain for PSK is minimized for $k_p = \pi(2n + 1)$, i.e., suppressed carrier. For high bit rates the SBS gain decreases linearly with $B/\Delta\nu_B$ (Fig. 8).

In wide-deviation FSK modulation, the laser frequency is modulated between two relatively widely-spaced frequencies ω_1 and ω_2—that is, the modulation depth $k_f = \omega_1 - \omega_2$ is at least several times the bit rate B. Consequently, the FSK spectrum is just the sum of two ASK spectra with $k_a = 1$ centered about ω_1 and ω_2. Therefore, the SBS gain for FSK is simply

$$g = g_B \left[\frac{1}{2} - \frac{B}{4\Delta\nu_B}(1 - e^{-\Delta\nu_B/B}) \right]. \qquad (16)$$

As shown in Fig. 8, the dependence of g on $B/\Delta\nu_B$ for FSK is the same as that for ASK. At high bit rates, the SBS gain decreases the cw value by a factor of four.

To summarize, SBS is a very strong nonlinear process that exhibits gain in the backward direction. This nonlinearity is most detrimental in systems employing narrow-linewidth lasers. In general, encoding pseudorandom data on the optical wave will reduce the effects of SBS. Maximum reduction occurs by using 100% modulation depth ($k_a = 1$) in ASK systems and $k_p = \pi(2n + 1)$ in PSK systems. The SBS gain decreases with increasing bit rates. In ASK and FSK systems, the maximum reduction is a factor of four. For high-bit-rate PSK systems, the SBS gain decreases linearly with B.

SBS has been observed in numerous experiments (Uesugi et al., 1981; Cotter, 1982) and as fiber loss was reduced, SBS thresholds approaching 2 mW have been observed at 1.52 μm in a 30-km-long fiber (Cotter, 1983). The effects on SBS thresholds of multiple frequencies were recently studied (Lichtman and Friesem, 1987; Aoki and Tajima, 1988) and the SBS gain reduction due to pump modulation predicted in Fig. 8 was confirmed (Lichtman et al., 1989; Aoki et al., 1988).

Backward SBS has recently been exploited to amplify laser signals in single-mode silica fibers (Olsson and van der Ziel, 1986a; Olsson and van der Ziel, 1986b). The pump light was injected counterpropagating to the signal light at the receiver end of the fiber. Gains as high as 4.3 dB per milliwatt of pump power were measured, with a total gain of 18 dB. Bit-error-rate measurements at 100 Mb/s showed no penalty due to the amplification process. One advantage of SBS amplification, as with SRS amplification, is that the transmission medium is also the amplifier. However, this kind of amplifier, unlike an SRS amplifier, requires narrow-linewidth pump lasers and precise control of the difference frequency between the pump and transmitter laser. Furthermore, it is difficult to amplify high-bit-rate (>1 Gb/s) signals using SBS because the pump spectrum needs to be significantly broadened (Olsson and van der Ziel, 1986b), leading to an unacceptable reduction in the available gain. However, for modest bit rates ($\lesssim 250$ Mb/s), SBS can be used as a novel channel selector and demodulator in a densely packed wavelength-multiplexed optical network (Chraplyvy and Tkach, 1986; Tkach et al., 1988, 1989). In this application, the narrow-band SBS gain provides amplification for the desired channel, leaving the neighboring channels unaffected. This amplified channel can then be detected with negligible interference from the other channels. To select a different channel, the gain region is tuned to the desired channel frequency by tuning the pump frequency. In effect the SBS process is a tunable narrowband optical amplifier.

VI. Four-Photon Mixing

The same nonlinearity that gives rise to the nonlinear refractive index also mediates the four-photon mixing process in single-mode fibers (Hill et al., 1978). The simplest embodiment of this effect is shown in Fig. 9. Two copropagating waves at frequencies f_1 and f_2 mix and generate sidebands at $2f_1 - f_2$ and $2f_2 - f_1$. These sidebands copropagate with the initial waves and grow at their expense. Similarly, three copropagating waves will generate nine new optical waves (Fig. 10) at frequencies $f_{ijk} = f_i + f_j - f_k$, where i, j, and k can be 1, 2, or 3. If the channels are equally spaced, some of the generated

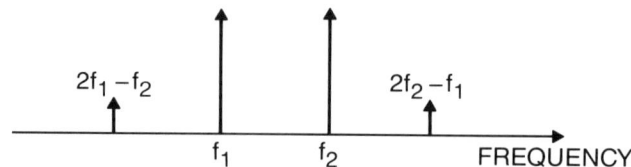

FIG. 9. Four-photon mixing with two injected waves at frequencies f_1 and f_2.

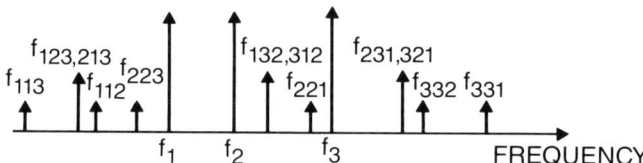

FIG. 10. Four-photon mixing with three injected waves at frequencies f_1, f_2, and f_3. The generated frequencies $f_{ijk} = f_i + f_j - f_k$.

waves will have the same frequencies as the injected waves. Clearly, the appearance of the additional waves as well as the depletion of the initial waves will degrade multichannel systems by crosstalk or excess attenuation.

The efficiency of four-photon mixing depends on the channel spacing and the fiber dispersion. Because of fiber chromatic dispersion, the interacting and generated waves have different group velocities. This destroys the phase matching of the interacting waves and lowers the efficiency of power generation at new frequencies (Hill et al., 1978; Shibata et al., 1987). The four-photon mixing efficiency decreases with increasing group velocity mismatch. Consequently, larger channel spacing and greater group velocity dispersion lead to lower efficiencies. The power $P_{ijk}(L)$ exiting the fiber generated at frequency f_{ijk} due to the interaction of channels at f_i, f_j, and f_k is (Hill et al., 1978; Shibata et al., 1987)

$$P_{ijk}(L) = \eta(1024\pi^6/n^4\lambda^2c^2)(6\chi_{1111})^2(L_e/A_e)^2 \times P_iP_jP_k \exp(-\alpha L), \quad (17)$$

where χ_{1111} is the third-order nonlinear susceptibility ($\chi_{1111} = 6 \times 10^{-15}$ cm^3/erg), and η is the efficiency of four-photon mixing. An explicit expression for η can be found in Shibata et al. (1987), but two specific examples in Fig. 11 are enough to provide some insight. The generation efficiency is plotted as a function of channel separation for two values of dispersion. The solid curve is the efficiency for the dispersion of a conventional single-mode fiber, 16 ps/nm–km. The dashed curve is the efficiency for dispersion-shifted fiber with a dispersion of 1 ps/nm–km. These plots show the frequency range over which the four photon mixing process is efficient. For example, in the conventional fiber only channels with separations less than 20 GHz will mix

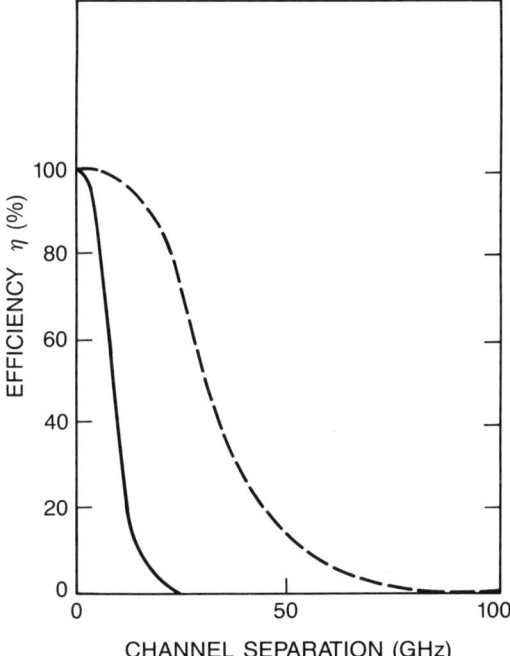

FIG. 11. Four-photon mixing efficiency as a function of channel separation at 1.55 μm. The solid curve represents standard single-mode fiber with dispersion of 16 ps/nm km. The dashed curve is for dispersion-shifted fiber with dispersion of 1 ps/nm km.

efficiently. On the other hand, in dispersion-shifted fibers, four-photon mixing efficiencies are greater than 20% for channel separations up to 50 GHz. Equation (17) and Fig. 11 can now be used to determine the four-photon mixing limitations in lightwave systems. With system parameters assumed previously, Fig. 12 shows the maximum power per channel that can be transmitted without degradation by four-photon mixing. The slight curvature at small channel number indicates that the four-photon interaction occurs between a channel and its two closest neighbors, as determined by Fig. 11.

Because the FPM nonlinearity couples nearby channels and does not have the frequency extent of SRS and CIP, statistical occurrence of marks in ASK systems need not be considered.

Several FPM experiments at different wavelengths have supported theoretical predictions. Two experiments (Shibata et al., 1986, 1987) used 0.8 μm-wavelength lasers where fiber dispersion is on the order of 100 ps/nm–km. More recently Shibata et al., 1988, experiments in dispersion-shifted fibers using 1.3-μm and 1.55-μm sources have confirmed the dependence of FPM efficiency on channel spacing and fiber dispersion.

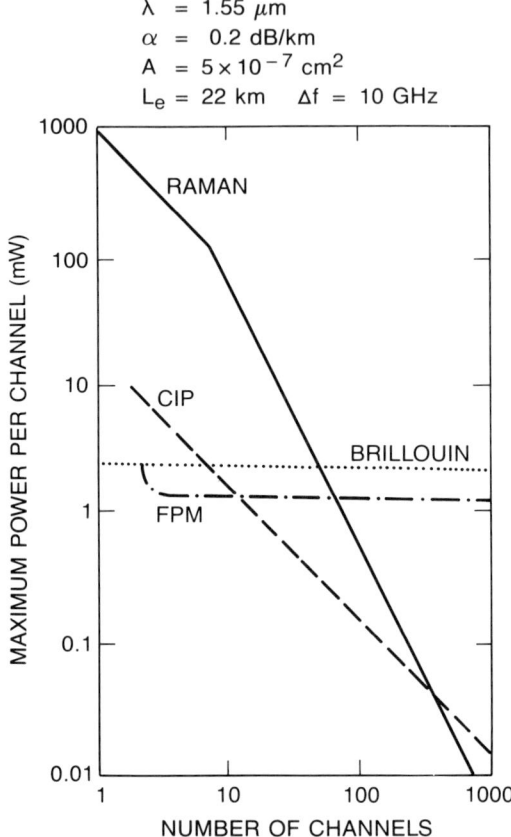

FIG. 12. Maximum power per channel, vs. number of channels, that ensures, SRS, CIP, SBS, and FPM degradations below 1 dB for all channels.

VII. Multiplexing Effects

Different methods of multiplexing (passive vs. frequency-selective) have a dramatic impact on the effects of optical nonlinearities. Passive multiplexing of N channels by a star coupler, for example, reduces the power per channel injected into the fiber by a factor of N. The higher the degree of multiplexing, the lower the power per channel injected into the fiber. (Multiplexing followed by semiconductor amplifiers will not change the situation, because the saturation power of a semiconductor amplifier is about the same as the output power of an injection laser.) Therefore, the power per channel injected into the fiber using passive multiplexing decreases with channel number, as shown in Fig. 13 for 10-mW and 50-mW laser transmitters (assuming no excess multiplexing loss). A particular nonlinearity will cause system degradation if the curve associated with that nonlinearity in Fig. 13 lies below the line representing the system transmitter power. For the case of frequency-selective multiplexing, the power per channel injected into the fiber will be independent of channel number. Consequently, such systems will be more susceptible to degradations by optical nonlinearities.

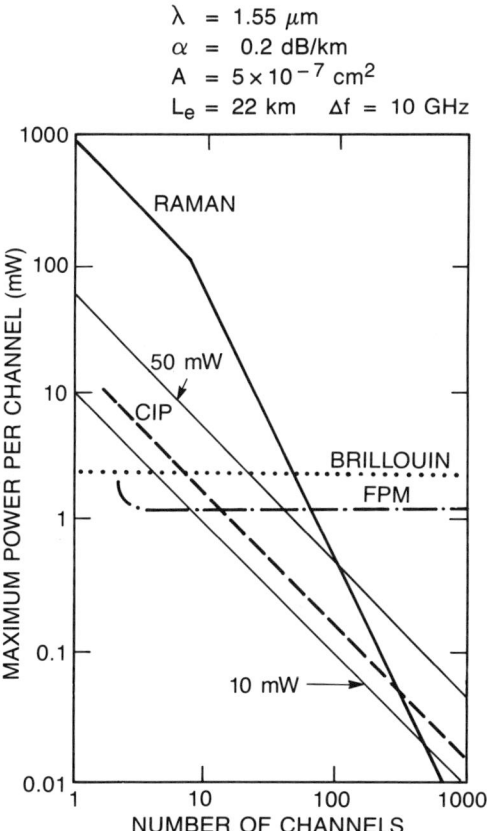

FIG. 13. Same as Fig. 12, but including curves for the power per channel injected into the fiber assuming lossless passive multiplexing for 10-mW and 50-mW lasers.

VIII. Scaling

The effect of changes in system parameters on the limitations due to optical nonlinearities is straightforward for SRS, CIP, and SBS. Scaling laws for four-photon mixing are complicated by the complex dependence of the mixing efficiency η on system paramaters (Shibata et al., 1987). The system parameters that affect the nonlinear optical effects are fiber attenuation coefficient α, fiber core area A_e, chromatic dispersion, channel separation, fiber polarization properties, and strength of nonlinear gain processes g.

SRS, CIP, and SBS scale with the ratio gL_e/bA_e. Changing fiber loss or fiber length changes L_e given by Eq. (3). Polarization-maintaining fibers will increase the nonlinear effect by a factor of two. Dependence on core area

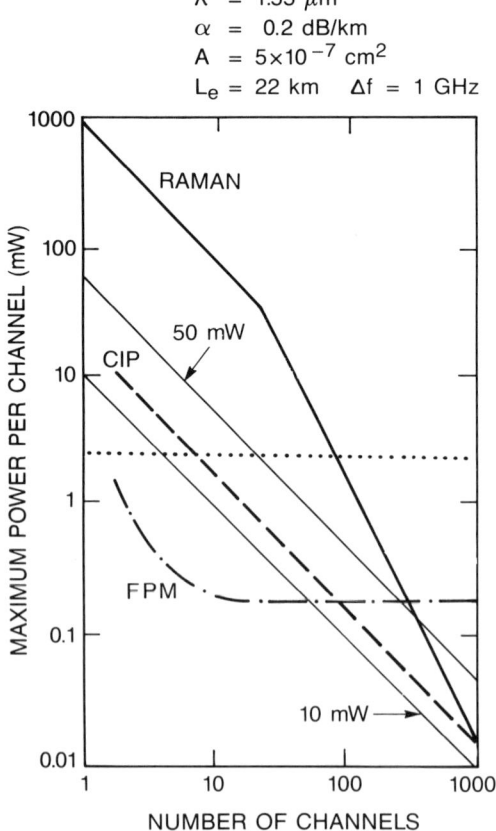

FIG. 14. Same as Fig. 13, but for 1-GHz channel spacing.

is obvious. Similarly, if the strength of the nonlinearity g is changed (for example, in the case of mid-infrared fibers with different material compositions), the effects of the nonlinearity will scale proportionately. Four-photon mixing depends on system parameters in a more complicated way and needs to be evaluated for each case. The dependence of nonlinear effects on channel separation is straightforward. For CIP and SBS, the effects are independent of channel spacing. For SRS, the effects are directly related to channel spacing (total occupied optical bandwidth, Eq. (5)). Decreasing channel spacing reduces SRS because of the nearly triangular shape of the Raman gain. The opposite is true for FPM. Decreasing channel spacing allows each channel to interact with more neighboring channels (Fig. 11), thereby increasing the nonlinear effects. Figures 14 and 15 are two examples of the

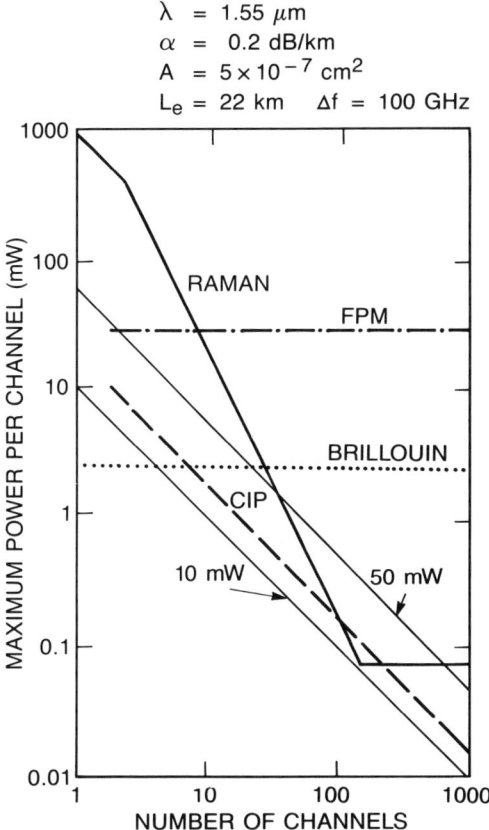

FIG. 15. Same as Fig. 13, but for 100-GHz channel spacing.

dependence of nonlinear effects on channel spacing. For 1-GHz channel separation (Fig. 14), the effects of SRS are diminished, whereas the effects of FPM are enhanced, compared to 10-GHz channel spacing. Note the added curvature in the FPM plot, indicating the increased extent of channel interaction. SBS and CIP curves are unaltered. For 100-GHz channel separation (Fig. 15), the effects of SRS are enhanced (total occupied bandwidth is larger). The break in the SRS curve at about channel number 150 indicates that the total occupied optical bandwidth is equal to the Raman width as approximated by a triangular profile. Adding more channels does not further degrade the system. The effects of FPM mixing are dramatically reduced because the FPM coefficient is small. Again, CIP and SBS are unaffected.

IX. Conclusion

Of the four nonlinearities, SRS is least likely to affect lightwave systems. Only in wavelength-multiplexed systems with hundreds of channels will SRS contribute to system degradations.

Carrier-induced phase noise is not a significant nonlinearity in PSK systems using external phase modulators. Directly phase-modulated lasers, however, have residual AM that can cause phase fluctuations detrimental to PSK systems. Even in passively multiplexed systems, lasers above about 20 mW in power will cause degradations. In frequency-selective-multiplexed systems, even low-power lasers (several milliwatts) will be unacceptable for systems with more than 10 channels.

The effects of SBS are directly related to transmitter power and independent of the number of channels. Consequently, SBS will degrade passively multiplexed systems with few channels more readily than systems with many channels. As usual, frequency-selective multiplexing exacerbates the effects of the nonlinearity. For typical high-speed, long-haul systems, transmitter powers exceeding 10 mW will degrade system performance.

Of all the nonlinearities, FPM is the most sensitive to system parameters. Not only does it depend on fiber length and core area, but it also depends on channel separation and fiber dispersion. To reduce the effects of FPM, channel separations should be greater than about 50 GHz and the wavelength region of minimum dispersion should be avoided. For standard fibers (dispersion zero at 1.3 μm) and channel separations of a few tens of gigahertz, FPM will degrade multiplexed systems for transmitter powers of a few milliwatts.

REFERENCES

Aoki, Y., and Tajima, K. (1987). "Dependence of the stimulated Brillouin scattering threshold in single-mode fibers on the number of longitudinal modes of a pump laser," *Tech. Digest, Conference on Lasers and Electrooptics*, April 26–May 1, Paper TuHH5, Baltimore, Maryland.

Aoki, Y., and Tajima, K. (1988). "Stimulated Brillouin scattering in a long single-mode fiber excited with a multimode pump laser," *J. Opt. Soc. B.*, **5**, 358.

Aoki, Y., Kishida, S., Honmou, H., Washio, K., and Sugimoto, M. (1983). "Efficient backward and forward pumping cw Raman amplification for InGaAsP laser light in silica fibers," *Electron. Lett.* **19**, 620.

Aoki, Y., Kishida, S., Washio, K., and Minenura, K. (1985). "Bit error rate evaluation of optical signals amplified via stimulated Raman process in optical fiber," *Tech. Digest, Conf. Opt. Fiber Commun.*, Feb. 11–13, paper TuO4, San Diego, California.

Aoki, Y., Tajima, K., and Mito, I. (1988). "Input power limits of single-mode optical fibers due to stimulated Brillouin scattering in optical communication systems," *J. Lightwave Technol.* **6**, 710.

Auyeung, J., and Yariv, A. (1978). "Spontaneous and stimulated Raman scattering in long low-loss fibers," *IEEE J. Quant. Electron.* **QE-14**, 347.

Bloembergen, N. (1967). "The stimulated Raman effect," *Am. J. Phys.* **35**, 989.

Chraplyvy, A. R. (1984). "Optical power limits in multichannel wavelength-division-multiplexed systems due to stimulated Roman scattering," *Electron. Lett.* **20**, 58.

Chraplyvy, A. R. and Henry, P. S. (1983). "Performance degradation due to stimulated Raman scattering in wavelength-division-multiplexed optical-fibre systems," *Electron. Lett.* **19**, 641.

Chraplyvy, A. R., and Stone, J. (1984a). "Synchronously pumped D_2 gas-in-glass fiber Raman laser operating at 1.56 μm," *Opt. Lett.* **9**, 241.

Chraplyvy, A. R., and Stone, J. (1984b). "Measurement of crossphase modulation in coherent wavelength-division multiplexing using injection lasers," *Electron. Lett.* **20**, 996.

Chraplyvy, A. R., and Stone, J. (1985). "Single-pass mode-locked or Q-switched pump operation of D_2 gas-in-glass fiber Raman lasers operating at 1.56 μm wavelength," *Opt. Lett.* **10**, 344.

Chraplyvy, A. R., and Tkach, R. W. (1986). "Narrowband tunable optical filter for channel selection in densely-packed WDM systems," *Electron. Lett.* **22**, 1084.

Chraplyvy, A. R., Stone, J., and Burrus, C. A. (1983). "Optical gain exceeding 35 dB at 1.56 μm due to stimulated Raman scattering by Molecular D_2 in a solid silica optical fiber," *Opt. Lett.* **8**, 415.

Chraplyvy, A. R., Marcuse, D. M., and Henry, P. S. (1984). "Carrier-induced phase noise in angle-modulated optical-fiber systems," *J. Lightwave Tech.* **LT-22**, 6.

Cohen, L. G., and Lin, Chinlon. (1978). "A universal fiber optic measurement system based on a near-IR fiber Raman laser," *IEEE J. Quantum Electron.* **QE-14**, 855.

Cotter, D. (1982). "Observation of stimulated Brillouin scattering in low-loss silica fibre at 1.3 μm," *Electron. Lett.* **18**, 495.

Cotter, D. (1983). "Optical nonlinearity in fibers: A new factor in systems design," *Br. Telecom. Technol. J.* **1**, 17.

Cotter, D. and Hill, A. M. (1984). "Stimulated Raman crosstalk in optical transmission: Effects of group velocity dispersion," *Electron Lett.* **20**, 185.

Desurvire, E., Papuchon, M., Pocholle, J. P., and Raffy, J. (1983). "High gain optical amplification of laser diode signal by Raman scattering in single-mode fibres," *Electron. Lett.* **19**, 751.

Hamilton, D. S., and Hellwarth, R. W. (1979). "Brillouin scattering measurements on optical glasses," *Phys. Rev. B.* **19**, 6583.

Hasegawa, A., and Tappert, F. (1973). "Transmission of stationary nonlinear optical pulses in dispersive dielectric fibers. 1. Anomalous dispersion," *Appl. Phys. Lett.* **23**, 142.

Hegarty J., Olsson, N. A., and McGlashan-Powell, M. (1985). "Measurement of the Raman crosstalk at 1.5 μm in a wavelength-division-multiplexed transmission system," *Electron. Lett.* **21**, 395.

Hegarty J., Olsson, N. A., and Goldner, L. (1986). "CW pumped Raman preamplifier in a 45-km long fibre transmission system operating at 1.5 μm and 1 Gbit/s," *Electron. Lett.* **21**, 290.

Hill, K. O., Johnson, D. C., Kawasaki, B. S., and MacDonald, R. I. (1978). "CW three-wave mixing in single-mode optical fibers," *J. Appl. Phys.* **49**, 5098.

Ikeda, M. (1981a). "Spectral power handling capability caused by stimulated Raman scattering effect in silica optical fibers," *Opt. Comm.* **37**, 388.

Ikeda, M. (1981b). "Stimulated Raman amplification characteristics in long span single-mode silica fibers," *Opt. Comm.* **39**, 148.

Ippen, E. P., and Stolen, R. H. (1972). "Stimulated Brillouin scattering in optical fibers," *Appl. Phys. Lett.* **21**, 539.

Johnson, D. C., Hill, K. O., and Kawasaki, B. S. (1977). *Radio Si.* **12**, 519.

Kaiser, W., and Maier, M. (1972). "Stimulated Rayleigh, Brillouin and Raman spectroscopy," in "Laser Handbook" (F. T. Arecchi and E. O. Schulz-Dubois, eds.), p. 1077. North-Holland, Amsterdam.

Koepf, G. A., Kalen, D. M., and Greene, K. H. (1982). "Raman amplification at 1.118 μm in single-mode fibre and its limitation by Brillouin scattering," *Electron. Lett.* **18**, 942.

Lichtman, E., and Friesem, A. A. (1987). "Stimulated Brillouin scattering excited by a multimode laser in single-mode optical fibers," *Opt. Comm.* **64**, 544.

Lichtman, E., Waarts, R. G., and Friesem, A. A. (1989). "Stimulated Brillouin scattering excited by a modulated pump wave in single-mode fibers," *J. Lightwave Technol.* **7**, 1.

Liu, P.-L., Ko, J.-S., Kaminow, I. P., Lee, T. P., and Burrus, C. A. (1983). "Steady-state intensity fluctuations, photon statistics and mode partitioning of Injection lasers," *Tech. Digest of Sixth Topical Meeting, Opt. Fiber Comm.*, Feb. 28–March 2, New Orleans, Los Angeles, Paper PD-3.

Mochizuki, K. (1985). "Optical fiber transmission systems using stimulated Raman scattering theory," *J. Lightwave Technol.* **LT-3**, 688.

Mollenauer, L. F., and Smith, K. (1988). "Demonstration of soliton transmission over more than 400 km in fiber with loss periodically compensated by Raman gain," *Opt. Lett.* **13**, 675.

Mollenauer, L. F., Stolen, R. H., and Gordon, J. P. (1980). "Experimental observation of picosecond pulse narrowing and solitons in optical fibers," *Phys. Rev. Lett.* **45**, 1095.

Nakazawa, M. (1985). "Highly efficient Raman amplification in a polarization-preserving optical fiber," *Appl. Phys. Lett.* **46**, 628.

Nakazawa, M., Tokuda, M., Negishi, Y., and Uchida, N. (1984). "Active transmission line: Light amplification by backward-stimulated Raman scattering in polarization-maintaining optical fiber," *J. Opt. Soc. Am. B.* **1**, 80.

Ohmori, Y., Sasaki, Y., Kawachi, M., and Edahiro, T. (1981). "Fibre-Length Dependence of Critical Power for Stimulated Raman Scattering", Electron. Lett. **17**, 593.

Olsson, N. A., and Hegarty, J. (1986). "Noise properties of a Raman amplifier," *J. Lightwave Technol.* **LT-44**, 396.

Olsson, N. A., and van der Ziel, J. P. (1986a). "Cancellation of fiber loss by semiconductor laser pumped Brillouin amplification at 1.5 μm," *Appl. Phys. Lett.* **48**, 1329.

Olsson, N. A., and van der Ziel, J. P. (1986b). "Fiber Brillouin amplifier with electronically controlled bandwidth," *Electron. Lett.* **22**, 488.

Olsson, N. A., Hegarty, J., Logan, R. A., Johnson, L. F., Walker, K. L., Cohen, L. G., Kasper, B. L., and Campbell, J. C. (1985). "68.3-km transmission with 1.37 Tbitkm/sec capacity using wavelength-division-multiplexing of ten single-frequency lasers at 1.5 μm," *Electron. Lett.* **21**, 105.

Prabhu, V. K. (1976). "PSK performance with imperfect carrier phase recovery," *IEEE Trans Aero. Elect. Syst.* **AES-12**, 275.

Shibata, N., Braun, R. P., and Waarts, R. G. (1986). "Crosstalk due to three-wave mixing process in a coherent single-mode transmission line." *Electron. Lett.* **22**, 675.

Shibata, N., Braun, R. P., and Waarts, R. G. (1987). "Phase-mismatch dependence of efficiency of wave generation through four-wave mixing in a single-mode optical fiber," *IEEE J. Quantum Electron.* **QE-23**, 1205.

Shibata, N., Azuma, Y., Tateda, M., and Nakamo, Y. (1988). "Experimental verification of efficiency of wave generation through four-wave mixing in low-loss dispersion-shifted single-mode optical fiber," *Electron. Lett.* **24**, 1528.

Shuker, R. and Gammon, R. W. (1970). "Raman-scattering selection-rule breaking and the density of states in amorphous materials," *Phys. Rev. Lett.* **25**, 222.

Smith, R. G. (1972). "Optical power handling capacity of low-loss optical fibers as determined by stimulated Raman and Brillouin scattering," *Appl. Opt.* **11**, 2489.

Stoicheff, B. P. (1963). "Characteristics of stimulated Raman radiation generated by coherent light," *Phys. Lett.* **7**, 186.

Stolen, R. H. (1979). "Nonlinear properties of optical fibers," *in* "Optical Fiber Telecommunications," (S. E. Miller and A. G. Chynoweth, ed.). Academic Press, New York.

Stolen, R. H. (1980). "Nonlinearity in fiber transmission," *Proc. IEEE* **68**, 1232.

Stolen, R. H. and Bjorkholm, J. E. (1982). "Parametric amplification and frequency conversion in optical fibers," *IEEE J. Quantum Electron.* **QE-18**, 1062–1072.

Stolen, R. H., and Ippen, E. P. (1973). "Raman gain in glass optical waveguides," *Appl. Phys. Lett.* **22**, 276.

Stolen, R. H., and Lin, Chinlon. (1978). "Self-phase modulation in silica optical fibers," *Phys. Rev. A* **17**, 1448.

Stolen, R. H., Lee, Clinton, and Jain, R. K. (1984). "Development of the stimulated Raman spectrum in single-mode silica fibers," *J. Opt. Soc. Am. B.* **1**, 652.

Stone, J., Chraplyvy, A. R., and Burrus, C. A. (1982). "Gas-in-glass—a new Raman-gain medium: Molecular hydrogen in solid-silica optical fibers," *Opt. Lett.* **7**, 297.

Stone, J., Chraplyvy, A. R., and Kasper, B. L. (1985). "Long-range 1.5 μm OTDR in a single-mode fibre using a D_2 gas-in-glass laser or semiconductor laser," *Electron. Lett.* **21**, 541.

Tang, C. L. (1966). "Saturation and spectral characteristics of the Stokes emission in the stimulated Brillouin process," *J. Appl. Phys.* **37**, 2945.

Tkach, R. W., Chraplyvy, A. R., and Derosier, R. M. (1988). "Optical demodulation and amplification of FSK signals using AlGaAs lasers," *Electron. Lett.* **24**, 260.

Tkach, R. W., Chraplyvy, A. R., and Derosier, R. M. (1989). "Performance of a WDM network based on stimulated Brillouin scattering," *IEEE Photonics Technol. Lett.* **1**, 111 (1989).

Tomita, A. (1983). "Crosstalk caused by stimulated Raman scattering in single-mode wavelength-division multiplexed systems," *Opt. Lett.* **8**, 412.

Uesugi, N., Ikeda, M., and Sasaki, Y. (1981). "Maximum single-frequency input power in a long optical fiber determined by stimulated Brillouin scattering," *Electron. Lett.* **17**, 379.

Vodhanel, R. S. (1989). "5 Gbit/s direct optical differential phase-shift keying modulation of a 1530-nm distributed-feedback laser," *IEEE Photonics Technol. Lett.* **1**, 218 (1989).

Yamamoto, Y., Saito, S., and Mukai, T. (1983). "AM and FM quantum noise in semiconductor lasers—Part II," *IEEE J. Quantum Electron.* **QE-19**, 47.

Yariv, A. (1975). "Quantum Electronics," 2nd ed. Wiley, New York.

Index

A

ac content, 35
Active feedback lightwave receiver circuits
  applications, 121
  basic nonintegrating high-sensitivity receiver, 121
  calculations of IC receiver sensitivities, 123
  comparison to previous designs, 121
  dynamic-range-extender/automatic-gain control (AGC), 121–123
  FET IC designs, 123
  gain elements, 122
  hybrid IC active-feedback receiver, 122
  IC receiver designs, 123
  ideal transimpedance amplifier, 122
  large-value feedback resistor, 122
  micro-FET feedback circuit, 121
  micro-FET feedback receiver, 122
  monolithic IC active feedback receiver, 122
  NMOS, CMOS, GaAs implementations, 123
  parasitic feedback capacitance, 122
  sensitivity, 121–122
  transimpedance receivers, 123
Active-feedback receivers, 80, 83
  dynamic-range-extender circuitry, 80
  IC versions, 81
AlGaAs
  gain-guided devices, 22
  and InGaAsP LEDs, 25–26
    applications in data link market, 26
    compared with lasers, 25–26
  optoelectronic devices, 26
    MM fibers, 26
    single-mode fiber, 26
    zero-chromatic dispersion, 26
  semiconductor laser, 189
    locked laser, 189–190
    longitudinal mode frequency, 190
  semiconductor laser amplifier, 186
    bandwidth-versus-gain relationship, 186
    coherence of phase-locked lasers, 186
    single mode operations, 186
  stripe geometry lasers' growth techniques, 22
    liquid phase epitaxy, 22
    oxide isolation, 22
    zinc diffusion, 22
Amplitude variations, 19
Analog transmission, 8–9, 45
  limits, 8
  signal-to-noise ratio, 45
  video applications, 45
Angle modulation, 151
  optical heterodyne, 151
  optical homodyne, 151
AT&T FT Series G 417 M/bs system, 57
AT&T SL TAT-8 trans-Atlantic subcable, 58
Audio and video coding bit rates, 8
Automatic bias control circuit, 38
Avalanche photodiode (APD) receivers, 87, 107–112
  avalanche multiplication noise, 110
  avalanche-noise amplitude distribution, 109
  device noise theory of McIntyre, 87
  electron and hole ionization coefficients, 108
  electron-initiated APD, 108
  heterostructure APDs, 110
  hole feedback, 108
  InGaAs/InP APD receiver sensitivities versus bit rate, 110–111
  leakage current, 111
  logic-zero bits, 108
  noiseless multiplication, 107–108
  optical receiver sensitivity calculations, 87
  photocurrent signal, 110
  photomultiplier, 108
  primary photocurrent, 107
    multiplied by impact ionization, 107
    noise, 107
  receiver noise theory of Smith and Personick, 87
  total equivalent mean-square primary photocurrent noise, 108
  tunneling current problem, 110
  wide-gap avalanche region, 110

B

Back facet monitoring, 49–53
  kinks, 49
  laser self-pulsations, 49
  low frequency noise, 49
  minimizing tap insertion loss, 53
  10-μm stripe lasers, 49
Biasing circuits for lasers, 31–40
  average power P_{AV}, 32
  biasing conditions, 33
    bias point I_B/I_{th}, 33, 38
    double feedback control circuits, 33–34
    highest and lowest extinction ratios, 33–34
    single control circuits, 37
  compared with LEDs, 31–32
  early circuit designs, 32
  extinction ratio ε, 32
  feedback strategies, 34–38
  fixed pulse current, 33
  jitter t_j, 32
  pattern dependence ΔP_p, 32
  slope change I_{th}, 34, 38
  timing delay t_d, 31–32, 34
Biasing circuits for LEDs, 29–31
  Boltzmann's constant, 31
  control with increasing temperature, 31
  current peaking and charge extraction circuits, 30–31
  and decrease in optical power, 29–32
  diffusion capacitance C_d, 30
  preshaped driving pulse, 30
  and pulse response time, 29–31
  reverse bias at the on–off transitions, 31
  rise and fall times, 31
  space charge capacitance C_s, 30
Binary PCM signals, 9
Biphase optical data transmission, 91
Bipolar technologies, 85
  heterojunction bipolar transistor, 85
BISDN (broadband integrated services digital network), 2
Bit rate × distance product, 2
Blue shift, 173, 177
  BH laser, 173
  BH type index-guided semiconductor lasers, 173
  effective refractive index, 173
  oscillating frequency, 173
Boltzmann's constant, 31

British Post Office (BPO), 116, 119–120
  encoding, 116, 120
  integrating receiver, 120
  limited equalization, 120
Burrus emitter, 20

C

Capacitive feedback receivers, 127–128
  capacitive feedback element, 127
  hybrid IC voltage gain, 127
  nonintegrating current-amplifier circuit, 127
  photodiode capacitance C_{pin}, 127
Carrier density modulation effect, 163–167
  comparison of frequency modulation and intensity modulation, 164
  Free Spectral Range (FSR), 165
  frequency characteristics, 164–165
  frequency deviation in AlGaAs CSP laser, 163
  frequency deviation in CSP laser, 164
  gain-guided lasers, 164
  index-guided lasers, 164
  relaxation oscillation, 163
  resonant enhancement, 163
  solid line curves, 163
Carrier-Induced Phase Modulation (CIP), 278–281
  chromatic dispersion, 281
  cross-phase modulation, 279–280
  dichroic mirror, 280
  intensity-dependent refractive index, 278
  nonlinear refractive index, 278–279
  phase noise, 278
  power fluctuations, 279
  self-reflexive interferometer, 280
  silica solitons, 281
  third-order electric susceptibility, 279
Channels in wavelength-multiplexed systems, 292
Characteristic temperature T_0, 21
Charge extraction circuits, 47
Chromatic dispersion, 10, 13, 227–228
Circuit and transistor technology implications, 84–85
Coherent laser sources
  FM and PM noise, 234–235
  frequency stabilization, 235
  modulation–demodulation schemes, 234

optical isolator, 238
oscillation wavelength, 234
semiconductor lasers, 235
    configuration for frequency stabilization, 236
    drawbacks as light sources, 235
    frequency stabilization, 237
    negative frequency feedback control, 237
    noise characteristics, 237
    optical isolators, 237
    oscillation frequency, 235
    quantum mechanical Langevin equation, 237
    quantum statistical treatment, 237
    spectral linewidths, 234, 235
Coherent modulation–demodulation, 218
    error-rate performance of various modulations, 218
    PSK homodyne detection, 218
Coherent optical fiber communication systems, 205–206, 233–234
    coherent optical carrier generation, 233
    FDM techniques, 206
    frequency-stabilized master oscillator, 233
    long-distance and high transmission lines, 205–206
    optical phase-locked loop (PLL), technique, 234
    present studies, 206
Coherent optical fiber transmission systems
    differences with current optical fiber transmission systems, 206–207
    fiber loss characteristics, 204
    first generation through fourth generation, 203–204
    lightwave intensity, 204
    long-distance high-speed transmission systems, 225–230
        dissolution of bandwidth limitation, 227–229
        improvement of launched power, 227
        improvement of modulation bandwidth, 229–230
        optical coherent modulation–demodulation techniques, 226
        regenerative repeater spacing, 225
    modulation and demodulation technology, 204
        DD direct detection, 204
        IM intensity modulation, 204
    semiconductor device technology, 204
        AlGaAs technology, 204
        Ge technology, 204
        InGaAsP technology, 204
        Si technology 204
        spectral coherence 204
Coherent optical modulation–demodulation methods, 206
Coherent systems, 227–228, 232
    chromatic dispersion, 227–228
Coherent transmission application modulation–demodulation technology, 238–244
Complementary distortion, 45
Complete receiver circuit noise expressions
    device figures of merit, 101–105
    FET, 101
    FET IC technologies, 101, 104
    FET transconductance and input capacitance, 101
    front-end figure of merit, 101
    high sensitivity pin-FET receiver, 103
    high sensitivity receivers and microwave or VHSIC technologies, 102
    1-μm gate-length GaAs MESFETs, 102
    1-μm gate silicon MOSFETs, 102
    optical receiver sensitivity, 102
    pin-photodiode, 101
    technology objectives, 102
    sensitivity calculations, 103
    stray-capacitance figure of merit, 102
    unity gain frequency of FET technology, 101
Controlled-dispersion fiber, 67
    Brillouin scattering, 67
    dispersion-flattened DF fibers, 67
    power-handling capability of SM silica fibers, 67
    wavelength-division-multiple (WDM), 67
Conventional and coherent systems, 218
    compared with direct detection, 219
    detectable power levels, 218–219
    nonreturn-to-zero (NRZ) rectanguler pulse, 218
Conventional optical fiber transmission systems
    amplitude-shift keying (ASK), 208
    avalanche photodiode (APD) p-i-n photodiode, 207
    components, 207

Conventional optical fiber transmission systems (*continued*)
  direct detection, 207
  frequency difference $|f_1 - f_0|$, 208–209
  frequency-shift keying (FSK), 208
  IM intensity modulation, 207
  intensity modulation scheme, 208
  intermediate frequency (IF) signal, 201
  light-emitting diode (LED), 207
  on–off keying (OOK), 208
  optical heterodyne detection, 208
  optical homodyne detection, 208
  phase-shift keying (PSK), 208
  semiconductor laser, 207
Conventional single-mode fiber, 211
  polarization controller, 211
Conventional systems, 227–228, 232
  chromatic dispersion, 227–228
Conventional transimpedance receiver, 113
  nonmultiplying pin photodiode, 113
  silicon APD detector, 113
Current peaking, 30–31, 47
  and charge extraction, 30–31
cw mode, 3

D

Data reference, 34–36
  feedback strategies, 34–36
Degradation of a short wavelength channel, 274–278
  color-center laser (FCL), 274
  long-range optical time-domain reflectometry, 278
  optical signal amplification, 274
  in single-mode fibers, 276–278
  as source of infrared light, 276–278
  and stimulated Raman scattering (SRS), 274–278
Demodulation, 241–244
  AM noise and SNR degradation, 241
  Costas-type loop, 243
  FM noise and SNR, 241
  heterodyne synchronous detection, 243
  high bias-current operation, 241
  optical heterodyne or homodyne receivers, 241–243
  optical PLLS, 242–243
  phase locking of local oscillator, 241
  shot-noise-limited detection, 241
  signal power attenuation and SNR, 241
  voltage-controlled oscillator (VCO), 242
Detectors
  finite modulation bandwidths, 7
  InGaAs avalanche photodiodes (APD), 8
  InGaAs PIN photodiode, 7
  lightwave scheme, 8
  noise generation limitations, 7
  silicon photodetectors, 7
  wavelength-dependent quantum efficiencies, 7
Differential, *see* External quantum efficiency
Digital bit-error rate, 82
Digital decision circuits
  asynchronous, 88–91
  automatic gain control (AGC)
  synchronous, 88–91
Digital receiver systems, 86–88
  associated waveforms, 87
  input optical signal power, 87
  non-return-to-zero (NRZ) signal format, 87
  optical fiber systems, 86
  signal-to-noise ratio, 87
  total noise power, 87
  video codec ICs, 86
Direct current modulation of lasers and LEDs, 3, 9, 19–26
  light emission in semiconductors, 20–23
  optical properties, 19–26
  pulse response characteristics, 19–26
Direct frequency modulation
  FM semiconductor laser, 180
  two-electrode laser, 180
  InGaAsP DFB laser, 181–186
    blue shift, 183–184
    electrodes, 183–184
    red shift, 183–184
    optical angular frequency deviation, 180–182
    parasitic intensity modulation, 185
    phase modulation, 151–152
Dispersion-limited fiber systems, 34
Distributed Bragg reflector (DBR), 227
Distributed feedback (DFD), 227
Distributed feedback laser
  Fabry–Perot structure, 14
  linewidth, 14
  perturbed waveguide, 14
Double feedback control

INDEX 301

circuits, 33–34
  bias point, 33
  feedback strategy, 34
  using front facet monitoring, 55–56
    constraints, 55
    double-t single-loop control, 55
    high-speed operational amplifier, 55
    monitoring extinction ratio, 55
    transimpedance amplifier, 55
Double heterostructure lasers
  carrier and photon confinement, 20
  external quantum efficiency v, 20–21
  Fabry–Perot cavity, 20
  high nonlinear electro-optic transfer functions, 20
  lasting threshold, I_{th}, 20
Drive current, 46
Dynamic range extenders, 128–132
  AGC circuits, 128–132
  dc bias source, 130
  equivalent circuit, 128
  input-shunt/voltage-gain-reduction AGC, 132
  linear drain-current-versus-drain-voltage-region, 128
  maximum input-voltage swing, 131
  variable input shunts, 128–130

E

Early voltage amplifier, 112
Edge emitter, 20
  coupling efficiency, 20
  integral lens, 20
  radiance values, 20
  tendency to lase, 20
Edge-emitting (EE) LEDs, 13
Edge enhancement, 47–48
Electrical-to-optical conversion, 3
  ICs in Si-bipolar, 3
  GaAs MESFET, 3
  Microwave interconnections, 3
  thermoelectric heat pump temperature control, 3
Electro-optic frequency modulation
  Franz–Keldish effect, 152
  fee-carrier absorption, 152
  injection current modulation, 152
  in semiconductor lasers, 152
  intracavity phase modulation, 152
Electro-optic waveguide modulators, 229

Equalization techniques, 115–116
  limiting gain, A, 116
  noise penalty, 115–116
Equivalent circuit models
  cascaded two-part model, 44
  current-controlled voltage source, 43
  damping mechanisms, 44
  effects of package parasitics, 43
  effects of space charge capacitance, 43
  intrinsic equivalent circuit, 43
  large signal model, 43
  normalized light response, 43
  photon density S, 41
  relaxation oscillation resonance frequency, 44
  single-mode rate equations, 41–42
  small-signal light response, 42
Etched mesa and buried heterostructure (EMBH), 60
External modulators
  directional-coupler type, 68
  loss type, 68
  Mach–Zehnder, 68
  total-internal-reflecting type, 68
External quantum efficiency, v, 20

F

Fabry–Perot cavity, 20
Facet monitoring, *see* Back facet monitoring; Front facet monitoring
Fast rise, t_r, and fall, t_f, times, 19
FDM systems, 232–233, 244
  application to local networks, 232–233
  chromatic dispersion, 244
  multiplexers/demultiplexers, 244
  optical interference, 244
Feedback-selective harmonic compensation, 45
FET technology, 80–82
Fiber amplifiers
  Erbium (Er)-doped, 249–251
  fiber-Brillouin, 249
  fiber-Raman, 249–251
Fiber and device technology
  direct intensity modulation, 205
    of light-emitting diodes (LEDs), 205
    of semiconductor lasers, 205
  direct power detection, 205
  repeater spacing, 205

Fiber-optic Raman amplification, 274–278
  copropogating, 275
  counterpropagating, 275
  pseudorandom bit stream, 276
  pump laser, 275
  soliton amplification, 275
  transmission medium, 275
  wavelength selective coupler, 275
Fiber perturbations, 246–247
  core-cladding interface irregularity, 246
  external fluctuation conditions, 247
  scattering center inhomogeneity, 246
Fibers, 4–5
Filtering and digital signal recovery circuits, 86
First-generation lightwave receivers, see Lightwave receivers
First-generation transmitters, 71
First-generation ∼0.8 μm, 10
Four photon mixing, 285–288
  dispersion-shifted fiber, 286–287
  fiber chromatic dispersion, 286
  group velocities, 286
Franz–Keldysh effect, 152
Frequency characteristics of direct frequency modulation, 161–163
  buried heterostructure (BH), 161
  channeled substrate planar (CSP), 161
  double-heterostructure AlGaAs semiconductor lasers, 161
  frequency deviation CSP, BH, and TJS, 161–162
  frequency deviation response of an InGaAsP laser diode, 162
  resonant phenomena, 161–162
  transverse-junction stripe (TJS) lasers, 161
Frequency characteristics of induced PM signal, 191–195
  brokenline curves, 193
  carrier refractive index change, 195
  closed-circle curves, 193
  frequency deviation of FM, 193
  locking half-bandwidth (LHB), 192, 194
  maximum phase deviation (MPD), 193–194
  modulation index, 192–193
  static phase shift, 192
  thermal refractive-index change, 194
  van der Pol equation, 191–192
Frequency modulation, 153–161, 197–199

birefringent crystal, 159
calculated gain spectra, 158
carrier density and temperature, 157
carrier modulation effect, 197
FM–AM conversion, 161
frequency deviations, 160–161
Halperin–Lax band tail, 158
heterodyne detection, 159
index, 160
injection-locked phase modulation, 197–198
MFD (maximum frequency deviation), 198
Michelson interferometer, 159
MPD (maximum phase deviation), 198
multielectrode semiconductor laser, 198
normalized frequency deviations, 157, 159
phase modulation, 198–199
power spectra, 154, 160
p-side-down mounted laser, 198
p-side-up mounted laser, 198
refractive-index-anomalous dispersion, 158–159
resonant peak, 197–198
scanning Fabry–Perot spectrometer, 159–160
small-signal van der Pol equation, 198
Frequency selectivity, 220–225
  homodyne receiver, 221
  incoming wave, 223
  optical heterodyne detection, 221
  response to multiplex signal, 221–222
  shot noise, 221
  superiority of coherent detection, 225
  thermal noise, 221
Front facet monitoring, 53–55
  advantages, 53
  coupling efficiency changes, 53
  minimizing tap insertion, 53
  tap current variation, 53–55
FT Series G 1.7 Gb/s 1.55-μm DFB WDM channel, 69–71
Full-width half-maximum (FWHM) optical spectral width, 13

G

GaAlAs, 20
Gain-guided lasers, 10, 21–23, 33–34, 48–56
  AlGaAs lasers, 22, 48–56
  bias points, 33–34
  stripe widths, 22

H

Heterodyne receivers, 211, 213, 215, 217, 241–244, 258
  beam splitter, 211
  detection schemes, 217
  modulation and demodulation, 241–243
  photomixing gain, 213, 215
  probability density functions, 216
High extinction ratio (on/off ratio), ε, 19
High impedance receiver, 112–113, *see also* Integrating receiver
Homodyne receiver, 211, 241–244, 258
  modulation and demodulation, 241–244
  optical phase-locked loop (PLL), 211
Hybrid IC active-feedback receivers, 132–135
  AGC/dynamic-range-extender circuitry, 133
  capacitive feedback, 132–133
  forward voltage amplifier, 132
  frequency response, 133–135
  loop stability considerations, 135
  photodiode capacitance feedback, 132

I

IC receivers, 85, 135–148
  active-feedback receivers, 135–143
    capacitive and nonlinear effects, 136
    gate bias source, 138
    Johnson noise, 137
    micro-FET feedback basic receiver, 135–136
    voltage amplifiers, 135–137
  design scaling laws, 143–144
    high bit-rates, 143–144
    parasitic feedback capacitance, 144
  sensitivity calculations, 144–148
    and bit-rate calculations, 144, 146
    comparison of GaAs and silicon FET IC technologies, 144–145
    gate input capacitance C_{FET}, 145
    mean-square noise, 147–148
    pin-photodiodes, 144–147
Index-guided lasers, 21–23, 33–34, 38, *see also* Narrow-spectrum index-guided lasers
  basic LF technique, 38
  bias point, 33–34
InGaAs avalanche photodiodes (APD), 8
InGaAs/InP APD-FET receiver measurements, 111–112
  avalanche multiplication of primary leakage current, 111–112
  theoretical germanium sensitivities, 112
InGaAsP, 20, 22, 24
InGaAs PIN photodiode, 7
Injection current frequency modulation, 152–153
  in AlGaAs and InGaAsP semiconductor lasers, 152
  refractive-index change, 152–153
  structure of oscillator and modulator, 152–153
Injection-locked oscillators, 227
Injection-locked semiconductor laser, 258
Input device noise theory for pin-FET receivers, 96–101
  circuit-equivalent input-noise power, 100
  filter frequency response function $F(f)$, 97
  filter frequency response shape $F^*(y)$, 97
  input resistor Johnson noise term, 100
  Johnson noise of the bias/feedback, 96
  mean-square FET drain noise current, 99
  mean-square input noise current, 99–100
  normalizing channel frequency response, 97
  root-mean-square noise current, 101
  Smith and Personick theory for noise, 96–101
Integral lens, 20
Integrating optical receivers, 80, 82, 113–121
  high impedance, 114
  photocurrent signal, 115
  sensitivities, 114–115
  transimpedance, 119–121
    bipolar junction transistor BJT cascade, 119
    BPO amplifier, 119
    encoding, 120
    FET input transistor, 119
    Ogawa amplifier, 119
Intensity modulation of sources, 9
I_{th} and optical pulse, 25

K

Kane function, 158
Kramers–Kronig integral of theoretical gain spectra, 158

L

Lasers, 2–3, 9–10, 14, 17–18, 19–44, 48–64, 132, 140–142
  bandwidths, 17–18
  circuits, 26–44, 140–142
    biasing, 31–40
    parasitic feedback capacitance, 141
    strategies, 26–44
    transimpedance AGC receiver circuit, 140–141
  diodes, 3
  direct current modulation, 9, 19–25
  gain-guided, 21–22, 48–56
  index-guided, 21–23, 33–34, 38, 58–60
  laser-chirping, 14
  long-wavelength laser diodes, 40
  pulse response, 23–25
  self-pulsations, 49
  transmitters, 48–64
    AlGaAs gain-guided lasers, 48–56
    1.3 μm InGaAsP multifrequency lasers, 56–64
Lasting threshold I_{th}, 20–21
LEDs (semiconductor light-emitting diodes), 2–3, 9, 19–25, 26–48, 132, 139–140
  bandwidths, 17–18
  biasing circuits, 29–31
  circuits, 26–44, 139–140
  direct current modulation, 9, 19–25
  pulse response, 23–25
    nonradiative t_{nr} lifetimes, 23
    normalized output power, $P(w)$, 23
    radiative t_r lifetimes, 23
  transmitters, 44–48
    1.3 μm IGaAsP sources, 45–48
    0.87 μm AlGaAs sources, 44–45
LF-modulated photodiode signal, 38
Light emission in semiconductors, 20–23
  direct band-gap semiconductor recombination, 20
  edge emitter, 20
  linear electro-optic transfer functions, 20
  surface emitter, 20
Lightwave
  communications, 1–3
  receivers, first and second generation, 112–121
    conventional transimpedance receiver, 113
    early voltage amplifier, 112
    high impedance receiver, 112
    integrating optical receivers, 113–116
    sensitivity versus dynamic-range tradeoff, 116
  scheme, 8
  system, 14–18
    lightwave window, 15–17
Linear amplifiers, 249
  Fabry–Perot cavity type, 249
  traveling wave type, 249
Linearization, 8, 45
Local oscillator, 243
Locked-laser-output, 186
Longitudinal modes, 10, 14
Long-span multigigabit systems, 14
Long-wavelength LED and laser sources, 17–18
Loop applications, 84
Low frequency (LF) signal, 38
Low source noise, ΔP_n, 19

M

Micro-FET feedback receiver, 123–126
  avoiding signal integration, 124–125
  circuit representation, 123–124
  current-to-voltage amplifier, 123
  dynamic-range-extender-circuit, 124
  input capacitance, 124
  parasitic shunt capacitance, 124–125
  receiver photocurrent response, 126
  special-design feedback micro FET (QF), 125
Microwave electronic circuits, 2
Modal noise, 22, 60–64
  BER fluctuations, 60
  channel substrate buried heterostructure (CSBH), 60
  effects of microwaves, 61–62
  in multimode fiber systems, 61–64
Mode partition noise, 14
Modulation, 2–10, 14–18, 26–29
  bandwidths, BW, 10
  circuits, 26–29
    common collector driver, 27–28
    common emitter saturating switch, 27
    current routing switch, 27
    Shottkey clamps, 27
    shunt driver, 27–28
    Si bipolar devices, 28–29

INDEX

formats, 8–9
  on–off keyed (OOK), 9
  pulse code modulated (PCM) format, 8
of the polarization vector, 4
properties of sources, 4–8
schemes, 2
rate, 3, 14–18
  lightwave window, 15–17
  limitations in lightwave system, 14–18
Modulation and demodulation technology, 204, 238–244
  amplitude modulation (AM), 204
  angle modulation, 238
  amplitude-shift keying (ASK), 204
  direct detection (DD), 204
  direct modulation of semiconductor lasers, 238
  external modulators, 238
  frequency modulation (FM), 204
  guided wave electro-optic modulation, 240
  heterodyne detection of optical signals, 204
  homodyne detection of optical signals, 204
  homodyne detection of optical signals, 204
  intensity modulation (IM), 204
  multiple quantum well (MQW), 240
Multifrequency lasers, 58
Multigigabit-per-second transmission systems, 66–71
  controlled-dispersion fiber, 67
  external modulators, 68–69
  long-haul trunking, 66
  present and future systems, 66–67
  submarine cable lightwave systems, 66
Multimode (MM) fibers, 5, 13, 34
Multiplexing effects, 288–290
  degradations by optical nonlinearities, 289
  passive versus frequency-selective, 288
  semiconductor amplifiers, 288–289

N

Narrow-spectrum index-guided lasers, 58–60
  back facet monitoring, 58
  etched mesa and buried heterostructure (EMBH), 60
  MM fiber applications, 58
  modal noise in MM fiber systems, 59–60
  negative eye degradation (NED), 60
  positive eye closure, 59–60

positive eye degradation (PED), 60
  temperature control, 58–59
  thermoelectric heat pump, 58–59
  SM fiber applications, 59
  spatial mode filtering, 60
Narrow-stripe lasers, 50–52
  parasitics and instabilities in gain-guiding lasers, 50–51
  severe pattern dependence ΔP_p, 50
  strong and weak pattern dependence, 50
Negative eye degradation (NED), 60
Nonfiber systems, 204–206
  coherent laser light, 204–205
  coherent optical modulation, 204–205
  demodulation techniques, 204–205
  frequency-division multiplexing (FDM), 205
Nonintegrating high-sensitivity wide-dynamic-range receiver circuits, 85
Nonlinear gain and system parameters, 268–269
  cross-sectional area, 268
  gain coefficient, 268
  optical nonlinearities, 269
  probe wave, 268
  pump, 268
Nonradiative recombination, 20
NRZ optical data transmission, 91

O

On–off keyed (OOK), PCM format, 9
Optical amplifiers, 230–232, 249–251
  Er-doped fiber amplifier, 249–251
  Fabry–Perot cavity-type amplifier, 249–251
  fiber Brillouin amplifiers, 249–251
  repeater systems, 230–232
  semiconductor amplifiers, 249
  traveling-wave type amplifier, 251
    beat noise, 232
    contrast to regenerative repeaters, 230
    fiber amplifiers, 230
    optical amplifier configuration, 231
    semiconductor amplifiers, 230
Optical CATV, 13
Optical depolarization, 4
Optical dispersion, 4
Optical feedback, 22
Optical feed forward, 45

Optical fibers, 82, 151, 244–249
  digital regenerators, 82
  polarization control and diversity techniques, 248
  polarization maintaining fibers, 245–248
  transmission systems, 151
Optical frequency modulation, 151-152
  acousto-optic effects, 152
  direct detection, 151
  electro-optic effects, 152
  heterodyne and homodyne detection, 151
  internal and external modulators, 152
Optical heterodyne and homodyne detection, 210–225
  advantages over direct detection, 211–225
  frequency selectivity, 211, 220–225
  receiver sensitivity, 212–215
Optical linearities, 267–268
  angle-modulated systems, 268
  cross-phase modulation, 267
  effects in fused silica fibers, 267
  four-photon mixing, 268
Optical loss, 4
Optical performance, 2
Optical phase modulation, 151, 153
Optical PLL technique, 242–243
  in-phase and quadrature (IQ) detection receiver, 243
  phase density receiver, 243
  stationary phase error, 243
  three-phase detection receiver, 243
Optical properties of sources for PCM transmitters, 18–26
  direct current modulation of lasers and LEDs, 19–25
  optical pulse requirements, 18–19
  source properties, 25–26
Optical pulses, 3, 18–19, 25–44
  circuit strategies, 26–44, *see also* Biasing circuits
  data rates at the gigabit per second range, 26
  time and temperature changes, 26
  width, 18
Optical receiver circuits, 79–80
  high-sensitivity integrating front ends, 80
  transistor noise physics influences, 79
Optical receiver noise power spectra, 90
Opto-electronic integrated circuit OEIC technology, 258

P

Pattern dependence, ΔP_p, 19
PCM transmitters, 26–44
  biasing circuits for lasers, 31–34
  biasing circuits for LEDS, 29–31
Phase modulation by injection locking, 186–189
  cavity (resonant) frequency, 186–187
  inherent injection locking phenomenon, 186
  maximum phase deviation (MPD), 188–189
  static phase shift, 187–188
Phase modulation measurement, 189–191
  Babinet–Soleil compensator, 190–191
  Mach–Zehnder interferometer, 190–191
  maximum phase deviation (MPD), 191
  phase modulation light (PM), 190
  sinusoidal signal, 191
Phase modulation for transmission applications, 195–197
  heterodyne and homodyne coherent optical
  transmission, 195–196
  phase shift, 195–196
  PM response, 195–196
Phase relationship between FM and AM signals, 171–174
  FM and AM modulated wave, 171
  MHz frequency deviation for CSP, BH, and TJS lasers, 171
  red shift, blue shift, 173
Photodetectors, 79, 81–82, 244
pin-FET receivers, 86, 105–107
  numerical sensitivity calculations, 105
  receiver front-end noise, 86
  sensitivity calculations for future IC receivers, 106
pin-photodiode receiver noise and sensitivity calculations, 91–95
  Gaussian noise discussion, 94
  open-loop noise calculations, 92
  receiver-circuit noise expressions, 92
  Smith and Personick expression for noise, 91
Pockels effects, *see* Optical phase modulation
Polarization control and density techniques, 248–249
Polarization maintaining fibers, 245–246
  bow-tie fibers, 246

eigenpolarization modes, 245
elliptical jacket, 246
modal birefringence, 245–246
PANDA, 246
Positive eye closure, 62
Positive eye degradation (PED), 60
Pulse code modulated (PCM) format, 8–9
Pulse response, 23–25, 46
  differences in LEDs and lasers, 23
  pulse modulation, 25
Pulse width variations, 19, 22

Q

Quantum mechanical Langevin equation, 237

R

Rayleigh scattering limit, 2
Receiver amplifier circuit, 85
Receiver and device requirements of lightwave systems, 83–85
  future optical-fiber applications, 84
  loop plant, 84
  receiver design goals, 83
Receiver sensitivity, 212–215, 219–220
  avalanche multiplication factor, 213
  calculations in coherent schemes, 220
  carrier-to-noise ratio (CNR), 213
  communication theory, 220
  multivalued system, 220
  shot noise, 212
  signal and noise power levels, 214
Red shift, 173, 177
Reduction of loop time delay, 237
Refractive-index change, 152–153
Regenerative repeater spacing versus data rate, 229
Relaxation oscillation resonance frequency, 44
Return-to-zero (RZ) optical data transmission, 91

S

Second generation 1.3 μm, 10–13
Semiconductor amplifiers, 249, 276, 288–289
  injection-locked oscillators, 249
  linear, 249
  multiplexing effects, 288–289
  and Raman amplification, 276
Semiconductor-laser linear amplifier, 227
Semiconductor lasers, 2, 10, 151, 157, 175–177, 205, 207, 229–230, 238–239, 241–243, 279
  AM noise, 241
  carrier distribution, 175–177
  carrier-induced phase modulation (CIP), 279
  direct frequency modulation, 151–152, 229–230
  direct IM feasibility, 207
  exhibiting interference fringes, 205
  high-bias current operation, 241
  local oscillator, 243
  modulation–demodulation technology, 238–239
  normalized oscillation frequency deviation, 157–159
Sensitivity versus dynamic-range tradeoff, 116
Separate-absorption-grading-multiplication avalanche photodiodes (SAGM APD), 8
7B/8B encoding technique, 116
Scaling 83, 290–292
Short-span subsciber loop systems, 13
Short-wavelength LEDs and lasers, 10
Shottky clamps, 27
Si bipolar devices, 28–29
Silicon avalanche photodetectors (APDs), 80–83
  first and second generation transmission systems, 82
  staircase APD proposal, 82
Single feedback loop strategies, 33–34
Single-frequency DFB lasers, 58, 67–68
  distributed feedback (DFB) lasers, 67
  injection lasers, 68
Single-frequency lasers, 14
Single-mode (SM) fiber, 5, 13
Sinusoidal response, 173–174
Slope change, 38
  low frequency (LF) signal, 38
  phase-sensitive detector (PSD), 38
Smith and Personick calculations of receiver noise, 82, 87, 91, 96–101
Snodgrass and Klinman commercial silicon bipolar receiver IC, 119

Solitons, 281
Source temperature control, 40–41
　thermistor, 40–41
　thermoelectric heat pump, 40–41
Source wavelength and linewidth, 10–14
　first generation, 10
　second generation, 10–13
　single-frequency lasers, 14
Staircase APD proposal, *see* Silicon avalanche photodetectors
Stimulated Brillouin scattering, 226–227, 281–285
　acoustic phonons, 281
　backward SBS, 285
Stimulated Raman scattering, 226–227, 269–278
　amplification of Raman-scattered light and degradations, 270–271
　frequency-shift-keyed systems (FSK), 274
　phase-shift-keyed (PSK) systems, 274
　shift-keyed systems (ASK), 274
Sunde's FSK, 209
Surface or Burrus emitter, 20, *see also* Edge emitter
System experiments, 251–257

T

10-μm and 5-μm stripe lasers, 49
Temperature-controlled laser transmitter, 26–27
Temperature modulation effect, 167–171
　laser chip layers, 167
　refractive-index change, 167
　sinusoidal current, 169
　thermal resistances, 169
　thermal response, 170
Thermistor, 40–41

Thermoelectric heat pump TEHP, 40–41, 58–59
Timing delay, 37–38
Transimpedance amplifier, 55, 82
Transimpedance optical receivers, 117–119
　feedback resistors, 118
　increasing sensitivity, 117
　signal integration, 119
Transimpedance receivers, 80, 112–113
Transistor technology, 79, 81
Transmitter design, 3–19
　modulation format, 8–9
　modulation rate, 14–18
　source wavelength and linewidth, 10–14
Transmitter design, 3–19
　modulation format, 8–9
　modulation rate, 14–18
　source wavelength and linewidth, 10–14
Transmitter testing, 64–66
　measurements, 65
　optical eye, 64
　reflection sensitivity tests, 65
Traveling-wave-type amplifiers, 227
Traveling-wave-type semiconductor amplifier, 230–232

V

Voltage-amplifier optical receiver, 113–114
　Johnson noise current, 113
　receiver circuit, 113
　sensitivities, 113–114

W

Wavelength-division-multiplex (WDM), 5
Wavelength-multiplexing experiment, 274
Wide-stripe gain-guided lasers, 49–50

Contents of Volume 1

1. *Modified Chemical Vapor Deposition*
   S. R. Nagel, J. B. MacChesney, and K. L. Walker
2. *Outside Vapor Deposition*
   Alan J. Morrow, Arnab Sarkar, and Peter C. Schultz
3. *Vapor-Phase Axial Deposition Method*
   Nobukazu Niizeki, Nobuo Inagaki, and Takao Edahiro
4. *Fiber Drawing and Strength Properties*
   F. V. DiMarcello, C. R. Kurkjian, and J. C. Williams
5. *Manufacturing of Optical Fibers*
   Donald P. Jablonowski, Charles W. Deneka, and Hiroshi Murata